Agri-Food Wastes and Biomass Valorization

Agri-Food Wastes and Biomass Valorization

Editors

Vassilis Athanasiadis
Dimitris Makris

Basel • Beijing • Wuhan • Barcelona • Belgrade • Novi Sad • Cluj • Manchester

Editors
Vassilis Athanasiadis
University of Thessaly
Karditsa, Greece

Dimitris Makris
University of Thessaly
Karditsa, Greece

Editorial Office
MDPI
St. Alban-Anlage 66
4052 Basel, Switzerland

This is a reprint of articles from the Special Issue published online in the open access journal *Waste* (ISSN 2813-0391) (available at: www.mdpi.com/journal/waste/special_issues/SB12N07HR9).

For citation purposes, cite each article independently as indicated on the article page online and as indicated below:

Lastname, A.A.; Lastname, B.B. Article Title. *Journal Name* **Year**, *Volume Number*, Page Range.

ISBN 978-3-0365-9539-9 (Hbk)
ISBN 978-3-0365-9538-2 (PDF)
doi.org/10.3390/books978-3-0365-9538-2

Contents

Review

Overview of the Biotransformation of Limonene and α-Pinene from Wood and Citrus Residues by Microorganisms

Adama Ndao * and Kokou Adjallé

Laboratoire de Biotechnologie Environnementale, Institut National de la Recherche Scientifique (INRS), Boulevard du Parc-Technologique, Quebec, QC G1P 4S5, Canada; kokou.adjalle@inrs.ca
* Correspondence: adama.ndao@inrs.ca

Abstract: This review provides an overview of the biotransformation of limonene and α-pinene, which are commonly found in wood residues and citrus fruit by-products, to produce high-value-added products. Essential oils derived from various plant parts contain monoterpene hydrocarbons, such as limonene and pinenes which are often considered waste due to their low sensory activity, poor water solubility, and tendency to autoxidize and polymerise. However, these terpene hydrocarbons serve as ideal starting materials for microbial transformations. Moreover, agro-industrial byproducts can be employed as nutrient and substrate sources, reducing fermentation costs, and enhancing industrial viability. Terpenes, being secondary metabolites of plants, are abundant in byproducts generated during fruit and plant processing. Microbial cells offer advantages over enzymes due to their higher stability, rapid growth rates, and genetic engineering potential. Fermentation parameters can be easily manipulated to enhance strain performance in large-scale processes. The economic advantages of biotransformation are highlighted by comparing the prices of substrates and products. For instance, R-limonene, priced at US\$ 34/L, can be transformed into carveol, valued at around US\$ 530/L. This review emphasises the potential of biotransformation to produce high-value products from limonene and α-pinene molecules, particularly present in wood residues and citrus fruit by-products. The utilisation of microbial transformations, along with agro-industrial byproducts, presents a promising approach to extract value from waste materials and enhance the sustainability of the antimicrobial, the fragrance and flavour industry.

Keywords: limonene; alpha-pinene; biotransformation; terpene; flavouring agent; solid-state fermentation

Citation: Ndao, A.; Adjallé, K. Overview of the Biotransformation of Limonene and α-Pinene from Wood and Citrus Residues by Microorganisms. *Waste* **2023**, *1*, 841–859. https://doi.org/10.3390/waste1040049

Academic Editors: Catherine N. Mulligan, Vassilis Athanasiadis and Dimitris P. Makris

Received: 21 August 2023
Revised: 11 September 2023
Accepted: 28 September 2023
Published: 4 October 2023

1. Introduction

Essential oils can be derived from various plant parts, including flowers, fruits, leaves, buds, seeds, twigs, bark, herbs, wood, and roots. These oils are predominantly composed of monoterpene hydrocarbons [1]. These terpene hydrocarbons, such as limonene, pinenes, and terpenes, serve as ideal starting materials for microbial transformations to obtained high value molecules such as carveol, terpineol, verbenol and verbenol. Microorganisms and enzymes can act as catalysts in two different processes: de novo synthesis and biotransformation of natural precursors. De novo synthesis involves utilising the entire metabolic pathway of microorganisms to produce a combination of flavour compounds, whereas biotransformation focuses on specific reactions that lead to the production of a major compound. In de novo synthesis, microorganisms metabolise carbohydrates, fats, and proteins, converting the breakdown products into flavour components [2,3]. However, this process yields only trace amounts of flavours and is not economically viable for industrial production due to low concentrations of the desired compounds [4]. In biotransformation, a precursor is introduced into the process, inducing the microorganism to follow a specific pathway and produce the final product through one or two chemical reactions [5]. Additionally, agro-industrial byproducts can be utilised as nutrient and substrate sources in fermentation processes, reducing fermentation costs and enabling industrial viability [6]. Terpenes are noteworthy substrates for biotransformation, as they are secondary metabolites of plants and can be found in byproducts generated during fruit and plant processing. Monoterpenes (such as limonene, α-pinene, and β-pinene) and sesquiterpenes (such as valencene and farnesene) are common volatile compounds found in various essential oils. Structurally, they are closely related to aroma compounds of significant interest in the industry, requiring only a few chemical reactions to obtain high-value products [7] (Table 1). From an economic perspective, the advantages of biotransformation become evident when comparing the prices of substrates and products. For example, in 2015, R-limonene has a reference price of US$34/L, while its oxygenated form, carveol, is priced around US$530/L [7,8]. Selecting the appropriate biocatalyst is another crucial step in natural aroma production. Microbial cells offer advantages over enzymes, as they possess higher stability and do not require additional cofactors for the reaction to occur. Microorganisms also exhibit rapid growth rates, and fermentation parameters can be easily manipulated, allowing for genetic engineering to enhance strain performance in large-scale processes [9]. The objective of this review is to provide approaches and strategies to improve the biotransformation of two of the most promising biomolecules found in agroforestry residual biomass. The review will discuss general knowledge of terpenes, recent advances of biotransformation of limonene and alpha-pinene, strategies to improve the biotransformation of these two molecules using solid-state fermentation and finally various extraction technology to recover the produce molecules.

Table 1. Chemical composition of balsam fir, black spruce, jack pine and citrus essential oils.

Essence	Parameters	Monoterpene	Oxygenated Monoterpene	Sesquiterpene	Esters	References
Balsam fir	Concentration	84.9%	1.3%	0.6%	9.1%	[10,11]
	major compound	α-Pinene β-pinene Limonene	α-Tepeniol	β-caraphyllene	Bornyl acetate	
Black spruce	Concentration	63.5%	2.6%			[10,11]
	major compound	α-Pinene β-pinene Y-3-carene	α-tepeniol 1,8-cineole	β-caraphyllene	Bornyl acetate	
Jack pine	Concentration	85.9%	2.7%	0.2%	8.2%	[10,11]
	major compound	α-pinene camphene	α-tepeniol 1,8-cineole	α-tepeniol 1,8-cineole	Bornyl acetate	
Citrus terpenes	major compounds	Limonene 95% α-pinene 2.5% camphene 2% β-myrcene 0.5%				[12]

2. A Brief Description of Terpenes

Conifers are the most abundant source of terpene compounds in nature [13]. In conifers, volatile compounds are contained in the oleoresin. This is a mixture of volatile monoterpenoids and sesquiterpenoids (turpentine), and non-volatile diterpenoids which represent the resin acids [14]. Monoterpenoids are low molecular weight compounds (136 Da for hydrocarbons and up to 200 Da for oxygenated derivatives) (Figure 1). Highly volatile, they are also odorants, naturally emitted by plants to attract pollinators or repel predators. They can be captured by processes such as hydrodistillation, to form extracts known as essential oils. Terpenes are often removed from essential oils as undesirable components, whereas synthetic oxy-functionalised derivatives find broad applications in flavours and fragrances industries. In addition to their perfuming power, these molecules respectively possess medicinal properties, which in synergy make essential oils renowned in aromatherapy. Volatile terpenoids also appear to be involved in protecting plants against heat and oxidative stress [14]. Their role as a communication molecule is thought to attract pollinating insects, thus aiding reproduction, but they also act as "signal" molecules to preserve surrounding trees in the event of an attack [15]. Despite the diversity of terpene compounds, they share a common biosynthetic pathway. This pathway comprises four stages [16]. Briefly, the first involves the formation of isopentenyl pyrophosphate (IPP), the C5 unit of biological isoprene. The main route for the synthesis of IPP and its allylic isomer, dimethylallyl pyrophosphates (DMAPP), is via the mevalonate pathway. A second synthesis pathway is also present, the methylerythritol phosphate (MEP/DOXP) pathway. In the second step, C5 units condense to generate three larger prenylpyrophosphates (GPP, FPP and GGPP). Subsequently, in the third step, these three pyrophosphates undergo a wide range of cyclisation and rearrangement to produce the parent carbon skeletons for each terpene class. The final stage encompasses various oxidation, reduction, conjugation and other transformations that convert the carbon skeletons into thousands of distinct terpene metabolites. Although several in vitro studies have shown that many monoterpenes and diterpenes have significant bactericidal effects, it is not always easy to achieve the same efficacy in vivo due to the multiple presence of molecules present in extracted oils. As a result, microbial biotransformation is a pertinent strategy to overcome difficulties and problems arising from the chemical synthesis, to have access to regiospecifc and stereospecifc compounds and explain the inactivity of essential oils or chemicals against some microorganisms.

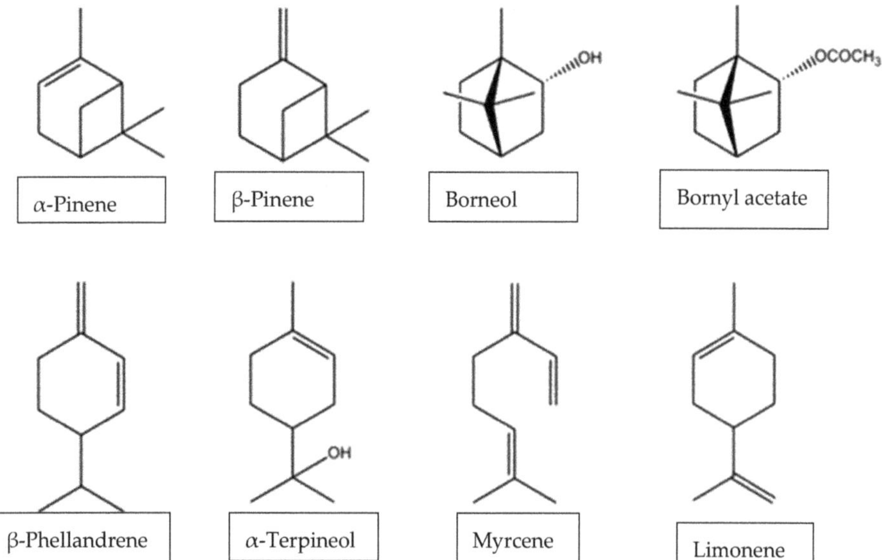

Figure 1. Some examples of monoterpenoides.

3. Recent Advances in Biotransformation of Terpenes: Study Case of Limonene and Alpha Pinene

Currently, the primary production of aroma compounds relies on chemical methods due to the challenges associated with direct extraction from natural sources. Natural extraction often results in complex mixtures with low concentrations and limited availability based on geography and seasonality [17]. Essential oils typically yield quantities ranging from 0.005% to 10%, with many falling below 1% [18]. The composition of essential oils varies, and they can contain over 100 different components, with major compounds often comprising more than 85% of the oil. The increasing focus on health, nutrition, and the desire to avoid synthetic chemicals in food has spurred the development of biotechnological processes to produce flavour compounds that can be classified as "natural" [19]. Some companies are actively working on biotechnological methods for aroma production. For instance, the Amyris Company utilises a genetically modified strain of Saccharomyces cerevisiae to produce β-farnesene from sugarcane. This sesquiterpene can also serve as a precursor to produce other aroma compounds in the fragrance industry [8]. Evolva and Isobionics companies have reported the biotechnological production of valencene, which can be further oxidised to nootkatone. These companies' market nootkatone as a flavouring agent and a potential natural insect repellent. Isobionics has developed efficient microbial strains for terpene biotransformation. Based on the above examples, bioprocessing of limonene and pinenes, mainly found in many essential oils and citrus fruits, offers significant potential for biotransformation [20].

3.1. Limonene Biotransformation

Limonene is the most abundant naturally occurring monoterpene found in orange peel oil, comprising up to 90% of its composition [21]. Due to its abundance and low cost, limonene has been extensively studied as a precursor to produce high-value derivatives, offering a potential strategy for enhancing the commercial value of agro-industrial residues like orange peel oil [22] (Figure 2). The applications of limonene as a flavour and fragrance additive in food products, cosmetics, household cleaning products, and textiles are well known [23]. Additionally, limonene finds utility as a solvent, a feedstock for fine chemicals, a precursor for polymeric biomaterials, and as an active ingredient in medicine [24]. Currently, most of the limonene is obtained from citrus rinds, a significant byproduct generated during fruit processing in citrus juice industries [25]. However, many citrus juice industries, particularly those in developing countries, lack the necessary infrastructure and technology for on-site recycling of large quantities of citrus waste [26]. Consequently, a substantial number of citrus rinds are disposed of in landfills rather than being utilised for limonene extraction. Limonene is extensively studied as a precursor in the biotechnological production of monoterpenoids, as it can be transformed into various value-added compounds such as carvone, carveol, perillyl alcohol, terpineols, menthol, and pinenes (Table 2). It serves as a cyclic monoterpene with multiple industrial applications as a bio-solvent, in surfactant formulations, as an antibacterial agent, and as a pesticide in nutrition, as well as a precursor for fine chemicals in the food, perfume, medicine, agrochemical, and oral hygiene industries [27]. Given the structural similarity between limonene and oxygenated monoterpenoids with pleasant fragrances, such as perillyl alcohol, carveol, carvone, menthol, and α-terpineol, limonene can be used as a precursor for the synthesis of these flavour compounds [24]. Several studies have reported the biotransformation of limonene into α-terpineol, using microorganisms such as *Cladosporium* sp., *Pseudomonas gladioli*, and an α-terpineol dehydratase isolated from the same strain [28]. Orange peel oil analysis has revealed that D-limonene is the major component, accounting for 96.1% of the total content [29]. The biotransformation of D-limonene to a-terpineol was conducted using a strain of *Penicillium digitatum* NRRL 1202. Experiments employed two distinct media: malt yeast broth (MYB) and malt extract broth (MEB). Among them, the MYB medium resulted in the most efficient bioconversion of D-limonene to a-terpineol. Optimisation of the bioconversion process using synthetic media supplemented with orange peel oil demonstrated increased yields of α-terpineol, reaching 79% at 3 h of fermentation and 95.5% after 7 h with the addition of MYB in the medium. This demonstrates that supplementation of synthetic medium (medium rich of reducing sugar, minerals and organic nitrogen) allows the microorganism to perform and increase the biotransformation of limonene. However, it should be noted that the use of expensive synthetic media and the concentration of orange peel oil were limitations of this study as the cost to scale-up the process was high because of the high cost of nutrients. Instead, the use of residual medium as wastewater coming from agroindustry could be a way to decrease the cost of the process and maintain high performance of biotransformation [29]. Molina et al. [7] investigated the bioconversion of limonene using two isolated strains, resulting in the production of high-value derivatives carvone and α-terpineol. The aspergillus LB-2038 strain achieved a higher concentration of carvone (47 mg/L), while the Penicillium LB-2025 strain stood out in α-terpineol production (10 mg/L). However, when 1% limonene supplementation was attempted, it affected initial biomass development and the microorganisms were unable to metabolise the substrates, indicating the toxicity of a high concentration of limonene to microbial cells. To mitigate the toxicity, a two-phase system with an organic phase containing limonene and an aqueous phase for the catalyst was implemented in this study. It is worth noting that comparing different strains for biotransformation results could have provided valuable insights [30]. Roy and Jayati [12] infected mandarin orange peels with Aspergillus oryzae and Penicillium species, resulting in GC-MS analysis showing a ratio of 10% beta pinene and 70% limonene for the infected orange peels, and 30% beta pinene, 27% beta-myrcene, and 32% limonene for the Penicillium-infected peel oil. The extraction of pure limonene oil with petroleum ether was performed in this study, but conducting a solid-state fermentation with different particle sizes of orange peels could have been an alternative approach to avoid early solvent addition and limonene toxicity. Furthermore, the optimisation of inoculum size and the use of multiple strains could have been beneficial for the optimisation of limonene biotransformation. Another study conducted by Bicas et al. [6] demonstrated that Pseudomonas Rhodesia and *Pseudomonas fluorescens* can metabolise limonene, with the latter exhibiting versatility and the ability to synthesise α-terpineol at high concentrations (11 g/L) [6]. The conversion of refining limonene to α-terpineol using fungal biocatalysts has been reported, with Penicillium and Fusarium oxysporum achieving yields of 3.45 g/L and 2.4 g/L, respectively. The conversion of limonene obtained from citrus extraction to carvone and carveol by Pseudomonas aeruginosa and *Rhodococcus opacus* resulted in yields of 0.63 g/L and 2.4 g/L, respectively [31]. From an economic perspective, limonene has a weighted average price of US\$ 34/L, while its oxygenated derivatives such as carveol, perillyl alcohol, and carvone have reference prices of US\$ 529/L, US\$ 405/L, and US\$ 350/L, respectively [8]. These monoterpenoids are widely used aroma compounds found in natural sources at low concentrations, making biotechnological approaches crucial for large-scale production. Therefore, the utilisation of limonene as a substrate to produce value-added derivatives through biotransformation is economically promising.

Table 2. Monoterpenes used as substrates in bioprocesses for the production of natural flavour compounds.

Terpene Substrate	Biotransformation Product	Strains	Chemical Name	Odour	Application	Reference
LIMONENE	α-Terpineol	*Pseudomonas fluorescens*	p-Menth-1-en-8-ol	lilac	Antioxidant activity, Anti-inflammatory activity, Antimicrobial activity: Activity against A.niger, Staphylococcus epidermidis, Cytostatic and cytocidal effects towards Goetrichum citri-aurantii.	[6]
	Carveol	*Rhodococcus opacus; Rhodococcus erythropolis* PWD8; *Pleurotus sapidus; Aspergillus cellulosae* M-77	p-Mentha-6,8-dien-2-ol	spearmint	Flavouring agent; food additive.	[31]
	Carvone	*Rhodococcus opacus; Rhodococcus eythropolis* PWD8; *Penicilium digitarium; Pleurotus sapidus*	p-mentha-1(6),8-dien-2-one	Mint aroma	Flavouring agent: flavour chewing gum and mint candies, provide aromas in personal-care products, air fresheners, and aromatherapy oils.	[31]
	Perillic acid or perillyl alcohol	*Pseudomonas putida* DSM 12264; *Aspergillus cellulosae* M-77; *Mycobacterium* sp. HXN-150	(4R)-4-prop-1-en-2-ylcyclohexene-1-carboxylic acid		Exhibit antimicrobial properties; Ingredients in cleaning products; Mosquito repellent when applied to the skin.	[31]

Table 2. *Cont.*

Terpene Substrate					
α-PINENE	Verbenol	*Aspergillus niger*	4,6,6-Trimethyl-bicyclo(3.1.1)hept-3-en-2-one	Balsamic aroma	used in fragrance formulation of soft drinks, soups used as important intermediates in cosmetics and pharmaceutical industries used as flavour in food, such as in meats, sausages and ice cream used as material to synthesise chemicals, such as citral.	[32]
	Verbenone		4,6,6-Trimethylbicyclo [3.1.1]heptan-3-one	minty spicy aroma	Used for insect control particularly against beetles as *Dendroctonus frontalis*; Used also in perfumery, aromatherapy, herbal teas and herbal remedies. The L-isomer is used as a cough suppressant under the name of levoverbenone. Verbenone may also have had antimicrobial properties.	

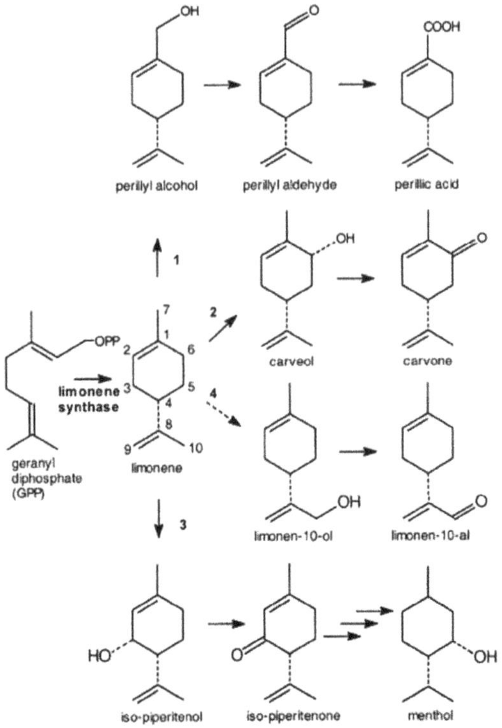

Figure 2. Biosynthetic pathways in plants starting from limonene.

3.2. Biotransformation of α-Pinene

One of the prominent antimicrobial monoterpenes found in various essential oils is α-pinene. This natural compound holds significant importance as it is widely utilised in the food and fragrance industries, pharmaceuticals, fine chemicals, and as a renewable fuel source [33]. Currently, industrial production of α-pinene primarily relies on tree tapping (gum turpentine) or as a byproduct of paper pulping (crude sulfate turpentine, CST). However, extracting α-pinene from trees is laborious, inefficient, and depletes natural resources due to its low content [34]. Hence, exploring sustainable technologies for the biotransformation of α-pinene would be a promising avenue (Figure 3). In a study conducted by [35], a cold-adapted fungus, *Chrysosporium pannorum*, was identified as a biocatalyst for the oxidation of refining α-pinene. Gas chromatography-mass spectrometry (GC-MS) analysis of the fermentation process revealed the production of two commercially valuable molecules, verbenol and verbenone, through the biotransformation of α-pinene. The oxidative activity of *C. pannorum* was observed over a wide temperature range (5–25 °C), with the optimum temperature being 10 °C to produce these bioactive compounds. By sequentially adding the substrate over a period of three days, the yield of verbenol and verbenone significantly increased to 722 mg/L and 176 mg/L, respectively, resulting in a twofold enhancement compared to a single supply of α-pinene. The total concentration of conversion products in the culture medium reached 1.33 g/L. Another study by [36] focused on the biotransformation of α-pinene to (+)-α-terpineol using *Candida tropicalis* MTCC 230. Under continuous agitation at 30 °C for 96 h, a concentration of 0.5 g/L of (+)-α-terpineol was obtained. The strain demonstrated efficient bioconversion capabilities, achieving a conversion rate of 77% and a production of 0.43 g/L under the specified conditions. Rottava et al. [37] reported the metabolism of α-pinene by the yeast *Hormonema* sp. UOFS Y-0067, isolated from pine tree samples. After 72 h, this yeast produced transverbenol (0.4 g/L) and verbenone (0.3 g/L) through α-pinene biotransformation. It is worth noting that all these experiments were conducted using pure or refine α-pinene. To the best of our knowledge, no research has been conducted on the biotransformation of α-pinene using residual wood, such as tree bark. Further investigations should include GC-MS analysis of different tree species or parts to identify sources with high α-pinene content. Based on the market value of the molecules obtained by the transformation of α-pinene, it would be important to study the technological feasibility of the biotransformation of α-pinene contained in wood residues rich in pinene. The key parameters for a good microbial fermentation in a semi-solid medium are widely studied in the literature. The particle size and specific surface area are key parameters to ensure a good contact surface between the microorganisms and the molecules to be biotransformed. Also, the particle size is an important factor in oxygen transfer during solid or semi-solid fermentation. Other important factors are the moisture content, the inoculum size and the fermentation temperature [38]. Regarding terpenes, the volatilisation of molecules and the presence of lignin can be factors slowing down the yield of molecules obtained. The following parts will discuss in detail the current practices to address these challenges.

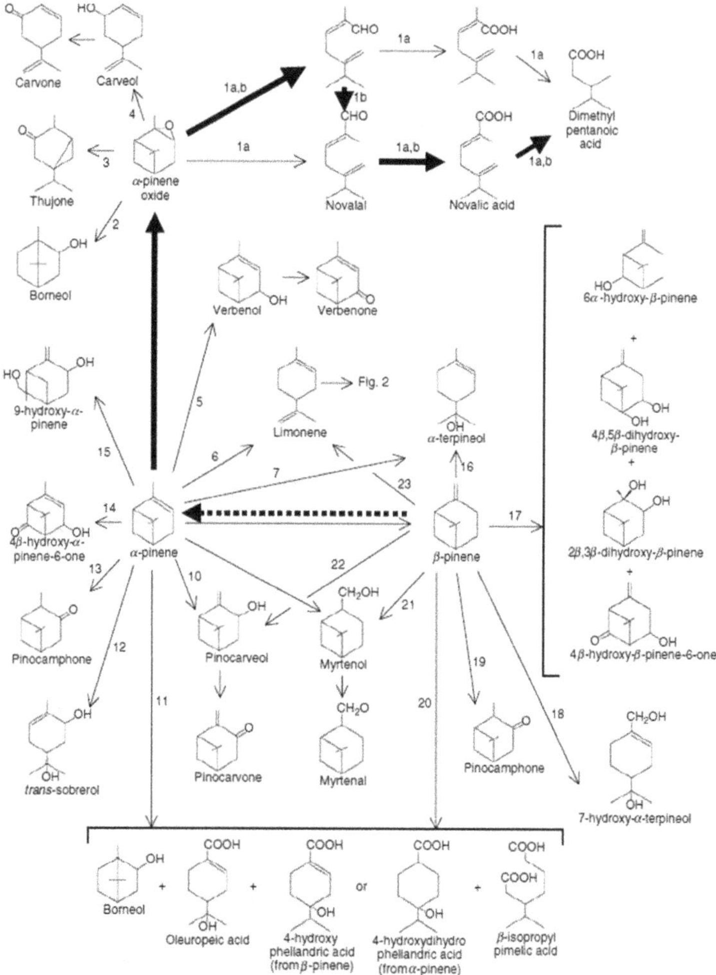

Figure 3. Degradation pathways of alpha pinene by Pseudomonas.

4. Avenues and Approaches for Improving the Biotransformation of Limonene and Alpha Pinene

4.1. Application of Solid-State Fermentation for Biotransformation from Filamentous Fungi

Solid-state fermentation corresponds to the growth of aerobic or anaerobic microorganisms on moist solid particles in the absence or near absence of free water [39]. These microorganisms grow on a solid matrix to which a liquid phase is bound, and a gaseous phase is trapped within or between these particles. Most solid-state fermentation processes involve filamentous fungi. However, there are a few processes that incorporate bacteria and yeast. Solid-state fermentation can involve a pure culture of organisms or the cultivation of several strains inoculated simultaneously or successively [40]. Current work on solid-state fermentation focuses on the application of these processes in the development of bioprocesses, i.e., bioremediation or the production of high value-added products. The latter are composed of biologically active secondary metabolites including antibiotics, aromatic compounds (limonene and pinene derivatives), alkaloids, biofertilizers, enzymes, and so on. In the case of the biotransformation of limonene or alpha pinene, it would be very interesting to carry out fermentation in a solid medium, as this brings the molecules present in the substrate into contact with the biotransforming strains. Also, the possibility of carrying out mixed cultures would enable them to produce different products and increase the yield of the process. Last but not least, this strategy avoids the need to use synthetic media to initiate the fermentation of the microorganisms and reduces or eliminates the use of solvents at the start of the process to extract the oil or juice containing the molecules to be biotransformed. Solid-state fermentation-based processes offer potential advantages in the biotransformation of bioactive molecules contained in forest and agricultural residues.

4.2. Selection of Microorganisms

Synthetic limonene or pinenes are often used instead of residues for biotransformation [41]. It would be of interest to use known strains to carry out solid or semi-solid fermentations of residual media containing limonene (citrus like orange, lemon) or pinene (white birch bark). This practice, rather than the use of oils, would minimise the toxicity of pinene or limonene at the start of fermentation. In addition, the use of solvents for juice extraction will be eliminated at the start of the process. Another possible step would be to isolate microbial strains from the residual wood and orange media. Indeed, the natural presence of strains in these residues testifies to their ability to withstand toxic doses of the molecules in question, as well as their capacity to metabolise them. Among strategies, bioprospecting is used by some authors in an attempt to identify new strains of microorganisms for the production of bioaromas [42]. Particularly, soil samples have been frequently used for the prospecting of microorganisms that are potential bio-transformers of terpene compounds. To illustrate this approach, some examples of microorganisms isolated from soil may be cited: Bacillus fusiformis, which was able to convert isoeugenol to vanillin (production of 8.10 g L1 after 72 h) [43]; Bacillus pumilus, which was also able to convert isoeugenol to vanillin (production of 3.75 g L1 after 150 h) [44]; Pseudomonas putida, which was able to convert isoeugenol to vanillic acid (98% of molar conversion after 40 min); Bacillus subtillis, which was able to convert isoeugenol to vanillin (production of 0. 9 g L1 after 48 h) [45] and *Chrysosporium pannorum*, which was able to convert a-pinene to verbenone and verbenol. Rottava et al. [37] have isolated strains from effluents of citrus industry as well as other samples (soil from the plantation of citrus, citrus fruits, and citrus leaves). Of the 405 strains isolated [37], eight were able to bioconvert limonene and fifteen converted pinene, generating α-terpineol as a product in both cases. The transformation of limonene to α-terpineol typically involves the oxidation of limonene. The first step is the hydroxylation of limonene, which can occur at different positions to yield a variety of alcohols. This is typically catalysed by cytochrome P450 monooxygenases and the formation of Limonene-1,2-diol: One of the primary alcohols formed from the hydroxylation of limonene is limonene-1,2-diol. Then the conversion to Terpineol: Limonene-1,2-diol can then be dehydrated to yield α-terpineol. This step can be enzymatically catalysed by dehydratase enzymes [46]. While the biotransformation of α-Pinene occurs typically with the oxidation of α-pinene to yield α-pinene oxide. This reaction is facilitated by the enzyme cytochrome P450 monooxygenase or similar enzymes present in Pseudomonas or other fungi. The α-pinene oxide can undergo a rearrangement to form generally verbenol. The verbenol formed can then be hydroxylated to produce terpineol. This hydroxylation can be mediated by specific hydroxylase enzymes present in fungi [46].

Bicas et al. [9] have obtained 248 microorganisms, of which seventy were developed in a medium containing limonene as the sole source of carbon. However, as reported by Molina et al. [30], there are several challenges to be overcome to enable the production of aromas by biotransformation, among them the high toxicity of both the substrate and the product and the low yields obtained. According to these authors, such difficulties can be overcome with the isolation and selection of strains as the studies mentioned above.

4.3. Pathways of Terpenes Biotransformation

When bacteria encounter monoterpenes, they must address the toxic effects associated with them [47]. To prevent the accumulation of monoterpenes in the cell and cytoplasmic membrane, bacteria modify their membrane lipids, transform monoterpenes, and employ active transport via efflux pumps [48]. At subtoxic concentrations, microorganisms utilise monoterpenes as their sole carbon and energy source. While numerous microbial cultures have reported monoterpene transformations over the past 50 years, the biochemical pathways involved have often been undisclosed. Furthermore, only a small fraction of the investigated strains has been deposited in culture collections. Without detailed knowledge of the genes or the availability of strains, the observations from biotransformation experiments hold limited value for future studies.

4.3.1. Bicyclic Monoterpenes

The isomers α-pinene and β-pinene (C10H16) are the primary components of wood resins, particularly conifers. *Pseudomonas Rhodesia* (CIP 107491) and *P. fluorescens* (NCIMB 11671) were found to grow on α-pinene as the sole carbon source. α-pinene undergoes oxidation to α-pinene oxide via a NADH-dependent α-pinene oxygenase and then experiences ring cleavage by a specific α-pinene oxidelyase, yielding isonovalal, which is subsequently isomerised to novalal [6]. The cleavage reaction of α-pinene oxide has also been observed in Nocardia sp. strain P18.3 [37]. An alternative degradation pathway for pinene via a monocyclic p-menthene derivative has been described in Pseudomonas sp. strain PIN [49]. Bacillus pallidus BR425 degrades α-pinene and β-pinene, yielding limonene and pinocarveol. While α-pinene is transformed into limonene and pinocarveol, β-pinene exclusively produces pinocarveol. Both intermediates can be further converted into carveol and carvone. Although the activity of specific monooxygenases has been suggested, experimental evidence is lacking [5]. Serratia marcescens utilises α-pinene as the sole carbon source, leading to the formation of trans-verbenol. In glucose and nitrogen-supplemented medium, this strain produces α-terpineol. The two oxidation products are considered dead-end products as they accumulate in cultures [6]. It should be noted that for many biotransformation studies, caution must be exercised as monoterpenes often contain impurities and oxidation products that can serve as substrates, resulting in traces of monoterpene and monoterpenoid transformation products that are not further metabolised.

4.3.2. Monocyclic Monoterpenes

Limonene (C10H16) is the most abundant monocyclic monoterpene and the second most abundant volatile organic compound (VOC) indoors, after toluene [23]. It is a major component of essential oils derived from citrus plants such as lemon and orange. Biotransformation of limonene start with activation of geranyldiphosphate by linalool synthase, which triggers ionization and rearrangement of gernayldiphosphate. The positively charged geranyl cation will be hydrolyzed by to form linalool and myrcene. For GPP to form limonene or terpineol, it must be activated in its cyclic form to give alpha terpinyl cation [50]. *Rhodococcus erythropolis* DCL14 transforms limonene into limonene-1,2-epoxide, which is then converted to limonene-1,2-diol by a limonene-1,2-monooxygenase and a limonene-1,2-epoxide hydrolase, respectively. A specific dehydrogenase forms the ketone, 1-hydroxy-2-oxolimonene, which is further oxidised to a lactone by a 1-hydroxy-2-oxolimonene 1,2-monooxygenase. Enzyme activities were only detected in limonene-induced cells, indicating tight regulation of limonene degradation. *R. erythropolis* DCL14 possesses a second pathway for limonene degradation in which limonene is hydroxylated to trans-carveol by a NADPH-dependent limonene 6-monooxygenase. *R. opacus* PWD4 follows the same pathway, converting limonene to trans-carveol. Biomass from a glucose-toluene chemostat culture of *R. opacus* PWD4 transformed limonene into trans-carveol, which was further oxidised to carvone by a trans-carveol dehydrogenase [31]. In *Pseudomonas gladioli*, studies on limonene metabolism identified α-terpineol and perillyl alcohol as major metabolites. However, none of the involved enzymes have been purified or further characterised. A α-terpineol dehydratase from *P. gladioli* was isolated and partially purified. The hydration reaction of limonene to form α-terpineol as the sole product was observed [51]. *Geobacillus stearothermophilus* (formerly *Bacillus*) has been shown to grow on limonene as the sole carbon source. The main transformation product of limonene in this case is perillyl alcohol, while α-terpineol and perillyl aldehyde are present in minor concentrations.

4.4. Improvements in Physical Pretreatment of Wood and Citrus Residues

4.4.1. Pretreatment and Conditioning

Pretreatment of forest and agricultural residual materials is an important link in the emerging circular economy in general and bioprocessing. Optimising this component will enable us to expand the range of high value-added products. Pretreatment methods can be thermal, chemical, physical or biological. Whatever the method, pretreatments aim to make biomass manageable during mechanical, microbiological or chemical transformations [52]. Pretreatment of forest and agricultural residues has an influence on the quantity and quality of terpenes extracted. Indeed, grinding is a technique that reduces the size of the biomass to improve the contact surface with steam during extraction. However, there may be losses in terms of terpenes yield due to adsorption on the mill walls or that may be volatilised in the open air due to reaching an equilibrium point with ambient temperature [53]. Nevertheless, the conditioning of fragmented forest and agricultural residues is accompanied by exothermic reactions. These reactions lead to material losses through fungal and bacterial attack that can reach around 25% and generate a reduction in energetic power [54]. Degradation phenomena generally depend on forest species, moisture content, particle size and storage time. The study by Douville et al. [54] showed that fungal attacks lead to weight losses in wood, as they degrade chemical compounds such as lignin and cellulose to consume their carbon molecules. Consequently, it is necessary to process the biomass immediately after harvesting and, to avoid the risk of bacterial or fungal degradation of the biomass and inhibit enzymatic activity after harvesting [55].

4.4.2. Thermal Pretreatment

According to several studies, torrefaction and pyrolysis are promising options for thermal pretreatment [56]. Torrefaction is a form of thermal pretreatment that produces a finished product characterised by low moisture and high calorific value compared with fresh biomass [57]. The biomass resulting from this process contains 70% of its initial weight and 90% of its energy content. Torrefaction is therefore a cost-effective and efficient method. The cost of torrefaction depends on plant costs (39%) and equipment costs (31%). However, pyrolysis is a form of thermochemical conversion that requires advanced technologies to produce a higher-quality end product [57].

4.4.3. Chemical Pretreatment

Chemical pretreatments use chemical reagents and solvents to modify or extract the major components of lignocellulosic biomass, such as cellulose, hemicellulose and lignin. This makes it possible to transform lignocellulosic biomass into interesting and environmentally friendly products [58]. The most commonly used processes for chemical pretreatment are alkaline pretreatment, acid dilution, steam explosion and the use of low-boiling organic solvents such as methanol and ethanol [59]. These methods aim to increase the internal surface area of the biomass and reduce the degree of polymerisation and crystallinity of the cellulose. For example, alkaline pretreatments degrade the lignin structure and break bonds with carbohydrate fractions [58]. Steam pretreatment, on the other hand, is energy-efficient and chemical-free, while avoiding dilution of the sugars. In fact, after steam pretreatment, the biomass slurry is often washed or diluted to remove the inhibitors formed during pretreatment (like furfural and HMF). This can result in a diluted sugar solution. The extent of dilution depends on the washing protocol and the intended downstream process. Also, as the biomass undergoes steam pretreatment, a portion of the hemicellulose fraction is solubilized, leading to the formation of sugars. The concentration of these sugars is affected by the severity of the pretreatment (combination of temperature, time, and pressure). In addition, this process can lead to condensation of soluble lignin, making BFR less digestible [60].

4.4.4. Biological Pre-Treatments

These are activities associated mainly with antifungal action, and have the power to degrade lignin, hemicellulose and cellulose. This process differs from the chemical and thermal processes and requires large amounts of energy [60]. In fact, biological pretreatment is an eco-responsible method that avoids toxic discharges into the environment. For example, white-rot fungi are responsible for lignin degradation and require less energy (oxygen, water and nutrients) during their reactions [61]. Two other enzymes derived from basidiomycetes: manganese peroxidase and laccase are responsible for hemicellulose and lignin degradation. On the one hand, these methods are relevant, as they improve productivity and avoid damaging the ecosystem through rejects. However, these methods require long reaction times to achieve relevant results. Admittedly, this is time-consuming for manufacturers [62]. On a large scale, biological pretreatment requires costly processes, as it requires a sterile environment. In addition, the microorganisms used during this process consume some of the biomass carbohydrates, reducing the efficiency of this method [61].

4.4.5. Extrusion

The lignocellulosic biomass consists mainly of lignin, cellulose, and hemicellulose, which form a strong and complex three-dimension structure [63]. To use lignocellulosic biomass as a feedstock for microbial biotransformation, it needs to be pretreated. The aim of this pretreatment is to break or weaken the lignocellulosic structure to extract/recover the compounds of interest. Usually, the target is to recover cellulose, hemicellulose and terpenes. But lignin blocks the access to those compounds [64]. Extrusion is a thermo-mechanical process that applies high shearing forces to a material (a solid or a semi-solid) for different purposes such as blending, compounding, mixing, disruption, compaction, pelletisation, etc. Therefore, extrusion is used in a wide range of domains: plastics, polymers, food, pharmaceuticals, ceramics, metal, biorefinery, paper industry, etc. They need to be moistened by a liquid. Adding water in non-optimal condition and ratios (solid/liquid) can cause biomass clogging problems inside the barrel due to evaporation and low viscosity. Compared to other physical pretreatments, extrusion is regarded as a low-energy consumption technology. Furthermore, extrusion does not need significant downstream operations because it does not produce inhibitors and does not use high water amounts as hydrothermal pretreatments do. The short pretreatment time is one of the main benefits of extrusion: while biological, chemical and some physico-chemical pretreatments may last from 30 min to several days, one extrusion pretreatment lasts a few minutes (usually around 2 or 3 min) [38].

5. Existing Technology for Terpenes Extraction

There are different types of extraction processes for volatile compounds such as terpenes which are based on essential oils extraction. These vary according to the needs of producers who are always seeking energy efficiency. Table 3 presents the principle and parameters affecting the extraction process (temperature, pressure, solvent usage), as well as the advantages and disadvantages of each extraction method in terms of yield, quality of essential oils, and environmental impact. The extraction of essential oils from plant tissues is ensured through the phenomenon of diffusion, which is a time-consuming process. The entrainment of volatile compounds depends on several factors, such as the quantity of the plant's lipid fraction, the degree of solubility of the volatile compounds, the gradients of total pressure, and the temperature gradients during the extraction process [65]. The extraction phenomenon follows the principle of Fick's law (Equation (1)). Diffusivity in Fick's law in Equation (1) is an important property that indicates the rate of mass transfer and it is useful for equipment design [66]:

$$\text{Fick's Law}: F = -D\frac{dN}{dx} \tag{1}$$

where F (mol.m^{-2}.s^{-1}) or (kg.m^{-2}.s^{-1}) is the mass flux of the solute, N (mol/m^3) or (kg/m^3) is the concentration of the solute in the solid particle, D (m^2.s^{-1}) is known as the diffusivity or diffusion coefficient for the solute in the solvent, and x (m) is the distance in the direction of the transfer.

5.1. Steam Distillation

This process involves an external steam source that diffuses through the raw material placed in the vessel, carrying away the volatile molecules. When the raw material is fragile, it is loaded onto plates (also called false bottoms) to create multiple stages. The advantage here is to maximise the exchange surface between the plant material and the steam, and also to avoid compaction [67]. For whole plants like lavender or branches such as fir, the raw material is loaded directly into the vessel. Steam distillation is the most commonly used method because it avoids overheating and reduces distillation times. Steam distillation is a gentler technique recommended for fragile raw materials such as flowers or leaves. However, it is not suitable for powders (such as ground wood or bark) as the steam tends to agglomerate them [68].

5.2. Solvent Extractions

Concretes, Resinoids, and Absolutes Extractions using nonpolar volatile solvents (hexane, diethyl ether, benzene, petroleum ether, dichloromethane, etc.) allow for the extraction of lipophilic molecules from the raw material. This includes volatile and non-volatile terpenoids, waxes, and fats. Extractions can be performed with hot solvents (as in the case of Soxhlet extraction) or at room temperature during cold maceration. In the essential oil industry, specific products are derived from these solvent extractions, namely concretes, resinoids, and absolutes [69]. The raw material loaded into tanks is "depleted" with the nonpolar solvent (often hexane). The solvent, enriched with lipophilic molecules, is evaporated at low temperatures. The resulting aromatic solid residue is called a concrete (or crude resinoid in the case of dry raw material or resin gum) [69]. To remove heavy molecules such as waxes and fats that do not contribute to the fragrance composition of the product, the concrete is extracted with ethanol at low temperatures under vigorous agitation. The resulting extract, after wax filtration, is called an absolute. These extractions of concretes, resinoids, and absolutes are used for fragile or low-yield raw materials (rose, jasmine, benzoin) and produce olfactively distinct products compared to essential oils obtained by hydrodistillation. They are primarily used in perfumery.

5.3. New Extraction Methods

5.3.1. Microwave-Assisted Extraction

Solvent-Free Microwave Extraction (SFME) allows for the extraction of essential oils from plants without the addition of water or steam, except for dry materials where a limited amount of water is added. The principle relies on the water naturally present in the raw material as a carrier for the volatile molecules, and microwave heating vaporise this water into steam which is carrying the essential oil. This technique is five to ten times faster than traditional distillation and allows extraction at temperatures below 100 °C, avoiding the thermal degradation of certain molecules [70]. Although more energy-efficient, this process is still less commonly used industrially compared to hydrodistillation [68].

5.3.2. CO_2 Extraction or Supercritical Fluid Extraction

Supercritical CO_2 is a fluid with high diffusion and density properties that can be used to extract volatile molecules from plants. The supercritical fluid passes through the raw material and collects the compounds. During expansion, the CO_2 transitions into the gas phase, allowing for its separation from the extract [71]. This gentle extraction process preserves the integrity of the molecules, often resulting in extracts that closely resemble the fresh plant in terms of fragrance. CO_2 extraction is advantageous due to its abundance, low cost, and non-toxic nature. However, this method requires significant investment in equipment and consumes a considerable amount of energy to reach the supercritical state (pressure conditions above 74 bars and a temperature of 31 °C) [72].

5.3.3. Subcritical Water Extraction

This process utilises hot water at its subcritical point, between 100 °C (boiling point) and 374.1 °C (critical point), maintained in a liquid state under high pressure (between 1 and 221 bars). In the subcritical state, water possesses different properties, allowing for the solubilisation of low-polarity molecules. Thus, the extraction of low-polarity or nonpolar volatile molecules, such as components of essential oils, is facilitated by reducing the polarity of water without the addition of organic solvents. This technique, which relies solely on water as the green solvent, is nevertheless costly in terms of equipment and energy [73].

Table 3. Various extraction methods for volatile molecules as terpenes.

Extraction Methods	Principle	Advantages and Disadvantages	Reference
Steam distillation	The forest or agricultural residue is in direct contact with steam, which is then condensed. Recovery takes place in a separator, where the volatile molecules are dispersed in water.	Méthode très simple Very simple method No energy expenditure	[74]
Solvent extraction	The volatile molecules are separated from the solvent by evaporation of the solvent at high temperatures.	Efficient, slow and costly method Requires high temperatures (degradation of some constituents of volatile molecules)	[75]
Hydrodistillation	The residual material is submerged in water, which is then heated to boiling point. After passing through the cooler, the mixture is collected in an essencier.	Efficient, but slow method for 100 g (4 h) High water consumption	[76]
Supercritical fluid extraction	Extraction requires a supercritical fluid (CO_2 in the presence of an organic solvent).	Efficient, low-cost method No oxidative degradation of lipids	[68]
Ultrasonic extraction	Sound waves exert vibrations on plant cell walls, improving extraction.	Reduced extraction time	[63]
Microwave extraction	The residual material is heated from the inside out, increasing the water pressure inside the cells and causing the cells to burst and spill their contents into the outside environment.	Environmental efficiency Fast method Saves time, water and residual solvent	[70]
Hydrodistillation combined with microwaves	For 100 g of plant material, this method requires power (1200 watts) and duration 15 min	Good yield, fast (75 min), low cost	[69]

6. Conclusions

Limonene is the most abundant naturally occurring monoterpene in orange peel oil, which has been extensively studied as a precursor for high-value derivatives, including flavour and fragrance additives, solvents, feedstocks for fine chemicals, and active ingredients in medicine. Various microorganisms, such as *Cladosporium* sp., *Pseudomonas gladioli*, and *Penicillium strains*, have been employed for the biotransformation of refine limonene into compounds like α-terpineol and carvone, showcasing the potential for large-scale production of value-added derivatives. Similarly, α-pinene, a prominent antimicrobial monoterpene found in essential oils, has garnered attention for its applications in food, fragrance, pharmaceutical, and fine chemical industries. The current industrial production of α-pinene from natural sources is laborious and inefficient, highlighting the need for sustainable biotransformation technologies. Studies have identified biocatalysts such as *Chrysosporium pannorum*, *Candida tropicalis*, and *Hormonema* sp. UOFS Y-0067 for the oxidation and transformation of α-pinene into commercially valuable compounds like verbenol and α-terpineol. The main biotransformation study of limonene and α-pinene currently relies on refine or pure precursors, which have limitations such as solvent utilisation, high toxicity effect, low concentrations of products obtained. However, biotechnological approaches offer a promising solution by utilising microorganisms that can metabolite terpenes found in residual wood and citrus. These molecules are abundant in these feedstocks and could be biotransform with solid-state fermentation and high value industrial by-products could be achieved by improving the technology with strategy as solid-state fermentation and extrusion of the lignocellulosic biomass. Further research is needed to explore alternative sources of α-pinene, optimise fermentation conditions of limonene, and identify potential microbes for its biotransformation.

Author Contributions: A.N.: Methodology, writing. K.A.: Reviewing and editing. All authors have read and agreed to the published version of the manuscript.

Funding: This research received no external funding.

Institutional Review Board Statement: Not applicable.

Informed Consent Statement: Not applicable.

Data Availability Statement: The authors confirm that the data supporting the findings of this study are available within the article.

Conflicts of Interest: The authors declare no conflict of interest.

References

1. Astani, A.; Reichling, J.; Schnitzler, P. Comparative study on the antiviral activity of selected monoterpenes derived from essential oils. *Phytother. Res. Int. J. Devoted Pharmacol. Toxicol. Eval. Nat. Prod. Deriv.* **2010**, *24*, 673–679. [CrossRef] [PubMed]
2. Braga, A.; Guerreiro, C.; Belo, I. Generation of flavors and fragrances through biotransformation and de novo synthesis. *Food Bioprocess Technol.* **2018**, *11*, 2217–2228. [CrossRef]
3. Panakkal, E.J.; Kitiborwornkul, N.; Sriariyanun, M.; Ratanapoompinyo, J.; Yasurin, P.; Asavasanti, S.; Rodiahwati, W.; Tantayotai, P. Production of food flavouring agents by enzymatic reaction and microbial fermentation. *Appl. Sci. Eng. Prog.* **2021**, *14*, 297–312. [CrossRef]
4. Gounaris, Y. Biotechnology for the production of essential oils, flavours and volatile isolates. A review. *Flavour Fragr. J.* **2010**, *25*, 367–386. [CrossRef]
5. Vespermann, K.A.; Paulino, B.N.; Barcelos, M.C.; Pessôa, M.G.; Pastore, G.M.; Molina, G. Biotransformation of α-and β-pinene into flavor compounds. *Appl. Microbiol. Biotechnol.* **2017**, *101*, 1805–1817. [CrossRef]
6. Bicas, J.L.; Dionisio, A.P.; Pastore, G.M. Bio-oxidation of terpenes: An approach for the flavor industry. *Chem. Rev.* **2009**, *109*, 4518–4531. [CrossRef]
7. Molina, G.; Pessôa, M.G.; Bicas, J.L.; Fontanille, P.; Larroche, C.; Pastore, G.M. Optimization of limonene biotransformation for the production of bulk amounts of α-terpineol. *Bioresour. Technol.* **2019**, *294*, 122180. [CrossRef]
8. de Oliveira Felipe, L.; de Oliveira, A.M.; Bicas, J.L. Bioaromas–perspectives for sustainable development. *Trends Food Sci. Technol.* **2017**, *62*, 141–153. [CrossRef]
9. Pessôa, M.G.; Vespermann, K.A.; Paulino, B.N.; Barcelos, M.C.; Pastore, G.M.; Molina, G. Newly isolated microorganisms with potential application in biotechnology. *Biotechnol. Adv.* **2019**, *37*, 319–339. [CrossRef]

10. Poaty, B.; Lahlah, J.; Porqueres, F.; Bouafif, H. Composition, antimicrobial and antioxidant activities of seven essential oils from the North American boreal forest. *World J. Microbiol. Biotechnol.* **2015**, *31*, 907–919. [CrossRef]

11. Teixeira, B.; Marques, A.; Ramos, C.; Neng, N.R.; Nogueira, J.M.; Saraiva, J.A.; Nunes, M.L. Chemical composition and antibacterial and antioxidant properties of commercial essential oils. *Ind. Crops Prod.* **2013**, *43*, 587–595. [CrossRef]

12. Roy, D.; Bhowal, J. Bioconversion of Mandarin Orange Peels by *Aspergillus oryzae* and *Penicillium* sp. In *Advances in Bioprocess Engineering and Technology: Select Proceedings ICABET 2020*; Springer: Berlin/Heidelberg, Germany, 2021; pp. 13–20.

13. Francezon, N.; Stevanovic, T. Chemical composition of essential oil and hydrosol from Picea mariana bark residue. *BioResources* **2017**, *12*, 2635–2645. [CrossRef]

14. Abbas, F.; Ke, Y.; Yu, R.; Yue, Y.; Amanullah, S.; Jahangir, M.M.; Fan, Y. Volatile terpenoids: Multiple functions, biosynthesis, modulation and manipulation by genetic engineering. *Planta* **2017**, *246*, 803–816. [CrossRef] [PubMed]

15. Chen, F.; Tholl, D.; Bohlmann, J.; Pichersky, E. The family of terpene synthases in plants: A mid-size family of genes for specialized metabolism that is highly diversified throughout the kingdom. *Plant J.* **2011**, *66*, 212–229. [CrossRef] [PubMed]

16. Karunanithi, P.S.; Zerbe, P. Terpene synthases as metabolic gatekeepers in the evolution of plant terpenoid chemical diversity. *Front. Plant Sci.* **2019**, *10*, 1166. [CrossRef] [PubMed]

17. Hadj Saadoun, J.; Bertani, G.; Levante, A.; Vezzosi, F.; Ricci, A.; Bernini, V.; Lazzi, C. Fermentation of agri-food waste: A promising route for the production of aroma compounds. *Foods* **2021**, *10*, 707. [CrossRef]

18. Wenda, S.; Illner, S.; Mell, A.; Kragl, U. Industrial biotechnology—The future of green chemistry? *Green Chem.* **2011**, *13*, 3007–3047. [CrossRef]

19. Pimentel, M.R.; Molina, G.; Dionísio, A.P.; Maróstica Junior, M.R.; Pastore, G.M. The use of endophytes to obtain bioactive compounds and their application in biotransformation process. *Biotechnol. Res. Int.* **2011**, *2011*, 576286. [CrossRef]

20. Sharma, P.; Vishvakarma, R.; Gautam, K.; Vimal, A.; Gaur, V.K.; Farooqui, A.; Varjani, S.; Younis, K. Valorization of citrus peel waste for the sustainable production of value-added products. *Bioresour. Technol.* **2022**, *351*, 127064. [CrossRef]

21. Vieira, A.J.; Beserra, F.P.; Souza, M.; Totti, B.; Rozza, A. Limonene: Aroma of innovation in health and disease. *Chem. Biol. Interact.* **2018**, *283*, 97–106. [CrossRef]

22. Maróstica Júnior, M.R.; Pastore, G.M. Biotransformação de limoneno: Uma revisão das principais rotas metabólicas. *Química Nova* **2007**, *30*, 382–387. [CrossRef]

23. Negro, V.; Mancini, G.; Ruggeri, B.; Fino, D. Citrus waste as feedstock for bio-based products recovery: Review on limonene case study and energy valorization. *Bioresour. Technol.* **2016**, *214*, 806–815. [CrossRef]

24. Ciriminna, R.; Lomeli-Rodriguez, M.; Cara, P.D.; Lopez-Sanchez, J.A.; Pagliaro, M. Limonene: A versatile chemical of the bioeconomy. *Chem. Commun.* **2014**, *50*, 15288–15296. [CrossRef]

25. John, I.; Muthukumar, K.; Arunagiri, A. A review on the potential of citrus waste for D-Limonene, pectin, and bioethanol production. *Int. J. Green Energy* **2017**, *14*, 599–612. [CrossRef]

26. Mahato, N.; Sharma, K.; Sinha, M.; Cho, M.H. Citrus waste derived nutra-/pharmaceuticals for health benefits: Current trends and future perspectives. *J. Funct. Foods* **2018**, *40*, 307–316. [CrossRef]

27. Siddiqui, S.A.; Pahmeyer, M.J.; Assadpour, E.; Jafari, S.M. Extraction and purification of d-limonene from orange peel wastes: Recent advances. *Ind. Crops Prod.* **2022**, *177*, 114484. [CrossRef]

28. Jongedijk, E.; Cankar, K.; Buchhaupt, M.; Schrader, J.; Bouwmeester, H.; Beekwilder, J. Biotechnological production of limonene in microorganisms. *Appl. Microbiol. Biotechnol.* **2016**, *100*, 2927–2938. [CrossRef] [PubMed]

29. Badee, A.; Helmy, S.A.; Morsy, N.F. Utilisation of orange peel in the production of α-terpineol by *Penicillium digitatum* (NRRL 1202). *Food Chem.* **2011**, *126*, 849–854. [CrossRef]

30. Molina, G.; Pinheiro, D.M.; Pimentel, M.R.; dos Ssanros, R.; Pastore, G.M. Monoterpene bioconversion for the production of aroma compounds by fungi isolated from Brazilian fruits. *Food Sci. Biotechnol.* **2013**, *22*, 999–1006. [CrossRef]

31. Duetz, W.A.; Fjällman, A.H.; Ren, S.; Jourdat, C.; Witholt, B. Biotransformation of D-limonene to (+) trans-carveol by toluene-grown *Rhodococcus opacus* PWD4 cells. *Appl. Environ. Microbiol.* **2001**, *67*, 2829–2832. [CrossRef]

32. Toniazzo, G.; de Oliveira, D.; Dariva, C.; Oestreicher, E.G.; Antunes, O.A. Biotransformation of (−) β-pinene by Aspergillus niger ATCC 9642. In *Twenty-Sixth Symposium on Biotechnology for Fuels and Chemicals*; Springer: Berlin/Heidelberg, Germany, 2005.

33. Allenspach, M.; Steuer, C. α-Pinene: A never-ending story. *Phytochemistry* **2021**, *190*, 112857. [CrossRef] [PubMed]

34. Karimkhani, M.M.; Nasrollahzadeh, M.; Maham, M.; Jamshidi, A.; Kharazmi, M.S.; Dehnad, D.; Jafari, S.M. Extraction and purification of α-pinene; a comprehensive review. *Crit. Rev. Food Sci. Nutr.* **2022**, 1–26. [CrossRef] [PubMed]

35. Trytek, M.; Jędrzejewski, K.; Fiedurek, J. Bioconversion of α-pinene by a novel cold-adapted fungus *Chrysosporium pannorum*. *J. Ind. Microbiol. Biotechnol.* **2015**, *42*, 181–188. [CrossRef] [PubMed]

36. Chatterjee, T.; De, B.; Bhattacharyya, D. *Microbial Oxidation of-Pinene to (+)-a-Terpineol by Candida tropicalis*; NISCAIR-CSIR: New Delhi, India, 1999.

37. Rottava, I.; Cortina, P.F.; Zanella, C.A.; Cansian, R.L.; Toniazzo, G.; Treichel, H.; Antunes, O.A.; Oestreicher, E.G.; de Oliveira, D. Microbial oxidation of (-)-α-pinene to verbenol production by newly isolated strains. *Appl. Biochem. Biotechnol.* **2010**, *162*, 2221–2231. [CrossRef]

38. Konan, D.; Koffi, E.; Ndao, A.; Peterson, E.C.; Rodrigue, D.; Adjallé, K. An Overview of Extrusion as a Pretreatment Method of Lignocellulosic Biomass. *Energies* **2022**, *15*, 3002. [CrossRef]

39. Sadh, P.K.; Duhan, S.; Duhan, J.S. Agro-industrial wastes and their utilization using solid state fermentation: A review. *Bioresour. Bioprocess.* **2018**, *5*, 1. [CrossRef]
40. Yafetto, L. Application of solid-state fermentation by microbial biotechnology for bioprocessing of agro-industrial wastes from 1970 to 2020: A review and bibliometric analysis. *Heliyon* **2022**, *8*, e09173. [CrossRef]
41. Bier, M.C.J.; Medeiros, A.B.P.; Soccol, C.R. Biotransformation of limonene by an endophytic fungus using synthetic and orange residue-based media. *Fungal Biol.* **2017**, *121*, 137–144. [CrossRef]
42. Baser, K.H.C.; Buchbauer, G. *Handbook of Essential Oils: Science, Technology, and Applications*; CRC Press: Boca Raton, FL, USA, 2015.
43. Zhao, L.-Q.; Sun, Z.-H.; Zheng, P.; He, J.-Y. Biotransformation of isoeugenol to vanillin by *Bacillus fusiformis* CGMCC1347 with the addition of resin HD-8. *Process Biochem.* **2006**, *41*, 1673–1676. [CrossRef]
44. Hua, D.; Ma, C.; Lin, S.; Song, L.; Deng, Z.; Maomy, Z.; Zhang, Z.; Yu, B.; Xu, P. Biotransformation of isoeugenol to vanillin by a newly isolated Bacillus pumilus strain: Identification of major metabolites. *J. Biotechnol.* **2007**, *130*, 463–470. [CrossRef]
45. Gallage, N.J.; Møller, B.L. Vanillin–bioconversion and bioengineering of the most popular plant flavor and its de novo biosynthesis in the vanilla orchid. *Mol. Plant* **2015**, *8*, 40–57. [CrossRef] [PubMed]
46. Kimura, M.; Ito, M. Bioconversion of essential oil components of *Perilla frutescens* by *Saccharomyces cerevisiae*. *J. Nat. Med.* **2020**, *74*, 189–199. [CrossRef] [PubMed]
47. Helfrich, E.J.; Lin, G.-M.; Voigt, C.A.; Clardy, J. Bacterial terpene biosynthesis: Challenges and opportunities for pathway engineering. *Beilstein J. Org. Chem.* **2019**, *15*, 2889–2906. [CrossRef] [PubMed]
48. Marmulla, R.; Harder, J. Microbial monoterpene transformations—A review. *Front. Microbiol.* **2014**, *5*, 346. [CrossRef]
49. Yoo, S.; Day, D. Bacterial metabolism of α-and β-pinene and related monoterpenes by *Pseudomonas* sp. strain PIN. *Process Biochem.* **2002**, *37*, 739–745. [CrossRef]
50. Sugiura, M.; Ito, S.; Saito, Y.; Niwa, Y.; Koltunow, A.M.; Sugimoto, O.; Sakai, H. Molecular cloning and characterization of a linalool synthase from lemon myrtle. *Biosci. Biotechnol. Biochem.* **2011**, *75*, 1245–1248. [CrossRef]
51. Sales, A.; Felipe, L.d.O.; Bicas, J.L. Production, properties, and applications of α-terpineol. *Food Bioprocess Technol.* **2020**, *13*, 1261–1279. [CrossRef]
52. Agbor, V.B.; Cicek, N.; Sparling, R.; Berlin, A.; Levin, D.B. Biomass pretreatment: Fundamentals toward application. *Biotechnol. Adv.* **2011**, *29*, 675–685. [CrossRef]
53. Tischer, B.; Vendruscolo, R.G.; Wagner, R.; Menezes, C.R.; Barin, C.S.; Giacomelli, S.R.; Budel, J.M.; Barin, J.S. Effect of grinding method on the analysis of essential oil from *Baccharis articulata* (Lam.) Pers. *Chem. Pap.* **2017**, *71*, 753–761. [CrossRef]
54. Douville, J.; David, J.; Lemieux, K.M.; Gaudreau, L.; Ramotar, D. The *Saccharomyces cerevisiae* phosphatase activator RRD1 is required to modulate gene expression in response to rapamycin exposure. *Genetics* **2006**, *172*, 1369–1372. [CrossRef]
55. Reinprecht, L.; Pánek, M. Ultrasonic technique for evaluation of bio-defects in wood: Part 1–Influence of the position, extent and degree of internal artificial rots. *Int. Wood Prod. J.* **2012**, *3*, 107–115. [CrossRef]
56. Jung, C.G. Voies de traitements de déchets solides: Valorisation matière et énergie. *Bull. Sci. Inst. Natl. Conserv. Nat.* **2013**, 50–54.
57. Senneca, O.; Cerciello, F.; Russo, C.; Wütscher, A.; Muhler, M.; Apicella, B. Thermal treatment of lignin, cellulose and hemicellulose in nitrogen and carbon dioxide. *Fuel* **2020**, *271*, 117656. [CrossRef]
58. Norrrahim, M.N.F.; Ilyas, R.A.; Nurazzi, N.M.; Rani, M.S.A.; Atikah, M.S.N.; Shazleen, S.S. Chemical pretreatment of lignocellulosic biomass for the production of bioproducts: An overview. *Appl. Sci. Eng. Prog.* **2021**, *14*, 588–605. [CrossRef]
59. Eloutassi, N.; Louaste, B.; Boudine, L.; Remmal, A. Hydrolyse physico-chimique et biologique de la biomasse ligno-cellulosique pour la production de bio-éthanol de deuxième génération. *Nat. Technol.* **2014**, *10*, 10–14.
60. Eloutassi, N.; Louaste, B.; Boudine, L.; Remmal, A. Valorisation de la biomasse lignocellulosique pour la production de bioéthanol de deuxième génération. *J. Renew. Energ.* **2014**, *17*, 600–609.
61. Motte, J.-C. Digestion Anaérobie par voie Sèche de Résidus Lignocellulosiques: Etude Dynamique des Relations Entre Paramètres de Procédés, Caractéristiques du Substrat et Écosystème Microbien. Ph.D. Thesis, University of Montpellier, University of Montpellier, Montpellier, France, 2013.
62. Chaturvedi, V.; Verma, P. An overview of key pretreatment processes employed for bioconversion of lignocellulosic biomass into biofuels and value added products. *3 Biotech* **2013**, *3*, 415–431. [CrossRef]
63. Zheng, J.; Rehmann, L. Extrusion pretreatment of lignocellulosic biomass: A review. *Int. J. Mol. Sci.* **2014**, *15*, 18967–18984. [CrossRef]
64. Duque, A.; Manzanares, P.; Ballesteros, M. Extrusion as a pretreatment for lignocellulosic biomass: Fundamentals and applications. *Renew. Energy* **2017**, *114*, 1427–1441. [CrossRef]
65. Djerrari, A. Influence du Mode D'extraction et des Conditions de Conservation sur la Composition des Huiles Essentielles de Thym et de Basilic. PhD Thesis, Université des Sciences et Techniques du Languedoc, Montpellier, France, 1986.
66. Chan, C.-H.; Yusoff, R.; Ngoh, G.-C. Modeling and kinetics study of conventional and assisted batch solvent extraction. *Chem. Eng. Res. Des.* **2014**, *92*, 1169–1186. [CrossRef]
67. Cassel, E.; Vargas, R.; Martinez, N.; Lorenzo, D.; Dellacassa, E. Steam distillation modeling for essential oil extraction process. *Ind. Crops Prod.* **2009**, *29*, 171–176. [CrossRef]
68. Boukhatem, M.N.; Ferhat, A.; Kameli, A. Méthodes d'extraction et de distillation des huiles essentielles: Revue de littérature. *Une* **2019**, *3*, 1653–1659.

69. Chemat, F.; Vian, M.A.; Fabiano-Tixier, A.-S.; Nutrizio, M.; Jambrak, A.R.; Munekata, P.E.; Lorenzo, J.M.; Barba, F.J.; Binello, A.; Cravotto, G. A review of sustainable and intensified techniques for extraction of food and natural products. *Green Chem.* **2020**, *22*, 2325–2353. [CrossRef]
70. Chan, C.-H.; Yusoff, R.; Ngoh, G.-C.; Kung, F.W.-L. Microwave-assisted extractions of active ingredients from plants. *J. Chromatogr. A* **2011**, *1218*, 6213–6225. [CrossRef]
71. Fornari, T.; Vicente, G.; Vázquez, E.; García-Risco, M.R.; Reglero, G. Isolation of essential oil from different plants and herbs by supercritical fluid extraction. *J. Chromatogr. A* **2012**, *1250*, 34–48. [CrossRef] [PubMed]
72. Yousefi, M.; Rahimi-Nasrabadi, M.; Pourmortazavi, S.M.; Wysokowski, M.; Jesionowski, T.; Ehrlich, H.; Mirsadeghi, S. Supercritical fluid extraction of essential oils. *TrAC Trends Anal. Chem.* **2019**, *118*, 182–193. [CrossRef]
73. Yang, Y.; Kayan, B.; Bozer, N.; Pate, B.; Baker, C.; Gizir, A.M. Terpene degradation and extraction from basil and oregano leaves using subcritical water. *J. Chromatogr. A* **2007**, *1152*, 262–267. [CrossRef]
74. Vinatoru, M. An overview of the ultrasonically assisted extraction of bioactive principles from herbs. *Ultrason. Sonochemistry* **2001**, *8*, 303–313. [CrossRef]
75. Hu, X.; Zhou, Q. Comparisons of microwave-assisted extraction, simultaneous distillation-solvent extraction, Soxhlet extraction and ultrasound probe for polycyclic musks in sediments: Recovery, repeatability, matrix effects and bioavailability. *Chromatographia* **2011**, *74*, 489–495. [CrossRef]
76. Lucchesi, M.-E. Extraction Sans Solvant Assistée par Micro-ondes Conception et Application à L'extraction des Huiles Essentielles. Ph.D. Thesis, Faculté des Sciences et Technologies, Université de la Réunion, Réunion, France, July 2005.

Article

Extraction of Bioactive Compounds via Solid-State Fermentation Using *Aspergillus niger* GH1 and *Saccharomyces cerevisiae* from Pomegranate Peel

Ana L. Izábal-Carvajal [1], Leonardo Sepúlveda [1], Mónica L. Chávez-González [1], Cristian Torres-León [2], Cristóbal N. Aguilar [1] and Juan A. Ascacio-Valdés [1,*]

[1] Bioprocesses & Bioproducts Group, Food Research Department, School of Chemistry, Autonomous University of Coahuila, Saltillo 25280, Coahuila, Mexico; anaizabal@uadec.edu.mx (A.L.I.-C.); leonardo_sepulveda@uadec.edu.mx (L.S.); monicachavez@uadec.edu.mx (M.L.C.-G.); cristobal.aguilar@uadec.edu.mx (C.N.A.)

[2] Research Center and Ethnobiological Garden (CIJE-UAdeC), Autonomous University of Coahuila, Viesca 27480, Coahuila, Mexico; ctorresleon@uadec.edu.mx

* Correspondence: alberto_ascaciovaldes@uadec.edu.mx; Tel.: +52-84-4415-5752; Fax: +52-84-4416-9213

Abstract: This study investigated the recovery of polyphenolic compounds such as punicalagin, punicalin, and ellagic acid via solid-state fermentation (SSF)-assisted extraction from pomegranate peel (*Punica granatum* L.) using *Aspergillus niger* GH1 and *Saccharomhyces cerevisiae*. Food processing has contributed to the increase in agroindustrial wastes, which has become a global concern due to environmental protection. However, these wastes can be valorized via the extraction of high-value components such as bioactive compounds. Ellagitannins extracted during the bioprocesses were identified via the HPLC–MS technique and quantified via total polyphenols (hydrolyzable and condensed assays). Enzymatic activities were tested. HPLC–MS analysis showed a decrease in the levels of punicalagin, the formation of punicaline, and the accumulation of ellagic acid during fermentation kinetics. The present study compares two different bioprocesses in order to obtain, from agroindustrial wastes, high-added-value compounds using SSF-.

Keywords: *Aspergillus niger*; ellagitannins; phenolic compounds; punica granatum; *saccharomyces cerevisiae*; solid-state fermentation

Citation: Izábal-Carvajal, A.L.; Sepúlveda, L.; Chávez-González, M.L.; Torres-León, C.; Aguilar, C.N.; Ascacio-Valdés, J.A. Extraction of Bioactive Compounds via Solid-State Fermentation Using *Aspergillus niger* GH1 and *Saccharomyces cerevisiae* from Pomegranate Peel. *Waste* **2023**, *1*, 806–814. https://doi.org/10.3390/waste1030047

Academic Editors: Catherine N. Mulligan, Dimitris P. Makris and Vassilis Athanasiadis

Received: 14 July 2023
Revised: 30 August 2023
Accepted: 5 September 2023
Published: 8 September 2023

1. Introduction

Pomegranate (*Punica granatum* L.) is a plant and fruit grown mainly in ancient Egypt, Italy, and Greece, though recently, it has spread to Asia, North Africa, and Europe [1]. The most important producers are India, Iran, Turkey China, the United States of America, Argentina, and Brazil, among others. [2]. Pomegranate has been investigated for its high content of polyphenolic compounds such as ellagitannins [3–6]. The antioxidative activity of pomegranate has been attributed to bioactive compounds that exhibit diverse medicinal properties and health benefit effects. Nowadays, pomegranate is recognized as an important fruit with a high percentage of antioxidant activity [7]. It has been reported that pomegranate peel (PP) has higher concentrations of bioactive compounds and exhibits a higher antioxidant activity than the rest of the fruit [8]. PP represents approximately 50% of the total weight of the fruit and is an important source of bioactive compounds such as ellagitannins [9]—mainly punicalin, pedunculagin, and punicalagin [5]. In 2017, a world production of 3.8 million tons of pomegranate peel was recorded [2]. The peel of the fruit is rich in phenolics acid such as gallic, ellagic, and caffeic acids. It has been reported that PP has a higher concentration of phenolic compounds than pomegranate juice and allows for a higher antioxidant activity [10]. Hydrolyzable tannins could be ellagitannins and gallotannins. The first ones are esters of hexahydroxydiphenic acid (HHDP) bound to a polyol and, when it is exposed to strong acids or bases, ester bonds hydrolyze and HHDP reorganizes into a dilactone, producing ellagic acid [11,12]. Ellagitannins can be found in grapes, rambutan, strawberries, cranberries, blueberries, guavas, and raspberries, among others [13]

Food processing has increased the by-products or agroindustrial wastes, which are produced along the food supply chain in stages such as agricultural production, manufacturing, processing, and distribution [14]. Recently, the use of agroindustrial wastes has become a global concern due to environmental protection; following this, residues can be valorized by extracting high-value compounds which can be reused as functional and nutraceutical ingredients; for example: proteins, polysaccharides, fibers, flavoring compounds, and phytochemicals [15].

Bioactive compounds have been recovered via various techniques and are mainly based on solvent extraction, supercritical fluid extraction, subcritical water extraction, enzymes, ultrasounds, and microwaves. These techniques allow for the recovery of specific bioactive compounds [14]. The extraction methods exhibit some disadvantages, such as their use of toxic solvents that damage the environment, their cost, their being difficult to implement, and their use high temperatures that can damage the bioactive compounds. Recently, ellagitannins have been extracted using emerging techniques to reduce the use of polluting solvents and obtain better extraction yields; for example: pressurized liquid extraction, ultrasound assisted extraction, microwave assisted simultaneous distillation, and supercritical fluid extraction [16].

SSF is a bioprocess whereby micro-organisms are allowed to grow in an environment without free water [17]. It has been demonstrated that this technique allows for higher yields of extraction and better product characteristics than submerged fermentation. SSF has great potential to recover bioactive compounds such as ellagitannins [18] since micro-organisms such as fungi naturally produce enzymes that degrade the cell wall, generating hydrolysis [19] and the mobilization of compounds towards the extraction solvent [20]. This bioprocess is low-cost and easy to implement since it requires small equipment and exhibits an important reduction in operating costs since it does not require sterilization, aeration, or agitation steps [17]. Several important factors must be considered in the development of SSF, the selection of the micro-organism strain, and the solid support. Bagasse, peels, seeds, and straw, among other examples, can be considered agroindustrial wastes. In the case of micro-organisms, filamentous fungi have been widely used since the technique mimics their natural habitat so they can synthetize enzymes and some other metabolites [21]. Moreover, the use of yeast has been recommended, which are able to grow over a material with low water activity and some bacteria.

Several in vivo and in vitro studies have reported that pomegranate presents antioxidant, antiviral, antibacterial, anti-inflammatory, antimicrobial, anticancer, and antiproliferative activity [7].

For this reason, PP is proposed, an agroindustrial waste that can be used due to its high content of bioactive compounds with highly relevant biological properties. In the present study, the SSF technique was applied to evaluate the ellagitannins extraction from PP using two different micro-organisms: *Aspergillus niger* GH1 and *Saccharomyces cerevisiae*.

2. Materials and Methods

Raw material: PP was collected from Cuatrociénegas, Coahuila, México. The peel was dehydrated for 48 h at 60 °C and grounded to powder (30-mesh particle size in an industrial homogenizer: 5 L; model LP12 Series 600-182, CDMX, México), stored in black plastics bags, and protected from light at room temperature.

Micro-organism and culture medium: In this study, we used the strains *Aspergillus niger* GH1 (collection DIA-UADEC, Saltillo, Coahuila, Mexico) and *Saccharomyces cerevisiae* 227 (collection of Instituto Tecnológico de Durango, México). Previously, the spores of *A. niger* GH1 were conserved in a cryoprotective solution at −55 °C (skimmed milk/glycerol 9:1). The strain GH1 of *Aspergillus niger* has been deposited in the Micoteca of the University of Minho, number MUM:23.16. Spores were reactivated in potato dextrose agar

(PDA-Bioxon) at 30 °C for five days and then were recovered using 0.01% Tween 80, and spores/mL were counted in a Neubauer chamber.

SSF with *Aspergillus niger* GH1: The SSF was performed on a polypropylene flask with pomegranate peel powder (1.5 g) as support/substrate. The conditions were inoculum 1×10^6 spores/g, 80% of humidity, and culture medium of $NaNO_3$ (15.6 g/L), KH_2PO_4 (3.04 g/L), KCl (1.52 g/L), and $MgSO_4$ (1.52 g/L). The reactors were incubated at 30 °C for 48 h, taking data every 6 h. All tests were conducted in triplicate. These conditions corresponded to those previously reported by Sepúlveda et al. (2012) [22].

SSF with *Saccharomyces cerevisiae*: The SSF was carried out in a polypropylene flask where the pomegranate peel powder (1.5 g) as support was inoculated with the yeast strain. The SSF conditions were temperature 25 °C, 70% humidity, inoculum 2×10^6, pH 5, peptone 10 g/L, extract of yeast 10 g/L, and NaCl 230 g/L. The reactors were incubated at 25 °C for 48 h, taking data every 6 h. All tests were conducted in triplicate. These conditions were previously reported by Moccia et al. (2012) [23].

Recovery fermentation extracts: All the fermentation extracts were recovered and filtered via manual compression every 6 h. Filtered extracts were protected from light at refrigerated at -20 °C until analytical determinations (total polyphenols, HPLC-ESI-MS analysis, and enzymatic activities).

Determination of total polyphenols: In both bioprocesses, condensed polyphenols present in PP fermentation extracts were determined via the HCl-butanol technique and ferric reagent [24]. Hydrolyzable polyphenols were determined via the Folin–Ciocalteu method [25]. The tests were conducted in triplicate, and the results were expressed as mg/g of filtered material. Total polyphenols are the result of adding the content of condensed and hydrolyzable polyphenols.

Analytical RP-HPLC–ESI-MS: Ellagitannins were quantified via HPLC–MS. The analysis via reverse phase-high performance liquid chromatography were performed on a Varian HPLC system with an autosampler (Varian ProStar 410, Palo Alto, CA, USA), a ternary pump (Varian ProStar 230I, Palo Alto, CA, USA), and a PDA detector (Varian ProStar 330, Palo Alto, CA, USA). All the samples (5 µL) were injected onto a Denali C18 column (150 mm × 2.1 mm, 3 µm, Grace, Palo Alto, CA, USA). The oven temperature was 30 °C. The eluents were formic acid (0.2%, v/v; solvent A) and acetonitrile (solvent B). Data were collected and processed using MS Workstation software (V 6.9). First, samples were analyzed in full-scan mode acquired in the m/z range of 50–2000 [26]. All the samples were filtered with 0.45 µm nylon membrane.

Enzymatic activities: The fermented extracts were used to determine the enzymatic activities reported as associated with the biotransformation of ellagitannins [27–32]. Spectrophotometric and chromatographic techniques determined the identification of these activities.

Ellagitannase [27]: TPP was used as substrate. We used a mixture of 1 mL of 1 mg/mL of TPP and 0.05 mL of enzymatic extract in citrate buffer 0.05 M at pH 5. It was incubated at 50 °C for 10 min. To stop the reaction, 0.02 mL of HCl 1.5 M was added. It was centrifuged at 6000 rpm for 30 min. The precipitated was suspended in 1.5 mL of ethanol and ultrasonicated for 30 min. The sample was filtered with a 0.45 µm nylon membrane and analyzed via HPLC. One enzyme activity unit was defined as the amount of enzyme that released 1 µmol of ellagic acid per min under the assay conditions.

Polyphenoloxidase [28]: A substrate solution was prepared with pyrocatechol 0.1 M in 0.05 M citrate buffer, pH 5, enzyme preparation, and 0.05 M citrate buffer, pH 5, and they were pre-incubated at 30 °C for 10 min before the assay. Then, 1.7 mL of buffer citrate and 0.5 mL of enzyme preparation were added into a test tube and incubated at 30 °C for 10 min. To stop the reaction, the test tube was boiled in a water bath for 2 min. The oxidized pyricatechol absorbance was read at 420 nm. The differences of absorbances were calculated via $A_{420} = A_{test} - A_{control}$. One unit of enzyme activity was defined as an increase in absorbance by 0.001 min^{-1}.

Cellulase [29]: CM cellulose (0.2 mL) and enzyme solution (0.05 mL) were incubated at 30 °C for 10 min. To stop the reaction, 2 mL of 0.3 M tricholoroacetic acid (TCA) was used. CM cellulose was calculated via the DNS method. The results were expressed as glucose concentration. One enzyme activity unit was defined as the amount of enzyme that released 1 μmol of glucose per min under the assay conditions.

β—glucosidase [30]: A standard reaction mixture was used, with 0.1 mL of 9 mM r-nitrophenol β-D-glucopyranoside (pNPG), 0.8 mL of 200 mM sodium acetate buffer (pH 4.6), and 0.1 mL of enzyme solution. The mixture was incubated for 15 min at 50 °C. The reaction was stopped with 1 mL of 0.1 M sodium carbonate. The released p-nitrophenol was read at 400 nm. One unit of enzyme was defined as the amount of enzyme that releases 1 μmol p-nitrophenol per min at pH 4.6 at 50 °C under the assay conditions.

Tannase [31]: Methylgalate was used as substrate, a reaction mixture (0.5 mL) was made of enzymatic extract and 0.01 M suspension of metylgalate 0.5 M with a citrate buffer at pH 5. The mixture was incubated at 30 °C for 10 min. The reaction was stopped by adding 0.2 mL of 0.5 N KOH. One enzyme activity unit was defined as the amount of enzyme that released 1 μmol of galic acid from methylgalate per min under the assay conditions.

Xylanase [32]: Xylan was used as a substrate. A total of 0.5 mL of enzyme solution and 0.5 mL of 0.5% (w/v) suspension of xylan with 0.05 M citrate buffer, pH 5.4, was used. The reaction mixture was incubated at 50 °C for 10 min. The reaction was stopped by adding 2 mL of 0.3 N TCA. One enzyme activity unit was defined as the amount of enzyme that released 1 μmol xylose from xylan per min under the assay conditions.

3. Results

3.1. SSF

A SSF kinetic was performed for 48 h, measuring fractions every 6 h, in order to obtain the highest concentration of polyphenols in SSF with A. niger GH1 and S. cerevisiae, using PP as the only carbon source. For comparative purposes, each fermented material was subjected to the same extraction procedure. The results show the polyphenolic content of the plant after fermentation. The maximum quantity of polyphenolic compounds was accumulated at 30 h with a value of 188.3 ± 9.5 mg/g of dry plant using *Aspergillus niger* GH1, and for *Saccharomyces cerevisiae*, the maximum value was 105.07 ± 8.4 mg/g of dry plant at 42 h of fermentation. The differences between the kinetic study with the fungus were statistically significant ($p < 0.05$).

3.2. RP-HPLC-ESI-MS Analysis

The recovered extracts were characterized via HPLC–MS. Table 1 reports the polyphenolic compounds found on the extracts of PP before and after (18 h) SSF with the *A. niger* GH1 and the *Saccharomyces cerevisiae*. Punicalagin is present in the unfermented material and on the yeast fermentation.

Table 1. Identification of ellagitannins-related compounds in the extraction of unfermented and fermented PP.

Retention Time (min)	[M-H]$^-$	Unfermented Material	Fugal Fermentation (18 h)	Yeast Fermentation (18 h)
3.350	350.8	1-Caffeoylquinic acid	1-Caffeoylquinic acid	1-Caffeoylquinic acid
4.554	352.9	3-Caffeoylquinic acid	3-Caffeoylquinic acid	3-Caffeoylquinic acid
5.249	480.9	-	HHDP-hexoside	-
7.179	330.8	Gallic acid 4-O-glucoside	Gallic acid 4-O-glucoside	Gallic acid 4-O-glucoside
9.696	780.8	Punicalin α	Punicalin α	Punicalin α
10.760	780.8	Punicalin β	Punicalin β	Punicalin β
12.952	782.7	Pedunculagin I	Pedunculagin I	Pedunculagin I

Table 1. *Cont.*

Retention Time (min)	[M-H]⁻	Unfermented Material	Fugal Fermentation (18 h)	Yeast Fermentation (18 h)
15.487	1082.6	Punicalagin	-	Punicalagin
20.204	782.8	Terflavin B	-	-
21.319	632.9	Galloyl-HHDP-hexoside	Galloyl-HHDP-hexoside	Galloyl-HHDP-hexoside
21.953	950.5	Granatin B	Granatin B	Granatin B
28.193	300.9	Ellagic acid	Ellagic acid	Ellagic acid

Figure 1 shows the results obtained via HPLC of the main quantified components during SSF with *Aspergillus niger* GH1 and *Saccharomyces cerevisiae* at 0- and 24-h using PP as support. Data show high initial levels of punicalagin and an accumulation of punicalin and ellagic acid. In Figure 2, the chromatograms of fungus and yeast fermentations can be observed, where punicalagin only appears during yeast fermentation, and ellagic acid appears in both bioprocesses.

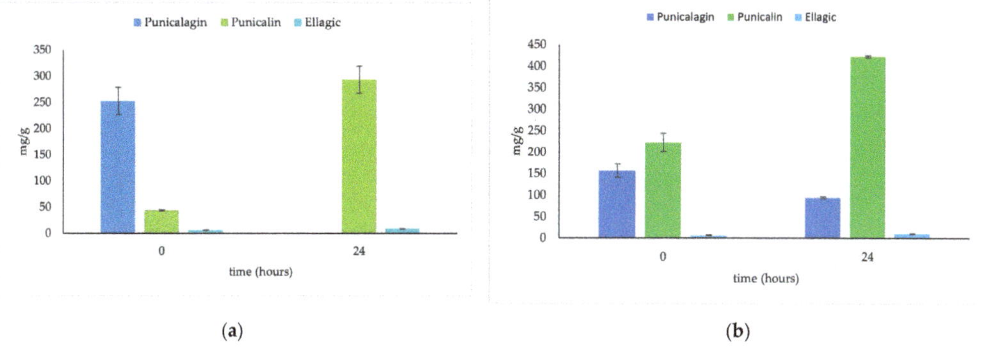

(a)

(b)

Figure 1. Quantitative analysis of the main ellagitannins identified in the fermented/unfermented PP extracts: (**a**) 0 and 24 h of SSF with *Aspergillus niger* GH1; (**b**) 0 and 24 h with *Saccharomyces cerevisiae*.

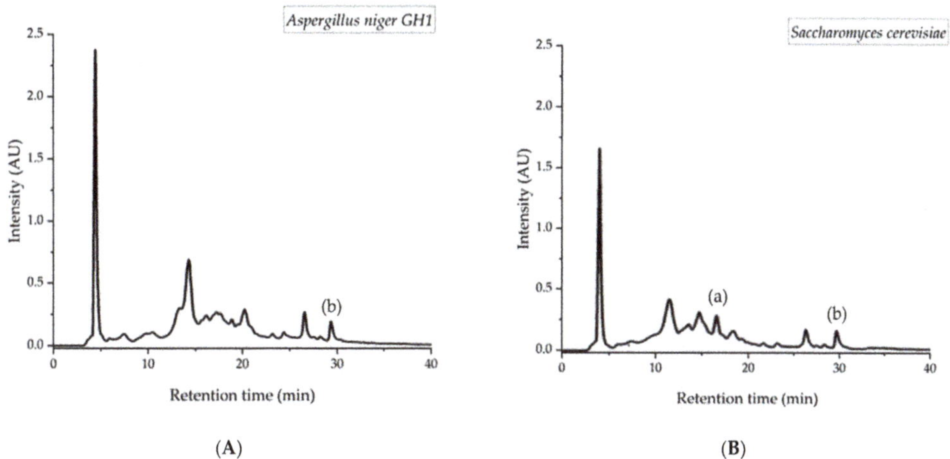

(**A**)

(**B**)

Figure 2. Chromatograms of the main ellagitannins extracted from PP in SSF. Compound (**A**): punicalagine; compound (**B**): ellagic acid; Bioprocess (a): yeast fermentation; Bioprocess (b): chromatograms of fungus.

3.3. Enzymatic Activities

The ellagitannins have important biological potential, so it is important to establish the enzymes involved during the biodegradation of these compounds to obtain some other molecules. Table 2 reports several enzymatic activities of assayed ellagitannase [27], polyphenoloxidase [28], cellulase [29], β—glucosidase [30], tannase [31], and xylanase [32]. The results suggest that PP ellagitannins are hydrolyzed by enzymes.

Table 2. Enzymatic activities evaluated.

	Fungal Fermentation		Yeast Fermentation	
Enzyme	**Max. Recorder Activity (U/L)**	**Time (Hours)**	**Max. Recorder Activity (U/L)**	**Time (Hours)**
β-Glucosidase	869.68 ± 60.64	12	1051.29 ± 281.79	36
Polyphenoloxidase	3435.42 ± 212.17	36	1576.60 ± 86.30	24
Cellulase	5051.62 ± 301.43	42	ND	-
Tannase	ND	-	ND	-
Xylanase	ND	-	ND	-
Ellagitannase	1184.12 ± 100.34	18	813.96 ± 140.37	24

ND: not detected.

4. Discussion

It has been reported that PP has numerous biological effects. It is an agroindustrial waste that can be used due its high content of bioactive compounds. SSF is an eco-friendly and low-cost technology. PP is a suitable support with which to perform an SSF, and the micro-organism was able to grow and invade the material.

SSF have been reported for the extraction of bioactive compounds, such as polyphenols, due to the ability of some micro-organisms to produce enzymes that allow for the bioavailability of compounds [11] and the mobilization of compounds towards the extraction solvent [20]. In this work, PP was used as a carbon and energy source for solid-state fermentation with fungus and yeast. The determination of total polyphenolic content involves the addition of total hydrolyzable polyphenols and total condensed polyphenols because the Folin–Ciocalteu reagent reacts mainly with the galloyl or HHDP groups, while the ferric reagent and hydrolysis with HCl-butanol act on compounds based on flavan-3-ol. The results show that the maximum accumulation time of polyphenolic compounds for *A. niger* GH1 was 188.3 mg/g, and for *S. cerevisiae*, it was 105.07 mg/g, indicating that SSF promotes the extraction of bioactive compounds from PP. There is a statistically significant difference between both micro-organisms, probably because they adapted differently to the support (PP) during fermentation.

Meanwhile, in order to learn the effect of SSF on the ellagitannins present in PP under the two evaluated conditions, the extracts were subjected to qualitative analysis via LC–MS. As expected, punicalagin is present in the unfermented material as well as in the material derived from *S. cerevisiae*. Punicalin, ellagic acid, and granatin B were found in all extracts. Moreover, some others ellagitannins were found on the evaluated material that corresponded to those reported as components of pomegranate ellagitannins generated via the hydrolysis of molecules with high weight [23]. PP is rich in phenolic acids such as gallic, ellagic, and caffeic acids. Ref. [33] reported that the most relevant bioactive compounds of pomegranate wine lees were hydrolyzable tannins: gallic acid, total punicalins, punicalagin, ellagic acid hexoside, and ellagic acid. Moreover, anthocyanins, flavonoids, and phenolic acids were detected. These results can be compared with those reported by [34], who studied the total polyphenolic content of pomegranate peel, where phenolic acids were the major phenolic compound, followed by flavonoids and tannins. Moreover, pomegranate peel has demonstrated a promising potential for biological activities, and it could be a source of functional food and nutraceuticals. Ref. [35] identified the phenolic compounds on ethanolic extracts of pomegranate peels via mass spectrometry (mainly punicalagin). The report shows that the punicalagin contained in the pomegranate peel improves the antimicrobial activity of the fruit.

As expected, the results suggest that PP wastes contain high levels of punicalagin. During the fermentation time with *A. niger* GH1, the punicalagin level drops from 0 to 24 h, though the levels of punicalin, an ellagic acid, increase. It has been reported that punicalagin leads to punicalin and the hexahydroxydiphenic acid group (HHDP), which, through spontaneous lactonization, converts to ellagic acid. On the other hand, for the *S. cerevisiae* fermented material, punicalagin is slowly reduced during the fermentation period, and an accumulation of punicalin and ellagic acid can be seen; this can be observed in Figure 2, where punicalagin can be detected under 18 h of bioprocessing with the yeast, whereas for the fungus, it is not detected. Ref. [23] reported that the pomegranate husk fermented with *Aspergillus* decreased the punicalagin levels in comparison to unfermented material. Moreover, with *saccharomyces*, the punicalagin was drastically reduced. Punicalagin is the main compound on PP and represents the hydrolase activity of the micro-organism as a pathway to accumulating ellagic acid.

In summary, punicalagin represents the main compound in the material, and the hydrolase activity of the micro-organism suggests a pathway for the formation of ellagic acid. The SSF with *A. niger* GH1 and *S. cerevisiae* promotes the hydrolysis of the ellagitannins to some other compounds, such as punicalin and ellagic acid, obtaining higher yields of extraction. The ellagic acid and punicalin are derived from the hydrolysis (partial or complete) of punicalagin [5].

Additionally, the biodegradation of pomegranate ellagitannins was evaluated. The maximum concentration of ellagic acid released was reached at 18 h of culture with a maximum value of 10.06 mg g^{-1}. Ref. [32] reported that xylanase and cellulase activities were responsible for the degradation of ellagitannins and the accumulation of ellagic acid. However, [27] mentioned that xylanase and cellulase were detected at 24 and 6 h, respectively, and reported that these enzymes are not directly related to the accumulation of ellagic acid. Xylanase and tannase activities were detected in neither the fungus nor the yeast fermentation. β-Glucosidase activity was detected at 18 h of culture with the fungus, and it is directly related to the accumulation of ellagic acid. As [30] mentioned, this enzyme is associated with ellagitannins degradation. Ref. [36] reported the production of ellagic acid by cultivating fungi in the solid state, using polyurethane foam (PUF) as a support, and an aqueous extract obtained from pomegranate peel as the only source of carbon and energy. The authors determined that *Aspergillus niger* GH1 consumed ellagitannins during the first 36 h of culture, with the maximum concentration of ellagic acid being reached at 48 h. The authors attributed the biodegradation of ellagitannins to a new tannase, which is probably different from acyl hydrolase.

Otherwise, ellagitannase activity was detected at 18 h of culture with the fungus and 24 h with the yeast. Ascacio-Valdés et al. (2014) reported that this enzyme is directly related with the biodegradation of ellagitannins present in pomegranate peel, such as punicalagin and punicalin. They proposed the accumulation of ellagic acid because this enzyme degraded the ester bonds between glycosides and the HHDP group.

5. Conclusions

Many agroindustrial wastes have been used as substrates for SSF, such as sugarcane bagasse, corn cobs, coconut husks, grape wastes, candelilla stalks, and rambutan, among others. PP is also an example of agroindustrial waste, and it is an important source of bioactive compounds such as ellagitannins (mainly punicalagin), which present high antioxidant activity. SSF has been reported as a technique by which to extract bioactive compounds such as polyphenols. In this study, we are reporting on the extraction of bioactive compounds via SSF with *A. niger* GH1 and *S. cerevisiae* using pomegranate peel as a substrate. These bioprocesses allow for the biodegradation of ellagitannins (mainly punicalagin) of high molecular weight, accumulating bioactive compounds such as punicalin and ellagic acid. The time of maximum accumulation of polyphenolic compounds with the fungus at 24 h was higher than that with the yeast, and this can be achieved due to *Aspergillus niger* GH1 being a filamentous fungus that has the ability of grow on the used support. Enzymatic activities, particularly those of ellagitannase, are responsible for ellagitannins degradation in both bioprocesses. The extraction of bioactive compounds has a relevant role in biological activities, and the components in question can be used as nutraceutical components in the food industry.

Author Contributions: Conceptualization, J.A.A.-V.; methodology, A.L.I.-C., L.S. and M.L.C.-G.; formal analysis, C.N.A. and C.T.-L.; investigation, J.A.A.-V. and C.N.A.; writing—original draft preparation, A.L.I.-C.; writing—review and editing, A.L.I.-C., J.A.A.-V. and L.S.; supervision, J.A.A.-V. and C.N.A. All authors have read and agreed to the published version of the manuscript.

Funding: This research received no external funding.

Institutional Review Board Statement: Not applicable.

Informed Consent Statement: Not applicable.

Data Availability Statement: Data sharing is not applicable to this article as no datasets were generated or analyzed during the current study.

Acknowledgments: Ana L. Izábal-Carvajal thanks CONAHCYT Mexico for scholarship support. The authors thank the School of Chemistry of the Autonomous University of Coahuila.

Conflicts of Interest: The authors declare no conflict of interest.

References

1. Evreinoff, V.A. Contribution à l'étude du Grenadier. *J. Agric. Trop Bot. Appl.* **1957**, *4*, 124–138. [CrossRef]
2. Kahramanoglu, I. Trends in Pomegranate Sector: Production, Postharvest Handling and Marketing. *Int. J. Agric. For. Life Sci.* **2019**, *3*, 239–246.
3. Bonzanini, F.; Bruni, R.; Palla, G.; Serlataite, N.; Caligiani, A. Identification and distribution of lignans in Punica granatum L. fruit endocarp, pulp, seeds, wood knots and commercial juices by GC-MS. *Food Chem.* **2009**, *117*, 745–749. [CrossRef]
4. García-Villalba, R.; Espín, J.C.; Aaby, K.; Alasalvar, C.; Heinonen, M.; Jacobs, G.; Voorspoels, S.; Koivumäki, T.; Kroon, P.A.; Pelvan, E.; et al. Validated Method for the Characterization and Quantification of Extractable and Nonextractable Ellagitannins after Acid Hydrolysis in Pomegranate Fruits, Juices, and Extracts. *J. Agric. Food Chem.* **2015**, *63*, 6555–6566. [CrossRef]
5. Seeram, N.; Lee, R.; Hardy, M.; Heber, D. Rapid large scale purification of ellagitannins from pomegranate husk, a by-product of the commercial juice industry. *Sep. Purif. Technol.* **2005**, *41*, 49–55. [CrossRef]
6. Pareek, S.; Valero, D.; Serrano, M. Postharvest biology and technology of pomegranate. *J. Sci. Food Agric.* **2015**, *95*, 2360–2379. [CrossRef]
7. Kalaycıoğlu, Z.; Erim, F.B. Total phenolic contents, antioxidant activities, and bioactive ingredients of juices from pomegranate cultivars worldwide. *Food Chem.* **2017**, *221*, 496–507. [CrossRef] [PubMed]
8. Akhtar, S.; Ismail, T.; Fraternale, D.; Sestili, P. Pomegranate peel and peel extracts: Chemistry and food features. *Food Chem.* **2015**, *174*, 417–425. [CrossRef]
9. Sreekumar, S.; Sithul, H.; Muraleedharan, P.; Azeez, J.M.; Sreeharshan, S. Pomegranate fruit as a rich source of biologically active compounds. *BioMed Res. Int.* **2014**, *2014*, 1–12. [CrossRef]
10. Kandylis, P.; Kokkinomagoulos, E. Food applications and potential health benefits of pomegranate and its derivatives. *Foods* **2020**, *9*, 122. [CrossRef]
11. Bala, I.; Bhardwaj, V.; Hariharan, S.; Kumar, M.N.V.R. Analytical methods for assay of ellagic acid and its solubility studies. *J. Pharm. Biomed. Anal.* **2006**, *40*, 206–210. [CrossRef] [PubMed]
12. Aguilera-Carbo, A.; Augur, C.; Prado-Barragan, L.A.; Favela-Torres, E.; Aguilar, C.N. Microbial production of ellagic acid and biodegradation of ellagitannins. *Appl. Microbiol. Biotechnol.* **2008**, *78*, 189–199. [CrossRef] [PubMed]
13. Banc, R.; Rusu, M.E.; Filip, L.; Popa, D.S. The Impact of Ellagitannins and Their Metabolites through Gut Microbiome on the Gut Health and Brain Wellness within the Gut–Brain Axis. *Foods* **2023**, *12*, 270. [CrossRef] [PubMed]
14. Kumar, K.; Yadav, A.N.; Kumar, V.; Vyas, P.; Dhaliwal, H.S. Food waste: A potential bioresource for extraction of nutraceuticals and bioactive compounds. *Bioresour. Bioprocess.* **2017**, *4*, 1–14. [CrossRef]
15. Baiano, A. Recovery of biomolecules from food wastes—A review. *Molecules* **2014**, *19*, 14821–14842. [CrossRef] [PubMed]
16. Domínguez-Rodríguez, G.; Marina, M.L.; Plaza, M. Strategies for the extraction and analysis of non-extractable polyphenols from plants. *J. Chromatogr. A* **2017**, *1514*, 1–15. [CrossRef] [PubMed]
17. Soccol, C.R.; da Costa, E.S.F.; Letti, L.A.J.; Karp, S.G.; Woiciechowski, A.L.; Vandenberghe, L.P.d.S. Recent developments and innovations in solid state fermentation. *Biotechnol. Res. Innov.* **2017**, *1*, 52–71. [CrossRef]

18. Martins, S.; Mussatto, S.I.; Martínez-Avila, G.; Montañez-Saenz, J.; Aguilar, C.N.; Teixeira, J.A. Bioactive phenolic compounds: Production and extraction by solid-state fermentation. A review. *Biotechnol. Adv.* **2011**, *29*, 365–373. [CrossRef]
19. Jamal, P.; Idris, Z.M.; Alam, M.Z. Effects of physicochemical parameters on the production of phenolic acids from palm oil mill effluent under liquid-state fermentation by Aspergillus niger IBS-103ZA. *Food Chem.* **2011**, *124*, 1595–1602. [CrossRef]
20. Torres-León, C.; Ramírez-Guzmán, N.; Ascacio-Valdés, J.; Serna-Cock, L.; Correia, M.T.d.S.; Contreras-Esquivel, J.C.; Aguilar, C.N. Solid-state fermentation with Aspergillus niger to enhance the phenolic contents and antioxidative activity of Mexican mango seed: A promising source of natural antioxidants. *LWT* **2019**, *112*, 108236. [CrossRef]
21. Farinas, C.S. Developments in solid-state fermentation for the production of biomass-degrading enzymes for the bioenergy sector. *Renew. Sustain. Energy Rev.* **2015**, *52*, 179–188. [CrossRef]
22. Sepúlveda, L.; Aguilera-Carbó, A.; Ascacio-Valdés, J.A.; Rodríguez-Herrera, R.; Martínez-Hernández, J.L.; Aguilar, C.N. Optimization of ellagic acid accumulation by Aspergillus niger GH1 in solid state culture using pomegranate shell powder as a support. *Process Biochem.* **2012**, *47*, 2199–2203. [CrossRef]
23. Moccia, F.; Flores-Gallegos, A.C.; Chávez-González, M.L.; Sepúlveda, L.; Marzorati, S.; Verotta, L.; Panzella, L.; Ascacio-Valdes, J.A.; Aguilar, C.N.; Napolitano, A. Ellagic acid recovery by solid state fermentation of pomegranate wastes by aspergillus Niger and saccharomyces cerevisiae: A comparison. *Molecules* **2019**, *24*, 3689. [CrossRef] [PubMed]
24. Swain, B.T.; Hillis, W.E. The phenolic constituents of pronus domestica. *J. Sci. Food Agric.* **1945**, *10*, 63–68. [CrossRef]
25. Makkar, H. *Measurement of Total Phenolics and Tannins Using Folin-Ciocalteu Method*; Springer: Dordrecht, The Netherlands; Berlin/Heidelberg, Germany, 2003.
26. Ascacio-Valdés, J.A.; Aguilera-Carbó, A.F.; Buenrostro, J.J.; Prado-Barragán, A.; Rodríguez-Herrera, R.; Aguilar, C.N. The complete biodegradation pathway of ellagitannins by Aspergillus niger in solid-state fermentation. *J. Basic Microbiol.* **2016**, *56*, 329–336. [CrossRef] [PubMed]
27. Ascacio-Valdés, J.A.; Buenrostro, J.J.; De la Cruz, R.; Sepúlveda, L.; Aguilera, A.F.; Prado, A.; Contreras, J.C.; Rodríguez, R.; Aguilar, C.N. Fungal biodegradation of pomegranate ellagitannins. *J. Basic Microbiol.* **2014**, *54*, 28–34. [CrossRef]
28. Shi, B.; He, Q.; Yao, K.; Huang, W.; Li, Q. Production of ellagic acid from degradation of valonea tannins by Aspergillus niger and Candida utilis. *J. Chem. Technol. Biotechnol.* **2005**, *80*, 1154–1159. [CrossRef]
29. Huang, W.; Niu, H.; Li, Z.; Lin, W.; Gong, G.; Wang, W. Effect of ellagitannin acyl hydrolase, xylanase and cellulase on ellagic acid production from cups extract of valonia acorns. *Process. Biochem.* **2007**, *42*, 1291–1295. [CrossRef]
30. Vattem, D.A.; Shetty, K. Ellagic acid production and phenolic antioxidant activity in cranberry pomace (*Vaccinium macrocarpon*) mediated by Lentinus edodes using a solid-state system. *Process. Biochem.* **2003**, *39*, 367–379. [CrossRef]
31. Sharma, S.; Bhat, T.K.; Dawra, R.K. A spectrophotometric method for assay of tannase using rhodanine. *Anal. Biochem.* **2000**, *279*, 85–89. [CrossRef]
32. Huang, W.; Niu, H.; Li, Z.; He, Y.; Gong, W.; Gong, G. Optimization of ellagic acid production from ellagitannins by co-culture and correlation between its yield and activities of relevant enzymes. *Bioresour. Technol.* **2008**, *99*, 769–775. [CrossRef] [PubMed]
33. Mena, P.; Ascacio-Valdés, J.A.; Gironés-Vilaplana, A.; Del Rio, D.; Moreno, D.A.; García-Viguera, C. Assessment of pomegranate wine lees as a valuable source for the recovery of (poly)phenolic compounds. *Food Chem.* **2014**, *145*, 327–334. [CrossRef] [PubMed]
34. Ambigaipalan, P.; De Camargo, A.C.; Shahidi, F. Phenolic Compounds of Pomegranate Byproducts (Outer Skin, Mesocarp, Divider Membrane) and Their Antioxidant Activities. *J. Agric. Food Chem.* **2016**, *64*, 6584–6604. [CrossRef]
35. Gosset-Erard, C.; Zhao, M.; Lordel-Madeleine, S.; Ennahar, S. Identification of punicalagin as the bioactive compound behind the antimicrobial activity of pomegranate (*Punica granatum* L.) peels. *Food Chem.* **2021**, *352*, 129396. [CrossRef] [PubMed]
36. Aguilera-Carbo, A.F.; Augur, C.; Prado-Barragan, L.A.; Aguilar, C.N.; Favela-Torres, E. Extraction and analysis of ellagic acid from novel complex sources. *Chem. Pap.* **2008**, *62*, 440–444. [CrossRef]

 *waste*

Article

Mechanical Properties of a Bio-Composite Produced from Two Biomaterials: Polylactic Acid and Brown Eggshell Waste Fillers

Duncan Cree [1,*], Stephen Owuamanam [1] and Majid Soleimani [2]

[1] Department of Mechanical Engineering, University of Saskatchewan, Saskatoon, SK S7N 5A9, Canada; stephen.owuamanam@usask.ca

[2] Department of Chemical and Biological Engineering, University of Saskatchewan, Saskatoon, SK S7N 5A9, Canada; mas233@mail.usask.ca

* Correspondence: duncan.cree@usask.ca; Tel.: +1-306-966-3244

Abstract: An option to reduce the exploitation and depletion of natural mineral resources is to repurpose current waste materials. Fillers are often added to polymers to improve the properties and lower the overall cost of the final product. Very few studies have assessed the use of waste brown eggshell powder (BESP) as filler in polylactic acid (PLA). The addition of mineral fillers in a polymer matrix can play an important role in the performance of a composite under load. Therefore, tailoring the amount of filler content can be a deciding factor as to which filler amount is best. The goal of this study was to investigate the effect of brown eggshells compared to conventional limestone (LS) powder on the mechanical properties of PLA composites. One-way analysis of variance (ANOVA) was used to carry out the statistical analysis on the average values of each composite mechanical property tested. Scanning electron microscopy (SEM) was used to view if there were any differences in the fractured surfaces. Overall, the LS performed marginally better than the BESP fillers. The highest ultimate tensile and ultimate flexural strengths for eggshell composites containing 32 μm fillers had values of 48 MPa (5–10 wt.% BESP) and 67 MPa (10 wt.%. BESP), respectively. Both the tensile and flexural modulus improved with filler contents and were highest at 20 wt.% with values of 4.5 GPa and 3.4 GPa, respectively. The Charpy impact strength decreased for all filler amounts. SEM micrographs identified changes in the fractured surfaces due to the additions of the filler materials. The ANOVA results showed statistically significant differences for the composite materials. After five weeks of soaking in distilled water, the composites containing 20 wt.% BESP fillers had the highest weight gain. The study demonstrated that waste brown eggshells in powdered form can be used as a filler in PLA composites.

Keywords: conventional limestone; waste brown eggshells; polylactic acid; mechanical properties; bio-composite

Citation: Cree, D.; Owuamanam, S.; Soleimani, M. Mechanical Properties of a Bio-Composite Produced from Two Biomaterials: Polylactic Acid and Brown Eggshell Waste Fillers. *Waste* **2023**, *1*, 740–760. https://doi.org/10.3390/waste1030044

Academic Editors: Catherine N. Mulligan, Vassilis Athanasiadis and Dimitris P. Makris

Received: 21 July 2023
Revised: 11 August 2023
Accepted: 22 August 2023
Published: 1 September 2023

1. Introduction

Polymers are used in many applications where synthetic thermoplastics are the most-popular and account for nearly 80% of all plastics around the world [1]. After their lifespan, developed countries tend to recycle them somewhat more than under-developed countries, where plastics are discarded to landfills. From 1950 to 2015, it is estimated that 6.3 billion metric tonnes (BMT) of plastic waste have been produced globally and about 60% or 4.9 BMT have been accumulating in landfills or the environment. If this waste trend continues, the amount of plastics discarded will reach an estimated 12 BMT by 2050 [2]. Recently, micro- and meso-plastics have been a growing problem in the marine environment [3]. According to Lebreton et al. [4], the Great Pacific Garbage Patch, a region in the North Pacific Ocean, has accumulated floating plastics in an area estimated to be 1.6 million km^2 [4]. Putting this into perspective, it is about 2.5-times the size of France (based on its land area of 643,801 km^2). Plastic degradation in the environment can take a long time. For example, depending on the type of plastic, it is estimated that plastic bottles can take 70–450 years to decompose, while plastic bags can take 10–1000 years [5].

Polylactic acid (PLA) is a bio-plastic product made from fermented plants such as corn, wheat, and sugarcane, which is derived from renewable and sustainable biomass sources. In order to lower the carbon footprint, PLA can be an alternative to petroleum-based synthetic thermoplastics. PLA has the largest market segment of all bioplastics produced. In 2022, the PLA global production was 20.7% or 460,000 tonnes and is estimated to reach 37.9% or about 2,390,000 tonnes by 2027 [6]. Presently, packaging relies heavily on synthetic polymers. A shift to bio-plastics could reduce contamination from these plastic products into the environment through biodegradation. PLA can be biodegraded by traditional composting [7] or mechanical [8] and chemical [9] recycling. A study showed that PLA has the ability to fully degrade into non-toxic products in a composting facility under specific conditions of temperature (30–60 °C), humidity (30–70%), and pH (8.5) over a period of 28–98 days [7]. Mechanical recycling consists of cleaning, drying, cutting, and pelletizing before reprocessing into new materials [8]. Chemical recycling involves the breakdown of PLA using solvents without affecting the monomer structure [9].

Lower-cost fillers such as calcium carbonate are used in polymer materials generally to reduce the overall cost of the polymers. Among other properties, fillers improve the tensile modulus and toughness of polymers. For instance, Zuiderduin et al. (2003) [10] added 0 to 60 wt.% of calcium carbonate to polypropylene (PP). Although the tensile strength decreased with the increase in filler loading, the tensile modulus and impact toughness increased within the tested temperature range of 20–70 °C compared to pure PP. The authors highlighted that a good dispersion of particulates within the matrix was important as agglomeration tended to have detrimental effects on the properties. Similarly, synthetic polymers containing calcium carbonate powder as fillers have been studied for polypropylene (PP) [11], LDPE [12–15], and HDPE [16,17] thermoplastic matrices.

The brown eggshell waste used in this study was acquired from the poultry industry, where they are composed of 96–97 wt.% calcium carbonate, a common ingredient found in geological formations of mineral limestone, with organic membranes as the balance [18]. Although both white and brown eggs are available on the market, the calcium carbonate content in the shells were found to be comparable [19]. Collecting eggshells from homes and restaurants would be a difficult task; however, there are large amounts of eggshell waste being produced every day by an industrial breaking plant process, which produces liquid eggs in bulk. There are many breaking plants located throughout the world. For instance, Egg Processing Innovations Cooperative is an egg-breaking plant located in Canada, which can process 180,000 dozen eggs per week. Newly emerging companies around the globe such as Eco-SHELL in Spain, Ateknea Solutions Hungary LLC in Hungary, and Bionovel Plus DOO in the Republic of Macedonia are advancing technology for extracting calcium carbonate powders from eggshells. Within a short time, these companies will want to find new niches for their powdered eggshells. The use of white eggshell powders in new materials has been suggested as alternative fillers in polymers, paints, ceramic tiles, mortar, and concrete [18]. A number of studies have been conducted on the use of white eggshells and seashells (e.g., bio-limestone) in synthetic thermoplastic polymers. For example, the mechanical properties have been evaluated for polypropylene (PP)/eggshells [20–24], LDPE/eggshells [25,26], LDPE/oyster shell [27] and for polyester (PET)/eggshells [28].

The prospect of using a green PLA polymer inspires the use of a sustainable filler material. Very few studies have been conducted on calcium carbonate fillers derived from eggshells for use in PLA composites. Hassan et al. (2014) [29] studied the effect of adding white eggshell powders to a biodegradable Bioplast thermoplastic polymer. The results showed that adding 1–2 wt.% eggshell filler improved the flexural strength and modulus, possibly due to the evenly dispersed particles throughout the matrix. Ashok et al. (2014) [30] investigated the addition of white eggshell fillers in PLA films. At all levels, the tensile strength and modulus increased, where the highest improvements were with 4 wt.% filler additions. The authors observed that the particles were uniformly distributed when less than 4 wt.% fillers were added, while at 5 wt.%, agglomeration occurred. More recently, Cree and Soleimani (2023) [31] investigated the addition of white eggshell powders to PLA. The tensile and flexural strengths peaked when 5 wt.% and 10 wt.% fillers were added, respectively, while the tensile and flexural modulus both improved as the filler loading increased. The dispersion of particles in a polymer matrix was noted to be difficult to achieve, which may affect the properties of the polymer composite materials.

In this work, calcium-carbonate-based filler; conventional limestone (LS), and brown eggshell powder (BESP) were added to a PLA matrix using an extrusion injection method. The brown chicken eggshells were initially crushed, washed, and dried prior to fine grinding, and both filler types were sieved to a particle size of 63 μm and 32 μm. The goal was to investigate the effect of adding various quantities of calcium carbonate powder to virgin PLA. The ultimate tensile, ultimate flexural, and impact toughness properties, fractured morphology, and water absorption, with and without calcium carbonate fillers, were studied. Statistical analysis of the mechanical property results was conducted using analysis of variance (ANOVA) to determine if a statistically significant difference existed between the average values of the composites containing filler loadings. After five weeks of soaking in distilled water, the moisture content and mass loss were measured. Water samples from each beaker were assessed for calcium carbonate contents and pH levels. The novelty of this study highlights the properties of PLA containing brown eggshells as a sustainable filler material.

2. Materials and Methods

2.1. Materials

The injection molding matrix was virgin PLA Prime Natural 4043D (NatureWork® LLC, Minnetonka, MN, USA) in pellet form. Conventional limestone was obtained from Imasco Minerals Inc. (Surrey, BC, Canada), while brown eggshells were obtained from Maple Lodge Hatcheries Ltd. (Stratford, ON, Canada).

2.2. Eggshell Preparation

The brown eggshells were dry-crushed in a steel drum, as shown in Figure 1a, which contained eleven 50.8 mm (2 in)-diameter steel balls (not visible). Prior to drying, the coarse eggshells were agitated and thoroughly washed several times to remove the organic membranes from the eggshells. A detailed explanation of the process can be obtained from a recent study [19]. To decrease the particle dimensions, a ball mill containing ceramic spheres was used, followed by rinsing with water and drying at 105 °C for 24 h. The ball mill jar with dimensions of 290 mm × 550 mm (diameter × length) is shown in Figure 1b and the pulverized brown eggshell powder (BESP) along with the ceramic balls ranging in diameter from 24–38 mm in Figure 1c. To evaluate the role of the filler particle size, the conventional limestone and eggshell powders were sieved to final mesh sizes of 230 (63 μm) and 450 (32 μm).

Figure 1. BESP preparation: (**a**) view inside steel drum, (**b**) ball mill, and (**c**) view inside ball mill.

2.3. Composite Formulations

A three-step process was used to fabricate the composite materials. First, the PLA pellets and calcium carbonate powders were mixed in a twin-screw extruder (SHJ-35, Nanjing Yougteng Chemical Equipment Co. Ltd., Nanjing, China). The extrusion was conducted at a temperature of 175 °C and a screw rotational speed of 16 rpm. Then, the filaments leaving the extruder were cooled in a water bath and further pelletized. The pellets were then dried at 80 °C for 4 h prior to injection molding. Finally, the composites were injection molded (Shen Zhou 2000, Zhangjiagang Shen Zhou Machinery Co., Shenzhen, China) with a temperature profile of 175, 180, 185, and 190 °C (feed zone to die zone). Compared to the virgin PLA (the control), the amount of filler loadings for each PLA composite formulation is shown in Table 1. The calcium carbonate fillers (LS and BESP) were used to replace a portion of the PLA polymer. The injection mold die was only able to produce tensile dog bone specimens and water absorption samples. The flexure and Charpy impact toughness specimens were made from the center of the dog bone samples cut to different lengths, but within ASTM standard sizes.

Table 1. PLA composites containing LS and BESP loadings for 63 μm and 32 μm particles.

	Calcium Carbonate Contents		
Sample	PLA (wt.%)	LS (wt.%)	BESP (wt.%)
Virgin PLA	100	0	0
LS-5	95	5	0
LS-10	90	10	0
LS-20	80	20	0
BESP-5	95	0	5
BESP-10	90	0	10
BESP-20	80	0	20

2.4. Mechanical Characterization

Tensile and flexural measurements were conducted using an Instron series 1137 (Instron, MA USA) instrument according to ASTM D638-14 and ASTM D790-17, respectively. The dimensions of the tensile specimens measured 200 mm × 12.74 mm × 3.25 mm, while the flexural specimens measured 65 mm × 12.74 mm × 3.25 mm (l × w × t). The tensile specimens were tested using a 10 kN load cell and a crosshead speed of 5 mm/min, while the three-point flexural test employed a 250 N load cell with a calculated crosshead speed of 1.30 mm/min. To retain a support span-to-depth ratio of 16:1, a 50 mm span length was used. For both tensile and flexural experiments, five specimens were tested for each composition at room temperature and averaged. Un-notched Charpy impact tests were conducted at room temperature using an Instron 450 MPX pendulum impact machine according to the guidance of ASTM D6110-18. The specimen's dimensions were 56 mm × 12.74 mm × 3.25 mm (l × w × t). At least ten samples were tested for each composite formulation and the mean values obtained. Prior to conducting the mechanical tests, all samples were conditioned at 23 ± 3 °C with a relative humidity of 50 ± 4% for about 24 h.

2.5. Scanning Electron Microscopy

The calcium carbonate particles and fractured surfaces of the tensile, flexure, and Charpy impact toughness samples were investigated using a JEOL JSM-6010 Scanning Electron Microscope (JEOL Ltd., Tokyo, Japan) in secondary electron (SEI) mode. All samples were initially gold coated to prevent electrostatic charging during viewing and observed at an operating voltage of 10–20 kV.

2.6. Water Uptake

The water uptake tests were conducted according to the recommended procedure outlined in ASTM D570-22. Test specimens measuring 57.4 mm × 36.4 mm × 2.7 mm (l × w × t) were performed at room temperature on a set of three samples placed in beakers filled with distilled water. Before beginning the experiments, all specimens were placed in an oven at 50 °C for 24 h. They were then removed, allowed to cool down in a desiccator, and weighed to obtain the initial dry weights. The weight gain percentage due to water absorption over a period of 5 weeks was initially measured after 24 h (1 day) followed by one-week intervals until saturation was reached using a digital balance (±0.1 mg). The specimens were removed one by one and quickly dried using a paper towel to remove excess surface water. The percentage of mass gain due to moisture uptake was measured by Equation (1), while the mass loss was evaluated by Equation (2).

$$\text{Moisture content (\%)} = [(m_2 - m_1)/m_1] \times 100\% \tag{1}$$

$$\text{Mass loss (\%)} = [(m_3 - m_1)/m_1] \times 100\% \tag{2}$$

where moisture content is the percentage mass gain, m_1 is the initial oven-dry mass, m_2 is the mass after every soaking interval, and m_3 is the mass after the 5-week period.

2.7. Leaching Measurements

To determine if calcium carbonate was coming out of the composite during the 5-week of water immersion, water samples from each uptake test were examined with an atomic absorption spectrophotometer (Model 240 Series AA, Agilent Technologies, Santa Clara, CA, USA).

2.8. pH Measurements

To determine if there were changes due to the degradation of the PLA and the leaching of the calcium carbonate fillers after a 5-week period in distilled water, the pH levels were measured with a pH meter (Orion Star™ A111 pH Benchtop Meter; Thermo Scientific Ward Hill, MA, USA) in triplicate. The distilled water was not refreshed during the 5-week period, where silicone caps were placed on the beakers to avoid evaporation. The pH values were measured at the end of the experiment, and the water temperature of the beakers was 20.6 °C.

2.9. Statistical Analysis

The experimental ultimate tensile strength/tensile modulus, ultimate flexural strength/ flexural modulus, and Charpy impact results for all formulations were compared individually using a one-way ANOVA F-test at a 95% confidence level since the variable changing was the filler quantity. The calculations were conducted with the Analysis ToolPak, a Microsoft Excel add-in program (Microsoft Corp., Redmond, WA, USA). If the F-critical value is less than the F-test value, the null hypothesis is rejected, indicating that the average was not the same for all groups. This suggests a statistically significant difference existed between the average values of each mechanical property for the different composite formulations. However, if the F-critical value approaches the F-test value, the null hypothesis cannot be rejected and considered as not being statistically significant.

3. Results and Discussion

3.1. Mechanical Properties

The mechanical properties evaluated in this study are given in Table 2. The (±) symbols in Table 2 and the error bars in Figures 2–4 represent the standard deviations.

Table 2. Mechanical properties of PLA and PLA/calcium carbonate composite materials.

Composite	Ultimate Tensile Strength (MPa)	Tensile Modulus (GPa)	Ultimate Flexural Strength (MPa)	Flexural Modulus (GPa)	Charpy Impact Strength (kJm^{-2})
Virgin PLA	50.9 ± 2.3	3.6 ± 0.10	78.9 ± 4.2	2.9 ± 0.03	17.3 ± 0.70
Filler size (63 μm)					
LS-5	50.1 ± 1.8	3.7 ± 0.18	52.6 ± 5.0	3.1 ± 0.05	13.2 ± 0.90
LS-10	46.6 ± 1.2	3.7 ± 0.28	76.9 ± 5.4	3.2 ± 0.09	11.1 ± 1.5
LS-20	43.9 ± 1.8	4.2 ± 0.35	71.4 ± 2.7	3.4 ± 0.04	10.3 ± 1.0
BESP-5	47.4 ± 2.0	3.6 ± 0.25	46.5 ± 5.5	3.1 ± 0.04	8.7 ± 1.3
BESP-10	33.4 ± 1.3	3.9 ± 0.15	63.7 ± 7.6	3.1 ± 0.20	6.9 ± 0.70
BESP-20	32.1 ± 0.60	4.0 ± 0.12	21.5 ± 6.1	3.4 ± 0.11	6.5 ± 1.1
Filler size (32 μm)					
LS-5	51.3 ± 0.70	3.4 ± 0.19	67.2 ± 1.6	3.1 ± 0.03	14.8 ± 1.0
LS-10	49.8 ± 0.30	3.9 ± 0.08	92.7 ± 1.2	3.3 ± 0.03	14.0 ± 1.2
LS-20	43.5 ± 1.9	4.4 ± 0.09	91.0 ± 2.1	3.5 ± 0.05	11.2 ± 1.3
BESP-5	48.1 ± 2.6	3.7 ± 0.31	60.3 ± 1.1	3.1 ± 0.04	10.4 ± 1.1
BESP-10	48.0 ± 3.2	3.9 ± 0.43	66.8 ± 1.6	3.3 ± 0.10	7.8 ± 1.5
BESP-20	42.7 ± 1.2	4.5 ± 0.19	44.7 ± 1.5	3.4 ± 0.20	7.2 ± 1.4

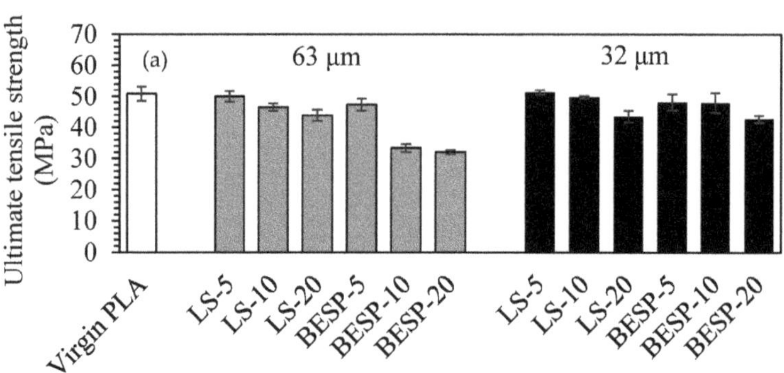

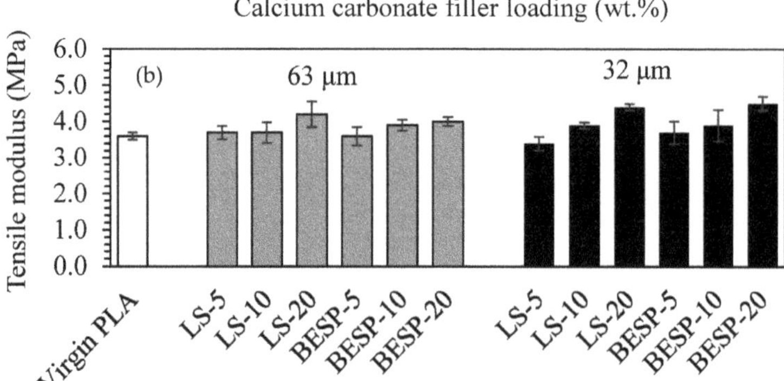

Figure 2. Tensile properties of virgin PLA and PLA composites with calcium carbonate fillers: (**a**) ultimate tensile strength and (**b**) tensile modulus.

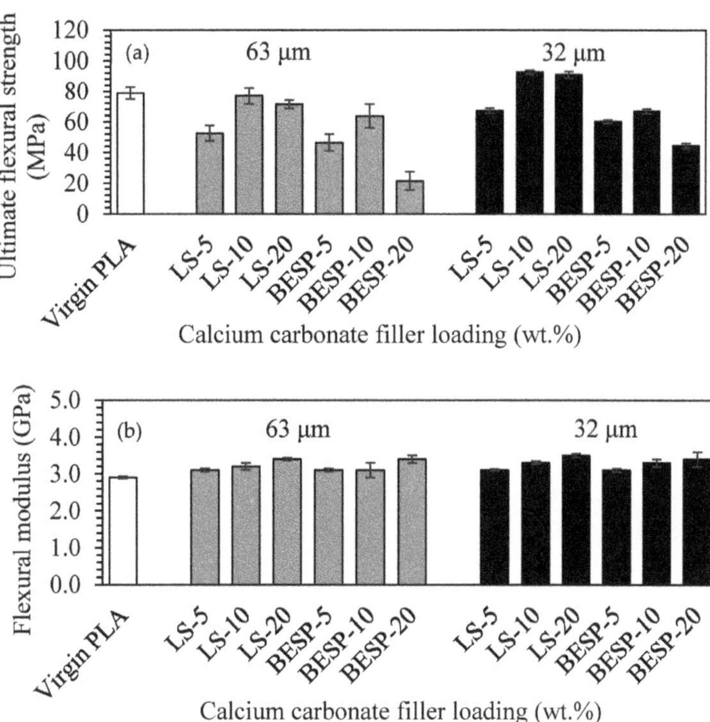

Figure 3. Flexural properties of virgin PLA and PLA composites with calcium carbonate fillers: (**a**) ultimate flexural strength and (**b**) flexural modulus.

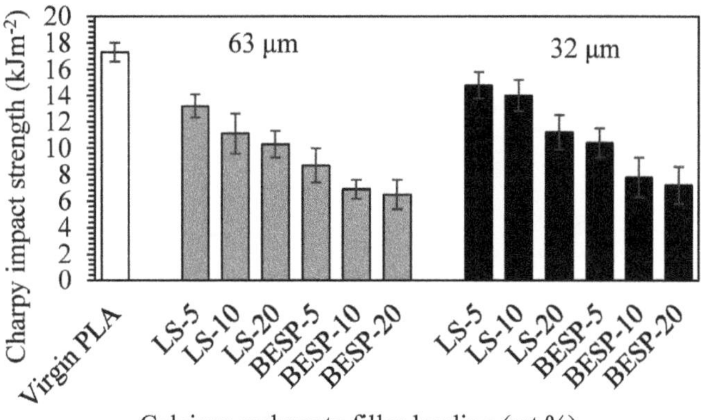

Figure 4. Charpy impact strength of virgin PLA and PLA composites with calcium carbonate fillers.

3.1.1. Tensile Properties

The tensile properties of the virgin PLA and PLA composites are summarized in Table 2 and shown in Figure 2 for two particle filler sizes (63 μm and 32 μm). The ultimate tensile strengths as a function of calcium carbonate filler loadings (Figure 2a) showed that composites containing 5 wt.% additions of LS and/or BESP for both particle sizes were similar to the control. When the loadings exceeded 5 wt.%, the strengths reduced below the control. This might be due to agglomeration of the fine filler powders in the PLA matrix, which lack dispersion at higher loadings. At 10 wt.%, the ultimate tensile strengths for LS and BESP (63 μm) reduced by 8% and 34%, respectively, while LS and BESP (32 μm) decreased by 2% and 6%, respectively. Composites containing 5–10 wt.% BESP (32 μm) were similar in ultimate tensile strength. The conventional-limestone-filled composites were not substantially affected; however, the addition of the brown eggshell fillers to the PLA showed a sharper decrease in strength. At 20 wt.% concentrations, considerable reductions in ultimate tensile strengths were observed. In general, the smaller 32 μm particles performed slightly better than the larger 63 μm particles. The performances of PLA composites are partly dependent on the interaction of the matrix with the filler. At low filler loadings, it is probable there will be more polymer-filler–particle interactions and lower filler agglomerations. The presence of rigid limestone fillers tends to inhibit polymer chain motion during the uniaxial tensile test. When the filler content increases, the dispersion of the particles becomes non-uniform and the likelihood of filler particles clustering with each other increases (e.g., particle-to-particle interactions). In polymer composites, the agglomeration and poor adhesion between particles leads to lower tensile strengths [32]. In contrast to the tensile modulus, the tensile strength is susceptible to how well the bond forms between the filler/matrix interfaces since a poor link will interrupt the stress transfer from the matrix to the fillers [22]. Physical and chemical surface modifications have been applied to inorganic fillers to improve compatibility between synthetic polymers. A common physical treatment is the use of stearic acid, a fatty acid that acts as a surfactant. Stearic acid acts to lower the surface tension between the hydrophobic polymer and hydrophilic calcium carbonate particles by coating the particles with a hydrophobic film. A chemical treatment such as a coupling agent reacts with the surface of the calcium carbonate and the polymer matrix via chemical covalent bonding. Silane, titanate, and zirconate have been used as couplings agents [33].

The tensile modulus as a function of calcium carbonate filler loadings (Figure 2b) showed a steady increase as the filler content changed from 5 to 20 wt.%. Due to the higher stiffness of the filler particles, compared to the more-pliable PLA matrix, both the calcium carbonate filler types were rigid and will have less deformation when a tensile load is applied. As the filler loadings increase, the tendency and amount of particles interacting with the PLA polymer chains amplifies, resulting in less movement of the overall number of polymer chains and, thus, higher tensile moduli [34]. The differences in tensile modulus may be due to how well the fillers are dispersed during the manufacturing process. A study on the addition of halloysite, a clay mineral, as a filler to polypropylene showed a better dispersion of fillers, led to a greater tensile modulus [35]. In another study, Moreno et al. (2015) [36] investigated injection-molded calcium carbonate/PLA composites by blending them with a twin-screw extruder. The research revealed that increasing the screw speed resulted in a higher shear rate, which tended to improve the dispersion of the fillers in the PLA matrix.

3.1.2. Flexural Properties

The flexural properties of the virgin PLA and PLA composites are presented in Table 2 and Figure 3 for the two particle sizes studied. The ultimate flexural strengths as a function of calcium carbonate filler loadings (Figure 3a) showed that conventional limestone fillers produced better ultimate flexural strengths than brown eggshell fillers for all filler loadings. The ultimate flexural strengths of composites containing 10–20 wt.% LS for both particle sizes were on par with or higher than the virgin PLA. Composites containing 5–10 wt.% BESP (32 μm) had the highest ultimate flexural strengths when the sustainable eggshell fillers were used. Particle dimensions are an important characteristic of fillers that influences the reinforcement effect of the PLA biopolymer. Small-sized particles have a high surface area, which provides a large surface region for contact between the matrix and the particles [11]. However, small particles also tend to assemble into larger agglomerates due to electrostatic forces, and this increases with greater additions of filler loadings. This phenomenon can describe the drop in ultimate flexural strength at 20 wt.% loadings for all composites. In contrast, large-sized particles do not agglomerate as much, but have a low surface area, which reduces the interaction between the matrix and filler when loads are applied to the composites. This factor is attributed to the lower strength of the composites containing 63 μm fillers compared to composites with 32 μm fillers.

The calcium carbonate fillers were able to provide a stiffening effect, as depicted by the flexural modulus results shown in Figure 3b. For both filler particle sizes, the flexural modulus increased with the increase of the filler content. At low loadings, the particles are anticipated to be well dispersed within the PLA matrix, resulting in larger inter-particle distances. As the filler content increases, the distances between the particles reduce, bringing particles closer to each other. This creates additional obstacles/restrictions for polymer chain mobility, which increase the flexural modulus. According to the results, it appears that the flexural modulus was less affected by the particle size and more by the amount of filler content.

3.1.3. Charpy Impact Properties

The Charpy impact strengths as a function of calcium carbonate filler loadings for the virgin PLA and PLA composites are provided in Table 2 and shown in Figure 4. The highest impact strength was observed for the virgin PLA. There was a gradual decrease of the impact strength with filler loading for all PLA composites; however, the composites with smaller particles sizes were slightly better. A study by Nakagawa and Sano (1985) [37] on polypropylene-filled calcium carbonate composites determined that smaller particles had better dispersion in the matrix and exhibited higher impact strengths. Toro et al. [38] studied the impact strength of polypropylene/eggshell composites. The authors noted that, although pure polypropylene samples experienced higher impact strengths, the addition of filler loadings reduced the total amount of polymer, which acted to decrease the toughness and impact strength of the composites. They also observed that composites with the highest tensile modulus (e.g., stiffness) exhibited the lowest impact properties due to the high stresses transferred from the polymer to the fillers. The impact strength could be affected by particle aggregation (e.g., particle-to-particle interactions), which leads to an inhomogeneous distribution of the fillers in the matrix. In addition, the forces holding particles together are adhesive forces such as weak secondary van der Waals forces. A study on the agglomeration tendency of eleven commercial calcium carbonate fillers was conducted ranging in particle sizes from 0.08 μm to 130 μm. The results indicated that the particle size was a controlling factor that determined the magnitude of the adhesive forces between particles where the aggregation of filler particles increased with decreasing particle size [39]. Although particle size is an important element, particle-to-particle aggregation has also been reported to be due to filler characteristics, specific surface area, composition, surface modification, surface tension, and process/manufacturing conditions [11,40].

The Charpy un-notched specimen test method measures the total energy needed for a crack to initiate and propagate. Un-notched testing can identify agglomeration in the composite material. Flaws in the form of large filler particles or large/small particles that agglomerate can be initiation points for cracks to generate and grow upon impact. Un-notched impact strengths obtained for polymers containing fillers has been reported to be sensitive to agglomerates [41]. In this study, the Charpy impact strengths decreased as the amounts of loadings increased, which suggests the un-notched samples were able to detect agglomerations in the polymer composite.

3.1.4. Scanning Electron Microscopy

Since the PLA composite's mechanical properties were generally better for the 32 μm fillers compared to the 63 μm fillers, SEM was conducted on the smaller particle sizes and fractured surfaces of composites containing these fillers. The particle microstructure for the powders of the conventional limestone (Figure 5a) and brown eggshells (Figure 5b) were similar in shape. The particles exhibited angular and irregular geometries for different sizes. The average particle size with the standard deviation (SD) for the 32 μm sieved

(No. 450 mesh) fillers was measured using the Image-J software (version 2.3.0/1.53f, NIH, Bethesda, Maryland, USA), where the LS and BESP were 26.9 ± 3.4 µm and 23.8 ± 7.6 µm, respectively, as shown in Figure 6.

Figure 5. Scanning electron micrographs showing 32 µm powder morphologies: (**a**) conventional limestone and (**b**) brown eggshells.

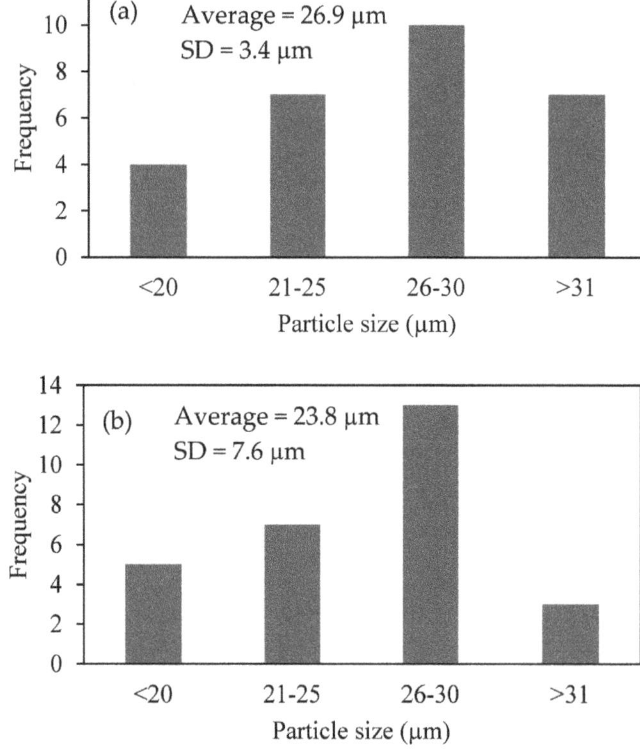

Figure 6. Size distribution of (**a**) conventional limestone and (**b**) brown eggshell fillers sieved to 32 µm.

The fractured surfaces of the tensile samples for the virgin PLA and the PLA composites containing the 32 μm sieved particles are shown in Figure 7. The morphology features of the virgin PLA (Figure 7a) were relatively smooth, revealing a brittle type of failure, which is characteristic of the PLA matrix. As the filler quantities increased, the surfaces were rougher, demonstrating ductile characteristics (Figure 7b–e), as given by the white ridges created during PLA stretching from the tensile test. From a visual observation, the virgin PLA (Figure 7a) and low filler content composites (Figure 7b,c) showed a dense fractured surface without any cavities, while at higher filler loadings (Figure 7d,e), some cavities were observed (red circles) due to pullout of the calcium carbonate fillers. This suggests the bonding between the filler and matrix was weak.

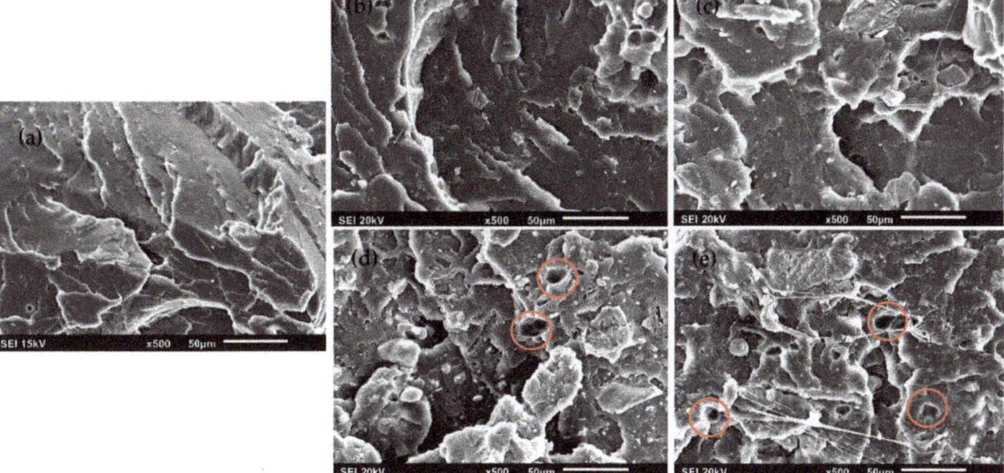

Figure 7. Scanning electron micrographs of tensile fractured surfaces: (**a**) virgin PLA, (**b**) PLA/LS-5, (**c**) PLA/BESP-5, (**d**) PLA/LS-20, and (**e**) PLA/BESP-20 (for 32 μm particles). Red circles show cavities in the matrix.

The SEM flexural fractured surfaces for the virgin PLA and the PLA composites containing 32 μm sieved particles are shown in Figure 8. Similar to the tensile fractured surfaces, the PLA without filler exhibited a brittle mode of failure, as depicted by the mostly flat regions in Figure 8a, while the filled PLA composites (Figure 8b–e) had irregular surfaces compared to the virgin PLA. When fillers were added (Figure 8b,c), the fractured surface after bending showed white fibrils, indicative of matrix plastic deformation. Void spaces were also concentrated around the stiff filler particles, as illustrated by the white regions, and this phenomenon was amplified at higher filler contents (Figure 8d,e). The particles appeared to be embedded in the fractured region of the PLA, suggesting synergy between the particles and matrix. However, during loading, the matrix deformed around the particles, indicating poor adhesion between the fillers and matrix.

The SEM Charpy impact fractured surfaces for the virgin PLA and PLA composites containing 32 μm particles are shown in Figure 9. The virgin PLA (Figure 9a) had flat areas associated with brittle fracture in addition to longitudinal patterns, which are linked to the crack propagation produced during impact loading. The PLA 5 wt.% composite fractured morphology (Figure 9b,c) appeared similar to the virgin PLA, which could explain the similar Charpy impact strength values associated with these materials, as echoed by the results in Figure 4. However, adding 20 wt.% loadings may be excessive, which caused embrittlement and reduced impact strength. By incorporating 20 wt.% fillers (Figure 9d,e), cavities appeared, possibly due to the dislodgement of filler particles in the matrix, as highlighted by the red circles. At greater loadings, there was no specific direction to the crack propagation, and this can be observed from the irregular broken surfaces. This was also observed in a polypropylene/calcium carbonate composite, where the impact strength decreased as the fractured surface became rougher [42].

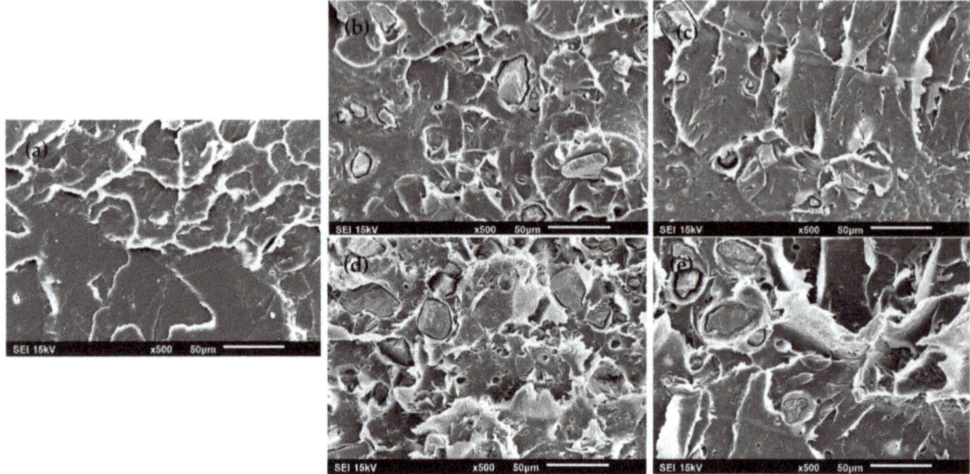

Figure 8. Scanning electron micrographs of flexural fractured surfaces: (**a**) virgin PLA, (**b**) PLA/LS-5, (**c**) PLA/BESP-5, (**d**) PLA/LS-20, and (**e**) PLA/BESP-20 (for 32 μm particles).

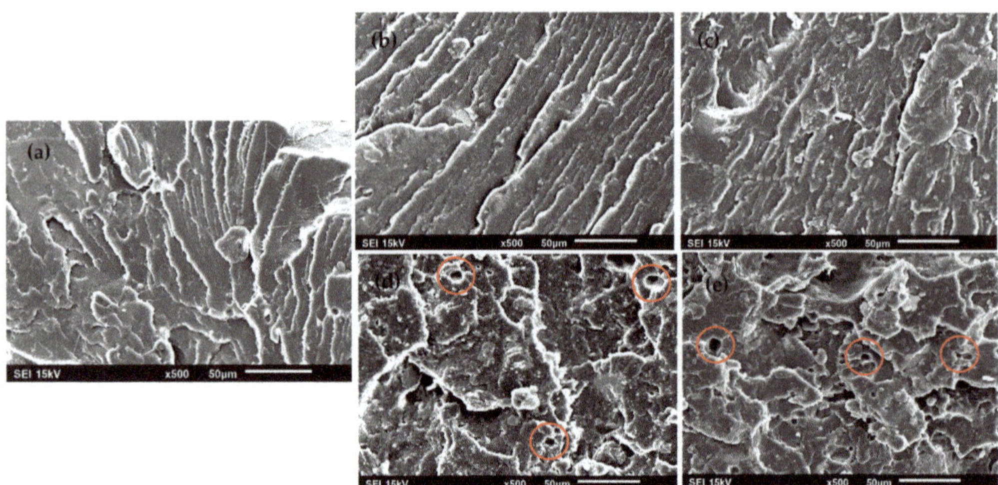

Figure 9. Scanning electron micrographs for Charpy impact fractured surfaces: (**a**) virgin PLA, (**b**) PLA/LS-5, (**c**) PLA/BESP-5, (**d**) PLA/LS-20, and (**e**) PLA/BESP-20 (for 32 μm particles). Red circles show cavities in the matrix.

3.1.5. Statistical Analysis

The tensile, flexural, and Charpy properties were assessed one-by-one for both particle sizes (63 μm and 32 μm), as given in Table 3. The mechanical property F-critical values were all less than the F-test values. Therefore, it can be concluded that a statistically significant difference existed between the average values of each property measured. The results of the

ANOVA analysis indicated that the addition of calcium carbonate filler materials (LS and BESP) to virgin PLA highly affected the tensile, flexural, and Charpy average properties.

Table 3. ANOVA mechanical property results of PLA containing calcium carbonate fillers (LS and BESP) with two different particle sizes of 63 µm and 32 µm.

Mechanical Property	Particle Size (µm)	Source of Variation	SS	df	MS	F-test	F-crit
Ultimate tensile strength (MPa)							
	63	BG	1431	6	238.6	84.49	2.508
		WG	67.77	24	2.824		
	32	BG	286.4	6	47.73	11.99	2.528
		WG	91.53	23	3.980		
Tensile modulus (GPa)							
	63	BG	1.527	6	0.2546	5.011	2.459
		WG	1.372	27	0.05080		
	32	BG	4.457	6	0.7429	12.54	2.490
		WG	1.482	25	0.0593		
Ultimate flexural strength (MPa)							
	63	BG	11,664	6	1944	72.61	2.549
		WG	589.0	22	26.77		
	32	BG	8255	6	1376	268.1	2.490
		WG	128.3	25	5.131		
Flexural modulus (GPa)							
	63	BG	0.9917	6	0.1653	42.92	2.549
		WG	0.0847	22	0.0039		
	32	BG	1.095	6	0.1825	21.42	2.490
		WG	0.2129	25	0.0085		
Charpy impact strength (KJ·m^{-2})							
	63	BG	858.6	6	143.1	129.4	2.246
		WG	69.68	63	1.106		
	32	BG	832.9	6	138.8	99.05	2.246
		WG	88.29	63	1.402		

BG, between groups; WG, within groups; SS, sum of squares; df, degree of freedom; MS, mean square.

3.1.6. Water Uptake

The water gains due to moisture content for the virgin PLA, as well as the PLA/LS and PLA/BESP composites containing 63 µm and 32 µm filler particles are shown in Figure 10. The water absorption weight gain for the virgin PLA at 2, 3, 4, and 5 weeks was 0.861%, 0.895%, 0.910%, and 0.910%, respectively. Similar results were obtained for an injection-molded PLA Ingeo Biopolymer 6100D polymer, where after 20 days (3 weeks) of immersion, the specimens reached their maximum capacity of about 1% of absorbed water [43]. The methyl side group of PLA provides hydrophobicity; however, PLA is not completely hydrophobic since water molecules are attracted to the polar oxygen linkages and end groups in the PLA molecule, which confers PLA's hydrophilicity [44]. The common consensus about PLA is that it is a polymer which attracts and absorbs moisture from the air over time. Following 5 weeks of water soaking, the composites containing LS and BESP particulates in amounts of 5, 10, and 20 wt.% had water weight gains of 0.910%, 0.983%, and 1.01% and 1.09%, 1.10%, and 1.13%, respectively for the 63 µm fillers. In the same way, composites containing 32 µm fillers had weight gains of 0.949%, 0.973%, and 1.09% and 1.10%, 1.15% and 1.26%, respectively. It was observed that smaller filler particle absorbed slightly more water than the larger 63 µm particles size. The physical absorption of water by fillers could be influenced by hydrophobicity/hydrophilicity, porosity, and surface area. Depending on the location of the sedimentary carbonate rock deposits, commercially available limestone can be of different purities. The majority of limestone contains calcium carbonate ($CaCO_3$) in the form of mineral calcite ranging in amounts of 85 wt.% (impure) to 98.5 wt.% (very high purity). The remaining are mineral impurities in the form of lime (CaO), magnesia (MgO), silica (SiO_2), and iron oxide (Fe_2O_3) [45]. Correspondingly, brown eggshells have been determined to comprise calcium carbonate with calcite being the principal mineral [19]. For both mineral limestone and eggshells, the calcite material has a rhombohedral structure based on the hexagonal unit cell [46]. At the microscopic level, the eggshell contains two layers of calcite. The outermost layer (~5–8 μm thick) consists of dense vertical calcite crystals. The layer below is the palisade layer and is about 200 μm thick. It includes columnar calcite crystals that are less dense than the vertical layer and positioned perpendicular to the surface of the eggshell [47]. Calcite has been reported to be highly hydrophilic [48], where water interacts with the surface of calcite by electrostatic interaction and by hydrogen bonding [49]. Limestone as a rock and eggshell as a bio-ceramic structure both have high permeability. For example, depending on the limestone geological formation and depth of recovery, it can have porosities between 8% and 52% [50]. In the literature, the reported surface porosity for chicken eggshells as a percentage value (to compare against limestone) is limited; however, eggshells are porous as conveyed by the number of pores on the surface of the eggshell being between 10,000 and 20,000 [51]. The architecture of an eggshell is interesting such that the microscopic pore structure functions to permit oxygen to diffuse inwards through the shell from the surroundings and allows water vapor and carbon dioxide created by the chicken embryo to diffuse outwards [52]. In contrast to the larger filler particles (e.g., 63 μm), the water weight gain increased for composites containing smaller particles (e.g., 32 μm) since they tend to have a greater surface area suggesting more water contact with the particles.

As was anticipated, the composites incorporating 32 μm particles had for the most part better mechanical properties than composites containing the larger 63 μm particles. Therefore, the PLA mass loss samples, as well as calcium carbonate contents and pH levels of the distilled water mediums were evaluated for the former.

3.1.7. Mass Loss

After 5 weeks immersed in distilled water, the mass loss of the virgin PLA decreased by 0.102% as shown in Figure 11. The PLA/LS and PLA/BESP composites with 5, 10, and 20 wt.% loadings displayed mass losses of 0.120%, 0.127%, and 0.131% and 0.143%, 0.207%, and 0.235%, respectively. The mass losses were more observable for the brown eggshell composites. In the same way, a study immersed PLA (Ingeo 3051D) in water at 23 °C and 51 °C for 30 days and observed a minor weight decrease of 0.07% and 0.21%, respectively [53]. The degradation of PLA when submitted to high moisture or submerged in water can breakdown the matrix by a hydrolysis (e.g., reaction with water) mechanism, which results in a mass loss [54] and the lowering of its molecular weight [55]. At ambient temperature, the hydrolytic degradation of PLA is marginally stable to slow but is accelerated as the temperature of the water increases. A previous study illustrated that the time for PLA (structure type not mentioned) to degrade varied with temperature. For instance, temperatures of 4, 13, 25, 30, 50, 60, and 70 °C had onset fragmentation (e.g., ester groups of polylactide chains in the presence of water) times of 64 months, 25 months, 6 months, 4.4 months, 1.5 months, 8.5 days and 1.8 days, respectively [56].

Mass losses can be attributed to the hydrolysis of the matrix compounded with the hydrolysis of the matrix/filler interface. The PLA Ingeo 4043D used in the present study is a highly amorphous bio-polyester with a D-isomer content of 4.3%, providing its lower crystallinity [57]. PLA with less than 1% D-isomer is used in injection molding, where high crystallization rates are required, while higher contents (4–8%) reducing the crystallization rates, which can be advantageous for injection molding of optically transparent applications [58]. More importantly, amorphous thermoplastics are ideal over semi-crystalline polymers for injection molding since they have a lower-dimensional shrinkage with little to no crystallization, which acts to reduce warpage [59]. However, amorphous PLA structures are less stable against hydrolysis as the structure allows water to enter more easily into the bulk of the sample compared to a more-stable semi-crystalline PLA, as its structure prevents water diffusion into the bulk of the material [60]. Over time, water will break down the high-molecular-weight PLA through hydrolysis. As the water molecules diffuse into the amorphous regions of the PLA, they react with ester bonds along the PLA backbone chain. This results in shorter, fragmented chains, which reduces the PLA's molecular weight. The hydrolysis process leads to ester bond cleavage occurring primarily in the amorphous regions, which produces water-soluble products such as lactic acid monomers and low-molecular-weight oligomer monomers. As time progresses, these products near the surface can leach out into the distilled water and PLA polymer degradation ensues [61]. Inorganic calcium carbonate is slightly soluble in pure water and has been reported to be between 6.8 mg/L [62] and 13 mg/L [63] at 25 °C. The solubility increases when the water contains carbon dioxide such as river water or rainwater found in nature. In addition, the solubility decreases with augmentation in temperature and increases with increasing mineral concentration [62]. As the composite absorbs water, the interface adhesion between the PLA matrix/industrial limestone and/or brown eggshell, fillers undergo hydraulic degradation, which promotes the detachment of filler particles [64]. It is hypothesized that the calcium-carbonate-based fillers are removed from the composite PLA surfaces as a solid material, which falls to the bottom of the beakers. A recent study attributed this phenomenon to osmosis degradation, where the calcium carbonate particles travel towards the external surfaces of the polymer composite and eventually are released into the aqueous media [65].

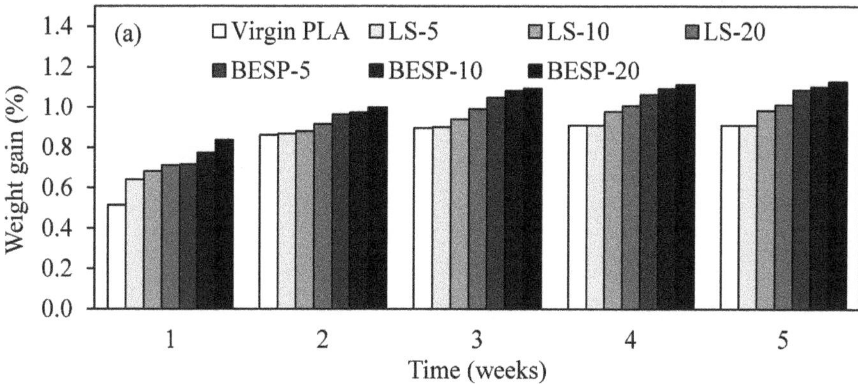

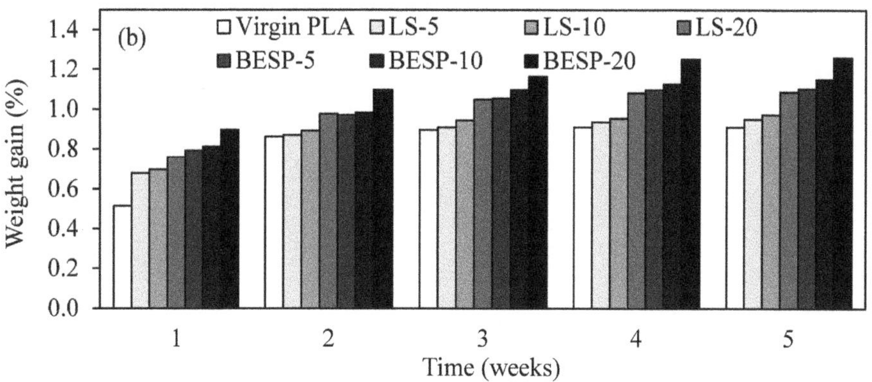

Figure 10. Weight gain due to moisture content for virgin PLA and PLA composites containing calcium carbonate fillers: (**a**) 63 μm particles and (**b**) 32 μm filler particles.

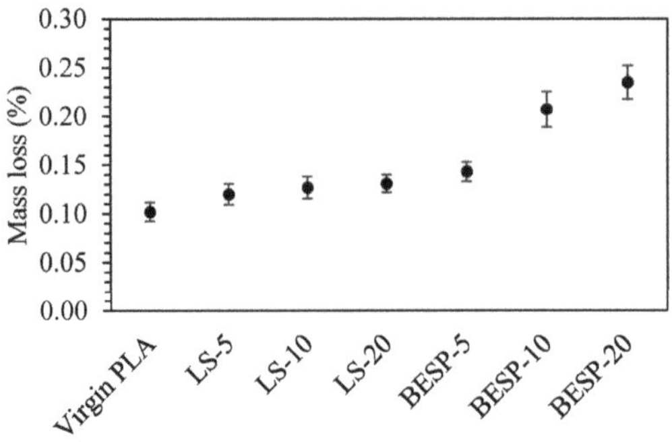

Figure 11. Mass loss after 5 weeks of the absorption test for virgin PLA and PLA composites containing 32 μm fillers at different loading concentrations.

3.1.8. Leaching Measurements

To determine the degree of leaching of the composites, if any, for both calcium-carbonate-based filler types (32 μm particles) and loadings, distilled water from individual beakers were tested after 5 weeks of water soaking, and the results are shown in Figure 12. An atomic absorption spectrophotometer was used to identify if calcium (Ca) was present in the water. The mass of calcium obtained was used to calculate the amount of $CaCO_3$ content in milligrams (mg) by multiplying the calcium content by a factor of 2.5. Distilled water and the virgin PLA water media were analyzed and did not contain any elemental calcium. Similarly, calcium traces in the PLA/LS composites were not detected. The PLA/BESP specimens with 5, 10 and 20 wt.% filler contents had 1.21, 2.05, and 4.68 mg of $CaCO_3$, respectively. The results convey that the conventional limestone did not leach out of the PLA matrix in contrast to the brown eggshell particulates.

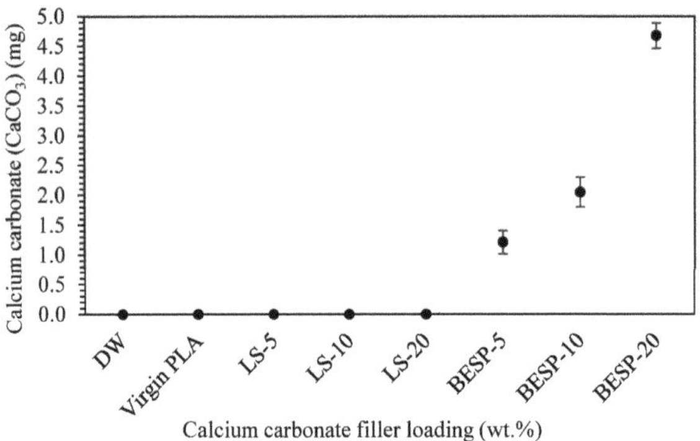

Figure 12. Calcium carbonate ($CaCO_3$) contents after 5 weeks in water for PLA composites containing 32 μm fillers at different loading concentrations.

Brown eggshells appeared to leach out more than mineral limestone, which can be due to the fact that there are amorphous regions of calcite within the crystalline calcite regions of the eggshell [66]. Amorphous regions are unstable and have a higher solubility in water than crystalline regions [67]. A "floating on water" test [68] was conducted where two beakers were filled with 100 mL of distilled water as shown in Figure 13a. Then, 5 g of the 32 μm-particle-size mineral limestone and brown eggshell powder were added in separate beakers (Figure 13b,c). The mixture was stirred at 1100 rpm for 5 min and allowed to settle for 5 days. From a visual observation, the majority of the mineral limestone beaded on top of the water, as depicted by the red arrow in Figure 13b, while a minor amount fell to the bottom of the beaker. In contrast, all the eggshell powder settled to the bottom of the beaker. The eggshell powder mixed better with the distilled water than the mineral limestone possibly due to the attached membrane on the eggshells. The eggshell membrane consists of proteins that are hydrophilic, therefore, they absorb water and support the sinking of the eggshell particles to the bottom of the beaker. When high-comminution-degree limestone powders are stirred, cohesive interactions between small particles can be responsible for the pasting of grains [68]. The higher solubility of brown eggshells in water may be attributed to the greater leaching in the PLA/BESP composites.

Figure 13. "Floating on water" test (**a**) prior to the addition of calcium carbonate, (**b**) side views after 5 days of rest for mineral limestone (**left**) and brown eggshells (**right**), and (**c**) top views after 5 days of rest for mineral limestone (**left**) and brown eggshells (**right**).

3.1.9. pH Measurements

Following the 5-week duration of water soaking, the leaching of PLA or calcium carbonate could cause the pH of the distilled water (DW) to change, as shown in Figure 14. As PLA breaks down in water, it creates an acidic medium due to carboxylic acid release [69]. In contrast, calcium carbonate offsets the acidic degradation products from PLA by increasing the pH [70]. The original distilled water had a measured pH of 6.17, and after the 5-week period, the pH value was 5.90. A reduction of the pH of the distilled water is indicative of the carbon dioxide (in the air) reaction with water to form carbonic acid. The virgin PLA water sample had a pH of 5.76, possibly due to marginal leaching of PLA. The PLA/LS and PLA/BESP composites had pH values of 5.83, 5.86, and

5.92 as well as 5.96, 6.05, and 6.33 for filler contents of 5, 10, and 20 wt.%, respectively. The PLA/LS composites had pH values in-line with the virgin PLA, while the pH values for the PLA/BESP composites rose slightly, which could be due to the leaching of the brown eggshell particles, as depicted in Figure 12.

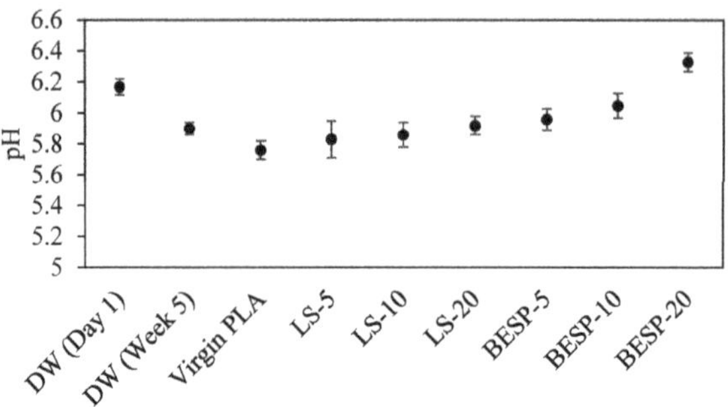

Figure 14. Variation of distilled water pH after 5 weeks for PLA composites containing 32 μm calcium carbonate fillers at different loading concentrations.

4. Conclusions

PLA composites containing either conventional limestone or brown eggshell fillers were prepared. For all mechanical properties, the additions of smaller particles were favored over larger particles, and conventional limestone performed marginally better than brown eggshells. The composites containing 5–10 wt.% brown eggshell fillers (32 μm) had the highest ultimate tensile strengths, while the additions of 10 wt.% brown eggshell fillers (32 μm) had the largest ultimate flexural strengths, but both properties were lower than the control. The tensile modulus and flexural modulus increased with filler loading; however, the Charpy impact strengths were lower than the control for both filler sizes of 63 μm and 32 μm. From scanning electron microscopy, changes in the fractured surfaces were observed from being relatively flat and smooth for virgin PLA to an increased roughness as the filler content increased. This suggests the additions of calcium carbonate fillers had the ability to alter the properties of the PLA. ANOVA showed that both filler size and filler contents had significant positive effects on the mechanical properties. The water uptake for the PLA/brown eggshell powder composites was greater than for the PLA/conventional limestone composites, which may be due to the porous architecture of the eggshell. Brown eggshell composites leached more than conventional limestone composites. The leaching was thought to be a result of the amorphous regions in the eggshells having a higher solubility than the conventional limestone. While the solubility of conventional limestone has been reported in the literature, there is a lack of solubility information for eggshells. The results indicated that waste brown eggshells are candidates for use as filler materials in polymer matrices.

Author Contributions: Conceptualization, D.C. and M.S.; methodology, D.C.; formal analysis, D.C., S.O. and M.S.; investigation, D.C.; resources, D.C.; data curation, D.C.; writing—original draft preparation, D.C.; writing—review and editing, D.C., S.O. and M.S.; supervision, D.C.; project administration, D.C.; funding acquisition, D.C. All authors have read and agreed to the published version of the manuscript.

Funding: The research was funded by the Natural Sciences and Engineering Research Council of Canada (NSERC) under the Discovery Grant (RGPIN-2020-06701). The APC was funded by MDPI.

Institutional Review Board Statement: Not applicable.

Informed Consent Statement: Not applicable.

Data Availability Statement: The data presented in this study are available upon request from the corresponding author. The data are not publicly available because the raw and processed data required to reproduce these findings cannot be shared at this time, as the data also form part of an ongoing study.

Acknowledgments: The authors would like to acknowledge Maple Lodge Hatcheries Ltd., Ontario, Canada, for the in-kind donation of the brown eggshells.

Conflicts of Interest: The authors declare no conflict of interest.

References

1. Al-Salem, S.M.; Lettieri, P.; Baeyens, J. Recycling and recovery routes of plastic solid waste (PSW): A review. *Waste Manag.* **2009**, *29*, 2625–2643. [CrossRef]
2. Geyer, R.; Jambeck, J.R.; Law, K.L. Production, use, and fate of all plastics ever made. *Sci. Adv.* **2017**, *3*, e1700782. [CrossRef]
3. Wright, S.L.; Thompson, R.C.; Galloway, T.S. The physical impacts of microplastics on marine organisms: A review. *Environ. Pollut.* **2013**, *178*, 483–492. [CrossRef] [PubMed]
4. Lebreton, L.; Slat, B.; Ferrari, F.; Sainte-Rose, B.; Aitken, J.; Marthouse, R.; Hajbane, S.; Cunsolo, S.; Schwarz, A.; Levivier, A.; et al. Evidence that the great pacific garbage patch is rapidly accumulating plastic. *Sci. Rep.* **2018**, *8*, 4666. [CrossRef]
5. Chamas, A.; Moon, H.; Zheng, J.; Qiu, Y.; Tabassum, T.; Jang, J.H.; Abu-Omar, M.; Scott, S.L.; Suh, S. Degradation rates of plastics in the environment. *ACS Sustain. Chem. Eng.* **2020**, *8*, 3494–3511. [CrossRef]
6. Market—European Bioplastics. 2022. Available online: https://www.european-bioplastics.org/market/ (accessed on 20 July 2023).
7. Emadian, S.M.; Onay, T.T.; Demirel, B. Biodegradation of bioplastics in natural environments. *Waste Manag.* **2017**, *59*, 526–536. [CrossRef]
8. Chariyachotilert, C.; Joshi, S.; Selke, S.E.; Auras, R. Assessment of the properties of poly (L-lactic acid) sheets produced with differing amounts of postconsumer recycled poly (L-lactic acid). *J. Plast. Film Sheeting* **2012**, *28*, 314–335. [CrossRef]
9. Iñiguez-Franco, F.; Auras, R.; Dolan, K.; Selke, S.; Holmes, D.; Rubino, M.; Soto-Valdez, H. Chemical recycling of poly (lactic acid) by water-ethanol solutions. *Polym. Degrad. Stab.* **2018**, *149*, 28–38. [CrossRef]
10. Zuiderduin, W.C.J.; Westzaan, C.; Huetink, J.; Gaymans, R.J. Toughening of polypropylene with calcium carbonate particles. *Polymer* **2003**, *44*, 261–275. [CrossRef]
11. Leong, Y.W.; Abu Bakar, M.B.; Ishak, Z.M.; Ariffin, A.; Pukanszky, B. Comparison of the mechanical properties and interfacial interactions between talc, kaolin, and calcium carbonate filled polypropylene composites. *J. Appl. Polym. Sci.* **2004**, *91*, 3315–3326. [CrossRef]
12. McNeill, I.C.; Mohammed, M.H. Thermal analysis and degradation mechanisms of blends of low density polyethylene, poly(ethyl acrylate) and ethylene ethyl acrylate copolymer with calcium carbonate. *Polym. Degrad. Stab.* **1995**, *49*, 263–273. [CrossRef]
13. Osman, M.A.; Atallah, A.; Suter, U.W. Influence of excessive filler coating on the tensile properties of LDPE–calcium carbonate composites. *Polymer* **2004**, *45*, 1177–1183. [CrossRef]
14. Wang, W.Y.; Zeng, X.F.; Wang, G.Q.; Chen, J.F. Preparation and characterization of calcium carbonate/low-density-polyethylene nanocomposites. *J. Appl. Polym. Sci.* **2007**, *106*, 1932–1938. [CrossRef]
15. Mantia, F.P.L.; Morreale, M.; Scaffaro, R.; Tulone, S. Rheological and mechanical behavior of LDPE/calcium carbonate nanocomposites and microcomposites. *J. Appl. Polym. Sci.* **2013**, *127*, 2544–2552. [CrossRef]
16. Bomal, Y.; Godard, P. Melt viscosity of calcium-carbonate-filled low density polyethylene: Influence of matrix-filler and particle-particle interactions. *Polym. Eng. Sci.* **1996**, *36*, 237–243. [CrossRef]
17. Teixeira, S.C.S.; Moreira, M.M.; Lima, A.P.; Santos, L.S.; Da Rocha, B.M.; De Lima, E.S.; da Costa, R.A.; da Silva, A.L.N.; Rocha, M.C.; Coutinho, F.M. Composites of high density polyethylene and different grades of calcium carbonate: Mechanical, rheological, thermal, and morphological properties. *J. Appl. Polym. Sci.* **2006**, *101*, 2559–2564. [CrossRef]
18. Cree, D.; Rutter, A. Sustainable bio-inspired limestone eggshell powder for potential industrialized applications. *ACS Sustain. Chem. Eng.* **2015**, *3*, 941–949. [CrossRef]
19. Pliya, P.; Cree, D. Limestone derived eggshell powder as a replacement in Portland cement mortar. *Constr. Build. Mater.* **2015**, *95*, 1–9. [CrossRef]
20. Toro, P.; Quijada, R.; Yazdani-Pedram, M.; Arias, J.L. Eggshell, a new bio-filler for polypropylene composites. *Mater. Lett.* **2007**, *61*, 4347–4350. [CrossRef]
21. Lin, Z.; Zhang, Z.; Mai, K. Preparation and properties of eggshell/β-polypropylene bio-composites. *J. Appl. Polym. Sci.* **2012**, *125*, 61–66. [CrossRef]
22. Ghabeer, T.; Dweiri, R.; Al-Khateeb, S. Thermal and mechanical characterization of polypropylene/eggshell biocomposites. *J. Reinf. Plast. Compos.* **2013**, *32*, 402–409. [CrossRef]
23. Feng, Y.; Ashok, B.; Madhukar, K.; Zhang, J.; Zhang, J.; Reddy, K.O.; Rajulu, A.V. Preparation and characterization of polypropylene carbonate bio-filler (eggshell powder) composite films. *Int. J. Polym. Anal. Charact.* **2014**, *19*, 637–647. [CrossRef]
24. Kumar, R.; Dhaliwal, J.S.; Kapur, G.S.; Shashikant. Mechanical properties of modified biofiller-polypropylene composites. *Polym. Compos.* **2014**, *35*, 708–714. [CrossRef]

25. Shuhadah, S.; Supri, A.G. LDPE-isophthalic acid modified egg shell powder composites (LDPE/ESPI). *J. Phys. Sci.* **2009**, *20*, 87–98.
26. Sivarao; Salleh, M.R.; Kamely, A.; Tajul, A.; Taufik, R.S. Mechanical properties modification of polyethylene (PE) for CaCO$_3$ particulated composites. *Adv. Mater. Res.* **2011**, *264*, 880–887.
27. Nwanonenyi, S.C.; Obidiegwu, M.U.; Onuchukwu, T.S.; Egbuna, I.C. Studies on the properties of linear low density polyethylene filled oyster shell powder. *Int. J. Eng. Sci.* **2013**, *2*, 42.
28. Hassan, S.B.; Aigbodion, V.S.; Patrick, S.N. Development of polyester/eggshell particulate composites. *Tribol. Ind.* **2012**, *34*, 217.
29. Hassan, T.A.; Rangari, V.K.; Jeelani, S. Value-added biopolymer nanocomposites from waste eggshell-based CaCO$_3$ nanoparticles as fillers. *ACS Sustain. Chem. Eng.* **2014**, *2*, 706–717. [CrossRef]
30. Ashok, B.; Naresh, S.; Reddy, K.O.; Madhukar, K.; Cai, J.; Zhang, L.; Rajulu, A.V. Tensile and thermal properties of poly (lactic acid)/eggshell powder composite films. *Int. J. Polym. Anal. Charact.* **2014**, *19*, 245–255. [CrossRef]
31. Cree, D.; Soleimani, M. Bio-Based White Eggshell as a Value-Added Filler in Poly (Lactic Acid) Composites. *J. Compos. Sci.* **2023**, *7*, 278. [CrossRef]
32. Liang, J.Z. Toughening and reinforcing in rigid inorganic particulate filled poly (propylene): A review. *J. Appl. Polym. Sci.* **2002**, *83*, 1547–1555. [CrossRef]
33. Rong, M.Z.; Zhang, M.Q.; Ruan, W.H. Surface modification of nanoscale fillers for improving properties of polymer nanocomposites: A review. *Mater. Sci. Technol.* **2006**, *22*, 787–796. [CrossRef]
34. Tan, M.A.; Yeoh, C.K.; Teh, P.L.; Rahim, N.A.A.; Song, C.C.; Voon, C.H. Effect of zinc oxide suspension on the overall filler content of the PLA/ZnO composites and cPLA/ZnO composites. *e-Polymers* **2023**, *23*, 20228113. [CrossRef]
35. Wei, T.; Jin, K.; Torkelson, J.M. Isolating the effect of polymer-grafted nanoparticle interactions with matrix polymer from dispersion on composite property enhancement: The example of polypropylene/halloysite nanocomposites. *Polymer* **2019**, *176*, 38–50. [CrossRef]
36. Moreno, J.F.; da Silva, A.L.N.; da Silva, A.H.M.D.F.T.; de Sousa, A.M.F. Preparation and characterization of composites based on poly (lactic acid) and CaCO$_3$ nanofiller. In *AIP Conference Proceedings*; AIP Publishing: Melville, NY, USA, 2015; Volume 1664. [CrossRef]
37. Nakagawa, H.; Sano, H. *Improvement of Impact Resistance of Calcium Carbonate Filled Polypropylene and Poly-Ethylene Block Copolymer*; Polymer Preprints, Division of Polymer Chemistry; American Chemical Society: Chicago, IL, USA, 1985; pp. 249–250.
38. Toro, P.; Quijada, R.; Arias, J.L.; Yazdani-Pedram, M. Mechanical and morphological studies of poly (propylene)-filled eggshell composites. *Macromol. Mater. Eng.* **2007**, *292*, 1027–1034. [CrossRef]
39. Pukánszky, B.; Fekete, E. Aggregation tendency of particulate fillers: Determination and consequences. *Polym. Polym. Compos.* **1998**, *6*, 313–322.
40. Lazzeri, A.; Thio, Y.S.; Cohen, R.E. Volume strain measurements on CaCO$_3$/polypropylene particulate composites: The effect of particle size. *J. Appl. Polym. Sci.* **2004**, *91*, 925–935. [CrossRef]
41. DeArmitt, C. Understanding filler interactions improves impact resistance. *Plast. Addit. Compd.* **2006**, *8*, 34–39. [CrossRef]
42. Zhu, Y.D.; Allen, G.C.; Jones, P.G.; Adams, J.M.; Gittins, D.I.; Heard, P.J.; Skuse, D.R. Dispersion characterisation of CaCO$_3$ particles in PP/CaCO$_3$ composites. *Compos. Part A Appl. Sci. Manuf.* **2014**, *60*, 38–43. [CrossRef]
43. Eselini, N.; Tirkes, S.; Akar, A.O.; Tayfun, U. Production and characterization of poly (lactic acid)-based biocomposites filled with basalt fiber and flax fiber hybrid. *J. Elastomers Plast.* **2020**, *52*, 701–716. [CrossRef]
44. Dugan, J.S. Novel properties of PLA fibers. *Int. Nonwovens J.* **2001**, *3*, 1558925001OS-01000308.
45. Mitchell, C. High purity limestone quest. *Ind. Miner.* **2011**, *531*, 48–51.
46. Alamillo-López, V.M.; Sánchez-Mendieta, V.; Olea-Mejía, O.F.; González-Pedroza, M.G.; Morales-Luckie, R.A. Efficient removal of heavy metals from aqueous solutions using a bionanocomposite of eggshell/Ag-Fe. *Catalysts* **2020**, *10*, 727. [CrossRef]
47. Hincke, M.T.; Nys, Y.; Gautron, J.; Mann, K.; Rodriguez-Navarro, A.B.; McKee, M.D. The eggshell: Structure, composition and mineralization. *Front. Biosci.-Landmark* **2012**, *17*, 1266–1280. [CrossRef]
48. Walker, R.A.; Wilson, K.; Lee, A.F.; Woodford, J.; Grassian, V.H.; Baltrusaitis, J.; Rubasinghege, G.; Cibin, G.; Dent, A. Preservation of York Minster historic limestone by hydrophobic surface coatings. *Sci. Rep.* **2012**, *2*, 880. [CrossRef]
49. Hakim, S.S.; Olsson, M.H.M.; Sørensen, H.O.; Bovet, N.; Bohr, J.; Feidenhans'l, R.; Stipp, S.L.S. Interactions of the calcite {10.4} surface with organic compounds: Structure and behaviour at mineral–organic interfaces. *Sci. Rep.* **2017**, *7*, 7592. [CrossRef] [PubMed]
50. Cassar, J. *Deterioration of the Globigerina Limestone of the Maltese Islands*; Special Publications; Geological Society: London, UK, 2002; Volume 205, pp. 33–49.
51. Baxter-Jones, C. Egg hygiene: Microbial contamination, significance and control. In *Avian Incubation, Proceedings of the Poultry Science Symposium, West of Scotland College, Auchincruive, Ayr, Scotland, 12–15 September 1989*; No. 22; Tullett, S.G., Ed.; Butterworth: London, UK, 1991; pp. 269–276.
52. Mueller, C.A.; Burggren, W.W.; Tazawa, H. The physiology of the avian embryo. In *Sturkie's Avian Physiology*; Academic Press: Cambridge, MA, USA, 2022; pp. 1015–1046.
53. Ndazi, B.S.; Karlsson, S. Characterization of hydrolytic degradation of polylactic acid/rice hulls composites in water at different temperatures. *Express Polym. Lett.* **2011**, *5*, 119–131. [CrossRef]

54. Casalini, T.; Rossi, F.; Castrovinci, A.; Perale, G. A perspective on polylactic acid-based polymers use for nanoparticles synthesis and applications. *Front. Bioeng. Biotechnol.* **2019**, *7*, 259. [CrossRef] [PubMed]
55. Taib, R.M.; Ramarad, S.; Ishak, Z.A.M.; Todo, M. Properties of kenaf fiber/polylactic acid biocomposites plasticized with polyethylene glycol. *Polym. Compos.* **2010**, *31*, 1213–1222. [CrossRef]
56. Lunt, J. Large-scale production, properties and commercial applications of polylactic acid polymers. *Polym. Degrad. Stab.* **1998**, *59*, 145–152. [CrossRef]
57. Ortenzi, M.A.; Gazzotti, S.; Marcos, B.; Antenucci, S.; Camazzola, S.; Piergiovanni, L.; Farina, H.; Di Silvestro, G.; Verotta, L. Synthesis of polylactic acid initiated through biobased antioxidants: Towards intrinsically active food packaging. *Polymers* **2020**, *12*, 1183. [CrossRef]
58. Wang, Y.; Xu, Y.; He, D.; Yao, W.; Liu, C.; Shen, C. "Nucleation density reduction" effect of biodegradable cellulose acetate butyrate on the crystallization of poly (lactic acid). *Mater. Lett.* **2014**, *128*, 85–88. [CrossRef]
59. Oliaei, E.; Heidari, B.S.; Davachi, S.M.; Bahrami, M.; Davoodi, S.; Hejazi, I.; Seyfi, J. Warpage and shrinkage optimization of injection-molded plastic spoon parts for biodegradable polymers using Taguchi, ANOVA and artificial neural network methods. *J. Mater. Sci. Technol.* **2016**, *32*, 710–720. [CrossRef]
60. Fukushima, K.; Feijoo, J.L.; Yang, M.C. Comparison of abiotic and biotic degradation of PDLLA, PCL and partially miscible PDLLA/PCL blend. *Eur. Polym. J.* **2013**, *49*, 706–717. [CrossRef]
61. Limsukon, W.; Auras, R.; Selke, S. Hydrolytic degradation and lifetime prediction of poly (lactic acid) modified with a multifunctional epoxy-based chain extender. *Polym. Test.* **2019**, *80*, 106108. [CrossRef]
62. Larson, T.E.; Sollo, F.W., Jr.; McGurk, F.F. Complexes affecting the solubility of calcium carbonate in water. In *ISWS Contract Report CR 145*; University of Illinois at Urbana-Champaign Water Resources Center: Urbana, IL, USA, 1973.
63. Tegethoff, F.W.; Rohleder, J.; Kroker, E. (Eds.) *Calcium Carbonate: From the Cretaceous Period into the 21st Century*; Springer Science & Business Media: Berlin/Heidelberg, Germany, 2001.
64. Tamboura, S.; Abdessalem, A.; Fitoussi, J.; Daly, H.B.; Tcharkhtchi, A. On the mechanical properties and damage mechanisms of short fibers reinforced composite submitted to hydrothermal aging: Application to sheet molding compound composite. *Eng. Fail. Anal.* **2022**, *131*, 105806. [CrossRef]
65. Abdessalem, A.; Tamboura, S.; Fitoussi, J.; Ben Daly, H.; Tcharkhtchi, A. Bi-phasic water diffusion in sheet molding compound composite. *J. Appl. Polym. Sci.* **2020**, *137*, 48381. [CrossRef]
66. Li, Y.; Li, Y.; Liu, S.; Tang, Y.; Mo, B.; Liao, H. New zonal structure and transition of the membrane to mammillae in the eggshell of chicken Gallus domesticus. *J. Struct. Biol.* **2018**, *203*, 162–169. [CrossRef]
67. Kendall, J. XCVI. The solubility of calcium carbonate in water. *Lond. Edinb. Dublin Philos. Mag. J. Sci.* **1912**, *23*, 958–976. [CrossRef]
68. Vogt, E. Effects of Commercial modifiers on flow properties of hydrophobized limestone powders. *Pol. J. Environ. Stud.* **2013**, *22*, 1213–1218.
69. Sabir, M.I.; Xu, X.; Li, L. A review on biodegradable polymeric materials for bone tissue engineering applications. *J. Mater. Sci.* **2009**, *44*, 5713–5724. [CrossRef]
70. Al-Shirawi, M.; Karimi, M.; Al-Maamari, R.S. Impact of carbonate surface mineralogy on wettability alteration using stearic acid. *J. Pet. Sci. Eng.* **2021**, *203*, 108674. [CrossRef]

waste

Article

Establishing Experimental Conditions to Produce Lignin-Degrading Enzymes on Wheat Bran by *Trametes versicolor* CM13 Using Solid State Fermentation

Paul W. Baker * and Adam Charlton

Biocomposites Centre, Bangor University, Deiniol Road, Bangor LL57 2UW, Wales, UK; adam.charlton@bangor.ac.uk
* Correspondence: paul.baker@bangor.ac.uk; Tel.: +44-(0)-124-838-2640

Abstract: Valorisation of wheat bran can be achieved by solid state fermentation (SSF), through application of this material as a growth substrate for a natural white rot fungal isolate, *Trametes versicolor* CM13, to produce lignin-degrading enzymes. One of the main challenges in optimising and upscaling (SSF) processes is the accurate adjustment and maintenance of moisture conditions. This factor was assessed in the scale up of microcosms and was evaluated over 28 days, under two slightly different moisture contents, reflecting minor differences in experimental conditions during set up and operation of the SSF process. In addition, the microcosms were processed differently from the initial trial using homogenisation of whole microcosms to create a homogeneous mixture prior to sampling. This appeared to result in less variation among the collected samples from the microcosms. Variation of measured parameters as a percentage of actual values measured ranged from 1.33% to 144% in the unmixed microcosms and from 0.77% to 36.0% in the pre-mixed microcosms. Decomposition in the more saturated microcosms progressed more quickly as hemicellulose content decreased and reached a steady state after 14 days, whereas hemicellulose content continued to decrease until 21 days in the less saturated microcosms. Lignin-degrading enzyme activities were not significantly different between either sets of experiments except for laccase on day 7. Laccase and manganese peroxidase activities were highest on day 21 and were similar in both sets of experiments. Enzyme activities on day 21 in the microcosms at moisture content of 42.9% and at 54.6% for laccase activities were 750 ± 30.5 and 820 ± 30.8 units, and for manganese peroxidase, activities were 23.3 ± 6.45 and 21.4 ± 21.4 units, respectively. These results revealed different decomposition rates during the early stage of solid-state fermentation as a function of the initial moisture content, whereas final enzyme activities and fibre content during the later stage were similar in microcosms having different moisture contents at the start.

Keywords: fungi; laccase; microcosm; homogenisation; hemicellulose; cellulose; fibre; biomass

Citation: Baker, P.W.; Charlton, A. Establishing Experimental Conditions to Produce Lignin-Degrading Enzymes on Wheat Bran by *Trametes versicolor* CM13 Using Solid State Fermentation. *Waste* **2023**, *1*, 711–723. https://doi.org/10.3390/waste1030042

Academic Editors: Vassilis Athanasiadis, Dimitris P. Makris and Catherine N. Mulligan

Received: 3 July 2023
Revised: 10 August 2023
Accepted: 15 August 2023
Published: 18 August 2023

1. Introduction

One potential application for agricultural crop waste is as a substrate for solid-state fermentation (SSF) with fungi, to produce industrial enzymes that have increasing uses and demand, particularly in the food industry. In terms of total global tonnage produced, wheat, rice and corn are the predominant crops globally, leading to large quantities of waste by-products [1]. Wheat is the major crop produced in Europe, and wheat bran is one of the by-products constituting 14–19% of the whole grain. Europe is estimated to produce 100 million tonnes of wheat bran annually [2] and some of this wheat bran could be used as a substrate for enzyme production using SSF with higher valorisation potential. SSF involves the growth of microorganisms, particularly fungi, on a moist, solid substrate but in the absence of free-flowing water [3]. The advantage of SSF, compared with the current, commercial processes for manufacturing enzymes using liquid (submerged) fermentation, is the reduced water input and the increased product output [4]. Submerged fermentation processes can give rise to dilute product streams and the removal of water is a costly process step that can be avoided using SSF [4]. Concentrated enzymes extracts were recovered from SSF during downstream processing with additional water. The majority of fungal SSF processes have focused on use of the Ascomycetes, particularly the *Aspergillus* and *Trichoderma* species, and while some studies described using Agaricomycetes, most of these were *Pleurotus* species [5]. Many of these studies may have used *Pleurotus* sp. because this fungus readily colonises and degrades a wide range of lignocellulose substrates. Consequently, another potential advantage of SSF when using these fungi is that the residual lignocellulosic material, after enzyme extraction, has potential application in downstream fermentation of the accessible sugars to a range of products, including biofuels and platform chemicals. In this study, *Trametes versicolor* CM13, a white rot fungus, and a natural isolate from an oak log, was used as an alternative to other fungi previously investigated, because it has been reported to exhibit high enzyme activity linked to the degradation of different wood species [6].

White rot fungi produce laccase, manganese peroxidase, lignin peroxidase and versatile peroxidase, which are the main groups of extracellular, lignin-degrading enzymes previously investigated, although some bacteria, e.g., *Pseudomonas* sp. and *Cupriavidus basilensis*, also possess peroxidases and laccases, resulting in the production of lipids and polyhydroxyalkanoate [7]. Lignin is a major component within woods and grasses and is a highly crosslinked, large molecular weight polymer, composed of three different monolignols (*p*-coumaryl alcohol, guaiacyl alcohol and sinapyl alcohol) and polyphenolic compounds [8]. The combinations of the lignin-degrading genes can vary significantly between different fungal species where, for example, *T. versicolor* was shown to possess 12 short manganese peroxidase genes and 10 lignin peroxidase genes in the whole genome [9]. A previous study reported that manganese peroxidase activity was higher compared with lignin peroxidase activity, which was often barely detectable [10]. Purified manganese peroxidase from *Ceriporiopsis subvermispora* showed significant degradation of a radioactively labelled lignin construct, linked to a polyethylene glycol backbone, and degradation increased as function of increasing enzyme concentration [11]. The role of manganese peroxidase in this process was further substantiated with the use of a recombinant manganese peroxidase, which revealed chemical changes in the aliphatic chains connected to the aromatic rings. In contrast, another study reported that the aromatic components within the lignin macromolecule were mineralised by lignin peroxidase [12]. Later studies have shown that higher levels of lignin degradation were achieved using a combination of both laccase and manganese peroxidase [13,14]. Consequently, manganese peroxidase has potential industrial application in degrading textile dyes [15] and lignin degradation for biofuel production in the pulp and paper industry [16,17].

One of the potential problems associated with SSF is the accurate adjustment of the moisture content at the start and minor changes in moisture content during fermentation that deviate from the optimal conditions for rapid fungal growth. Most lignocellulosic substrates have a heterogenous composition in terms of cell types and sizes that will affect water uptake. Furthermore, maintaining the same moisture content is impossible when using forced aeration because this caused evaporation, while, conversely, biomass decomposition leads to increased saturation. Consequently, moisture content has been highlighted as one of the most important parameters affecting the efficiency and potential upscaling of fungal SSF processes [18]. Therefore, the aim of this study was to examine the effect of minor differences in the initial moisture content on the outcome of the lignin-degrading enzymes, along with fibre decomposition.

2. Materials and Methods

2.1. Particle Analysis of Wheat Bran

Triplicate samples of wheat bran (20 g) were separated through overlapping 3 cm diameter sieves on the Octogon 200 sieve shaker (Endecotts, London, UK) for 10 min at an amplitude of 2.7 mm at 3000 min^{-1} and 50 Hz. The material collected in each fraction was weighed.

2.2. Analysis of Growth of T. versicolor CM13 on Wheat Bran over Time

Microcosms containing wheat bran (20 g) were adjusted with deionised water (15 mL), autoclaved, and then inoculated with spores collected from *T. versicolor* CM13 growing on malt extract agar for one month. Sterile deionised water (10 mL) was added to the plates and the surface was scrapped aseptically with a loop. The spore suspension was pipetted from the plates into 30 mL sterile deionised water and mixed thoroughly. Microscopy revealed a spore population of about 1×10^7 cells per ml with only a few fragments of aerial mycelium being present. This suspension (3 mL) was added to each microcosm with mixing so that the final moisture content was 47.4%. All microcosms were incubated at 22 °C and when initial growth was observed, duplicate microcosms were destructively sampled by mixing the entire contents with a spatula and collecting two samples (5 g). One of these samples was used for moisture analysis and the resulting dried material was then used for fibre analysis, while the other sample was used for enzyme extraction and pH measurement.

2.3. Effect of Enzyme Activities of T. versicolor CM13 at Two Different Moisture Contents

The experiment was repeated as described in the previous section, except that a larger quantity of material was used to reduce differences caused by rates of evaporation. Two sets of microcosms were prepared with wheat bran (100 g) containing deionised water, one with a final moisture content of 42.9% and another with 54.6%. The moisture content at 42.9% represented the lowest possible moisture content where fungal growth occurred and the moisture content at 54.6% was sufficiently high enough to be statistically different from the moisture content at 42.9% at $p < 0.05$. Spores were collected from the surface of *T. versicolor* CM13 growing on malt extract agar for one month in sterile distilled water, and the spores were inoculated into the microcosms with mixing. The microcosms were destructively sampled in triplicate on days 7, 14 and 21, as described in the previous section, by mixing the degraded substrate with a sterile spatula. The entire microcosms were then homogenised (Waring blender BB255SK, Conair Corporation, Stamford, CT, USA) for 60–150 s until the clumps of mycelia were no longer visible. Samples (5 g) were removed from each microcosm for moisture analysis, fibre analysis, extraction of enzymes and pH measurements.

2.4. Analysis of Samples

On the sampling day, the contents of the microcosm were mixed using a spatula and 5 g was removed to determine moisture analysis, which was dried at 103 °C until the rate of moisture loss was less than 20 mg/min. The dried material (0.5 ± 0.025 g) was also used in stepwise fibre analysis using neutral detergent fibre (NDF) and acid detergent fibre (ADF) solutions in the Fibre analyser (Ankom, New York, NY, USA) machine to determine non-fibre, hemicellulose and cellulose contents, using a reported method [13]. The insoluble lignin content in the bags was determined after ADF extraction using 72% (v/v) sulphuric acid (500 mL) in the Daisy machine (Ankom, New York, NY, USA) by rotating at room temperature (3 h). The bags were washed repeatedly in 2 L tap water until pH 7. Ash contents were determined by heating 0.5 g of the original material before fibre analysis in crucibles in a furnace (600 °C for 4 h). The calculated ash contents in each fibre were subtracted from each fibre fraction assuming equal distribution of ash among the different types of fibre.

Klason lignin contents were determined on a dried sample (0.3 g) using 72% sulphuric acid (5 mL), which was incubated at 25 °C for 3 h and was regularly stirred. Deionised water (240 mL) was added to the suspension, which was then autoclaved (121 °C, for 1 h). The hot suspensions were filtered through pre-weighed glass microfibre filters (Whatman, Little Chalfont, Buckinghamshire, UK) along with additional hot deionised water to remove any remaining soluble compounds. The filters were oven dried (103 °C) overnight and weighed and the lignin content was determined by subtracting the weight of the filter. The ash contents were subtracted from these values.

Another sample (5 g) was resuspended in deionised water (100 mL), blended (1 L Waring blender) and filtered through cotton wool. The pH in this filtrate was measured. A sample was removed from the crude filtrate and filtered through a 0.2 μm polycarbonate filter to remove any remaining solids that could increase the background colour of the colorimetric assays.

Laccase activity was determined using 400 μL 3 mM ABTS (2,2′-azino-bis (3-ethylbenzo thiazoline-6-sulfonic acid), 400 μL 1 mM sodium acetate buffer, pH 4.5 and 400 μL of sample and the absorbance values of the samples were measured at 420 nm. The control sample consisted of 400 μL deionised water instead of the sample. Laccase activity was calculated based on the molar extinction coefficient of ABTS being ε_{420}, 36,000 $M^{-1} \cdot cm^{-1}$ [19].

Manganese peroxide was determined in 1.1 mL of 50 mM sodium succinate (pH 4.5), 50 mM sodium lactate (pH 4.5) using 0.1 mM $MnSO_4$, 0.1 mM phenol red with 50 μL of sample [20]. Activity was initiated with 50 μM hydrogen peroxide (prepared as 20 μL of 0.43%) and at the end of 30 min incubation, 57.5 μL of 10% sodium hydroxide was added. The absorbance values of the samples were measured at 610 nm and calculated using the molar extinction coefficient (ε_{610}, 30,737 $M^{-1} \cdot cm^{-1}$). An experimental control was performed using deionised water instead of the sample.

Another assay to measure decolourisation activity was performed with Remazol brilliant blue (800 μL, 0.16 mM) in water with a 400 μL sample and hydrogen peroxide (20 μL, 38 mM). Decolourisation of this azo dye is indicative of the combined activities of laccase and manganese peroxidase, where one or the other enzyme compensates for the other enzyme when the activity of one enzyme is higher than the other [21]. The absorbance values of the samples were measured at 592 nm and activity was calculated based on the molar extinction coefficient of Remazol brilliant blue being ε_{592}, 6170 $M^{-1} \cdot cm^{-1}$ [22]. The sample changed the absorbance value of the assay and, consequently, it was necessary to include the sample within the control. Experimental controls were performed using 400 μL of the samples and deionised water (20 μL), instead of hydrogen peroxide.

Lignin peroxidase activity was determined in 125 mM sodium tartrate (pH 3) using 0.16 mM azure B and 5 mM hydrogen peroxidase based on a similar method previously described [23]. The absorbance values of the samples were measured in a microplate reader at 651 nm (ε_{651}, 48,800 $M^{-1} \cdot cm^{-1}$). The same method of experimental controls was performed as with Remazol brilliant blue using the sample instead of water.

All enzyme activities were determined using pre-warmed reagents incubated (30 °C) with horizontal shaking (600 rpm). At the end of the incubation (30 min), 100 μL of each sample was measured in a microplate reader (BioTek Epoch, Agilent, Santa Clara, CA, USA) at the respective absorbance value. Enzyme activities were calculated using the molar extinction coefficient of the coloured compound being measured. Enzyme units are μM of the coloured compound formed or decolourised per min. The pH values were measured in the remaining filtrates.

2.5. Statistical Analysis

Growth studies of *T. versicolor* CM13 on wheat bran over time and under different moisture conditions were conducted using duplicate and triplicate microcosms, respectively, and single samples were analysed from each microcosm. Data analysis was conducted using ANOVA (analysis of variance) in SPSS version 27 with Tukey's post hoc test. The averages and standard deviations are shown in the figures and tables, while superscript letters represent significant differences. Those with the same letters are similar and those with different letters are significantly different at $p < 0.05$.

3. Results and Discussion

3.1. Growth of T. versicolor on Wheat Bran over Time

In relation to the size distribution of the wheat bran, it was determined that 91% of particles ranged from 0.5 to 2 mm (Figure 1), and this can have a significant impact on the efficiency of SSF processes due to increased uptake of water by the growth substrate as a function of reducing particle size. For example, the thermo-mechanical pre-treatment of *Miscanthus*, reported previously, using continuous pressurised disc refining resulted in the formation of smaller sized particles (<250 µm) compared to the original chopped material (>1 mm), and lower moisture contents were required to achieve saturation in the refined material [11]. The potential benefits of lower moisture contents could lead to increased gaseous exchange and more rapid dissipation of heat, which are critical factors effecting SSF efficiency. Wheat bran is composed of three different tissues, the outer pericarp, an intermediate strip and the aleurone layer possessing elastic properties [24]. In addition, a minor proportion of the seed starch (~19%) becomes associated with wheat bran due to the imperfect detachment of the bran layer from the grain [2]. Consequently, it assumed that any moisture is either directly absorbed due to the water binding capacity of the fibres or forms a thin film around the wheat bran particles, which may limit the amount of water absorbed by the bran. This is supported by previous studies which reported that the moisture content for the optimal growth on wheat bran of *Trichoderma reesei* QM9414 was ~46% [25] and for generating high cellulase activity from a mutagenic strain of *Aspergillus* sp. was 33% [26]. These values are considerably lower than the atypical low moisture content of 67%, which only supported the growth of only one species of white rot fungi, *Pleurotus erygnii*, on wheat straw [27]. This study also reported that the addition of 30% wheat bran to wheat straw reduced the moisture content to 44%, but this level of moisture continued to support the growth of *P. erygnii*.

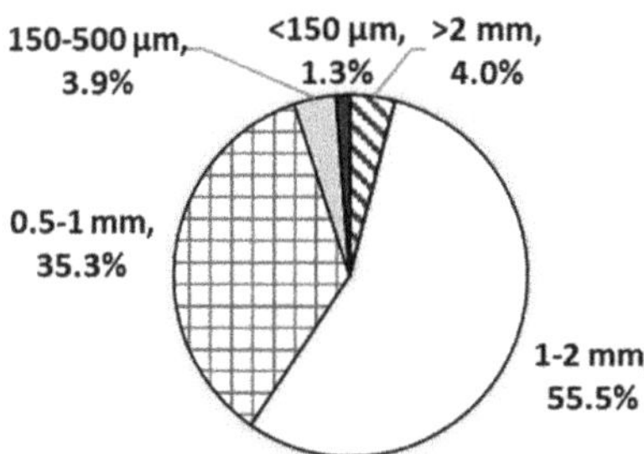

Figure 1. Percentage distribution of particle sizes of dry wheat bran before inoculation with *T. versicolor* CM13 spores which was determined using a sieve shaker.

The percentage distribution of hemicellulose, cellulose and lignin in wheat bran were 22.7%, 6.8% and 3.2%, respectively, with the remaining 67.3% composed of the non-fibre component. These values compare well with previously reported values [28]. During SSF, fungal degradation caused the non-fibre content to decrease on days 25 to 37 compared with days 0 to 13, which was statistically significant (Figure 2), possibly indicating that the presence of starch [2], pectins and soluble acetylated hemicelluloses [29] that contributed to fungal growth. Arabinoxylans form about 50% of the dry weight present in wheat bran [21], which is higher than the total fibre content, indicating that a significant proportion of these are soluble. In addition, the hemicellulose content showed a similar decrease over time and was significantly lower on days 13 and days 25 to 37 compared with day 7, whilst the cellulose content significantly increased on days 25 to 37 compared with days 0 to 13. These results indicate that the fungus was producing hemicellulases rather than cellulases, which was indicated by the decreasing quantities of hemicellulose lost through decomposition whereas cellulose content increased. Furthermore, the quantities of hemicellulose in wheat bran were higher than cellulose that may facilitate the fungus to produce hemicellulases.

Similar patterns have been observed with mixed substrates of wheat straw and wheat bran appearing to suppress cellulose degradation [27,30]. However, fungal degradation of wheat straw without wheat bran showed significant cellulose decomposition. Similarly, the insoluble lignin content was significantly higher on days 13 and 37 compared with days 0 and 7. Consequently, the lignin content showed a decrease only during the initial stage of decomposition. Lignin is closely associated with hemicellulose and cellulose, and, presumably, the degradation of lignin with increasing lignin-degrading enzyme activity is accompanied with increased hemicellulose decomposition. A previous study highlighted that the addition of increasing quantities of wheat bran (2% to 10%) to wheat straw had no effect on delignification, possibly due to the higher nitrogen content associated with wheat bran which might be expected to lower the rate of delignification [31]. The ash content appeared to increase with incubation time and was higher on day 33 compared with days 0–18, indicating a loss in biomass. The increase in ash content resulted from decomposition of organic material, rather than inorganic components which presumably remain intact during fungal growth.

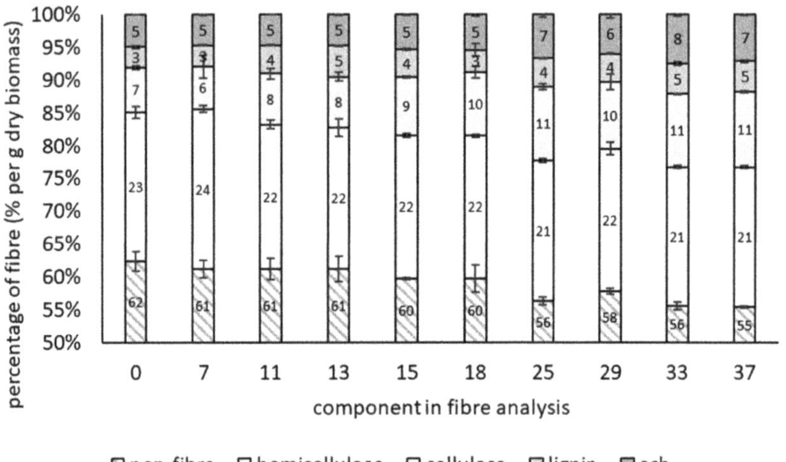

Figure 2. Bars represent the proportions of fibre content along with standard deviations during fungal degradation by *T. versicolor* growing on wheat bran. The decrease in non-fibre content may be due to soluble hemicelluloses acting as nutrients for the fungus.

During the SSF experiments, the pH values decreased over time and were lowest between days 11 and 18 (Figure 3), and this was probably due to organic acid production which was observed by some strains of filamentous fungi [32]. The moisture contents, manganese peroxidase activities and Klason lignin contents (Figures 3 and 4) showed no significant differences over time. These results showed that moisture content was kept constant for the duration of the experiment. However, it was expected that manganese peroxidase activities and Klason lignin content would have changed over the experiment. This is because there appeared to be an increase in manganese peroxidase activities on day 25 and Klason lignin contents appeared to decrease on day 15, which is linked to peak laccase activities. However, the lack of significant differences may be due to the high errors associated with some of the data. Klason lignin values were relatively high, and this may be due to the presence of cutin [33]. Laccase activity was significantly higher on days 13 to 33 compared with days 7, 11 and 37. Laccase activities appeared to increase and decrease over this period perhaps reflecting the expression of redundant genes at different stages during the growth cycle [34]. Laccase activity initially peaked on day 15 and there were two phases of manganese peroxidase activity which peaked on days 11 and 25, while. A similar trend was observed with the growth of *Inonotus obliquus* on wheat bran, although higher manganese peroxidase activity was determined during the first phase [35]. A previous study has shown that the occurrence and length of expression of lignin-degrading enzymes was dissimilar among different species of white rot fungi, when growing on wheat straw [10].

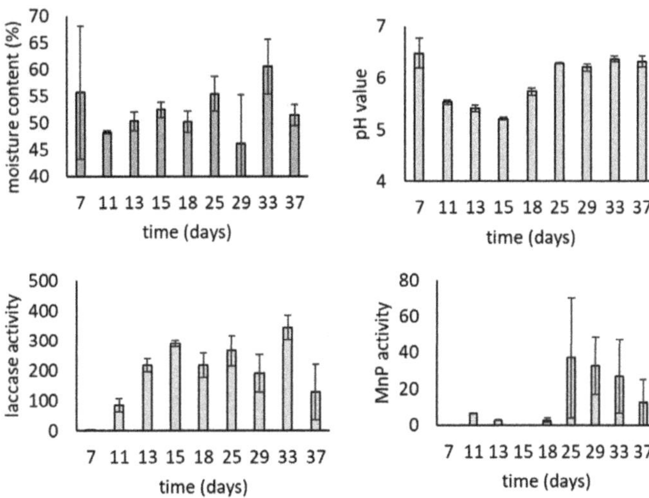

Figure 3. The moisture contents of duplicate microcosms destructively sampled after initial colonisation of wheat bran by *T. versicolor* CM13. The changes in pH possibly relate to the production of organic acids formed by this fungus. Laccase activities increased early and appeared to fluctuate during the experiment whereas manganese peroxidase activities (MnP) were mostly determined later in the experiment.

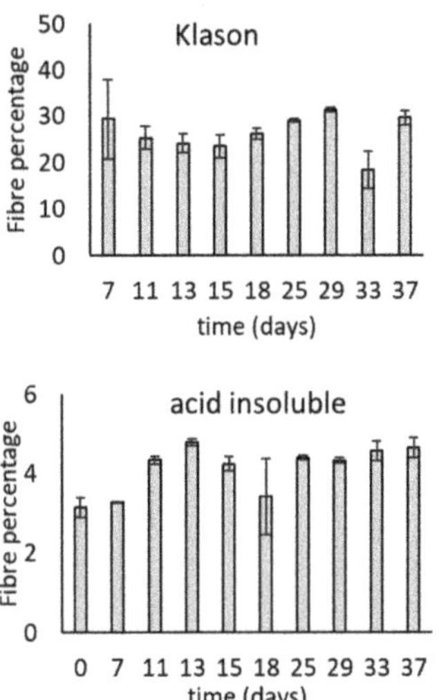

Figure 4. Quantities of Klason lignin (total lignin) and acid insoluble lignin (after removal of hemicellulose and cellulose) in wheat bran degraded by *T. versicolor* CM13.

3.2. Growth of T. versicolor CM13 on Wheat Bran under Two Different Moisture Conditions

Visual examination of the microcosms clearly showed complete colonisation of the microcosms, but some regions showed extensive mycelial growth often near the glass vessel surface. A previous study using scanning electron microscopy reported a network of fungal mycelial clumps (9–12 µm diameter) of *T. versicolor* within a network of growing mycelia on semi-solid medium containing suspended wheat bran [36]. Consequently, samples randomly removed from the microcosms may differ significantly in fungal biomass contents due to this irregular growth of mycelia. Furthermore, it was evident that these mycelial clumps often remained intact during the homogenisation extraction step to recover the enzymes and, therefore, another homogenisation step was incorporated beforehand to process the whole microcosm rather than a small sub-sample.

The use of homogenisation (high shear mixing) to redistribute the fungal mycelium more evenly in the whole microcosm, resulted in lower standard deviations (Table 1), following measurements from a sample of this material. Standard deviations associated with the smaller microcosms (20 g) were greater where sampling was performed without prior homogenization. The effects of evaporation were less pronounced in the larger microcosms due to the larger quantity of biomass.

Table 1. Standard deviations of SSF experimental parameters in the microcosms as a proportion of actual measured values (represented as percentages).

Measured Parameter	Time	Saturation	
Microcosm size	20 g	20 g	100 g
Prior homogenisation	no	no	yes
Moisture content	7.90	15.1	2.25
Biomass loss	ND	19.8	7.53
pH	1.33	3.37	0.77
Laccase	32.7	36.1	6.13
Manganese peroxidase	34.7	98.4	36.0
Remazol Brilliant Blue activity	ND	144	14.8

The moisture contents between the microcosms at 42.9% and at 54.6% were sufficiently close to each other to mimic minor variations around a midpoint of 48.8% that could occur when establishing the microcosms and during operation yet were also significantly different at the start of the experiment (Table 2). The moisture contents of both sets of microcosms at the start of the experiment were within the range for optimal cellulase production by *Aspergillus* sp., growing on wheat and reported in a previous study [26]. Both sets of microcosms differed from each other by 11.8% and both showed a significant increase in moisture content over time, which corresponded with significant biomass decomposition. The final moisture contents of both sets of microcosms differed from each other by 6.1% at the end of the experiment but these results were not significantly different, unlike the moisture contents of both sets of microcosms at the start. Therefore, moisture contents appeared to converge over time during fungal decomposition. However, these microcosms received passive aeration and the effect of forced aeration would undoubtedly have resulted in a different outcome. The moisture content in the more saturated microcosms appeared to remain unchanged from day 14 onwards and this was reflected by the slow increase in biomass loss, pH changes, and laccase and lignin-degrading activities. This indicated that rapid fungal colonisation occurred within 14 days and, thereafter, entered a stationary phase where there was limited growth. In contrast, the moisture content of the less saturated microcosms showed a slower rate of increase that continued until day 21, at which point the moisture contents of both sets of microcosms were not significantly different. Fungal decomposition as shown by biomass loss, pH changes and hemicellulose decomposition revealed that the pattern was similar in the less saturated microcosms as in the more saturated microcosms except there was a lag phase of 7 days in the less saturated microcosms. A slower rate of decomposition occurred from day 14 in the more saturated microcosms that was indicative of stationary phase growth. It was apparent that the stationary phase was achieved on day 21 in the less saturated microcosms. Fungal decomposition during the stationary phase progresses more slowly and, therefore, any differences between the microcosms will become minimal.

Table 2. Physical parameters and lignin enzyme activities of wheat bran colonised by *T. versicolor* CM13 at different moisture contents.

	Day	Low	High
Moisture content	0	42.9 ± 0.18 [a]	54.6 ± 0.80 [b]
	7	54.4 ± 3.87 [b]	57.0 ± 1.73 [bc]
	14	58.9 ± 1.61 [b]	67.5 ± 1.61 [d]
	21	62.6 ± 1.40 [cd]	67.7 ± 0.95 [d]
Biomass loss	7	27.7 ± 6.06 [a]	32.9 ± 5.28 [a]
	14	26.5 ± 2.79 [a]	43.7 ± 3.29 [b]
	21	45.1 ± 0.83 [b]	46.7 ± 0.81 [b]
pH	0	6.22 ± 0.03 [ab]	6.25 ± 0.01 [b]
	7	5.55 ± 0.11 [a]	5.96 ± 0.09 [b]
	14	5.80 ± 0.04 [c]	6.10 ± 0.01 [bc]
	21	6.22 ± 0.02 [ab]	6.27 ± 0.04 [a]
Laccase	0	0.12 ± 0.00 [a]	0.58 ± 0.59 [a]
	7	366 ± 25.6 [b]	628 ± 27.7 [c]
	14	529 ± 87.5 [c]	777 ± 25.2 [d]
	21	750 ± 30.5 [d]	820 ± 30.8 [d]
Manganese peroxidase	7	1.69 ± 0.23 [a]	1.23 ± 0.83 [a]
	14	3.41 ± 1.47 [a]	6.85 ± 3.46 [a]
	21	23.3 ± 6.45 [b]	21.4 ± 2.54 [b]
Lignin peroxidase	7	ND	ND
	14	0.47 ± 0.29 [a]	0.55 ± 0.36 [ab]
	21	0.00 ± 0.66 [ab]	0.00 ± 4.16 [ab]
Lignin-degrading activity	7	0.00 ± 54.66 [a]	107 ± 114 [a]
	14	632 ± 14.5 [b]	792 ± 43.7 [b]
	21	72.4 ± 39.1 [b]	102 ± 10.1 [b]

Numbers with the same letters denotes similarity and those with different letters are significantly different at $p < 0.5$.

In general, the manganese peroxidase activities showed a significant rate of increase that were not significantly different from each other at the end of the experiment, but both had significantly increased compared with activities measured on days 7 and 14. In another study, a comparison was made in the growth of *Pleurotus pulmonarius* on wheat bran at two different moisture contents, which revealed higher manganese peroxidase activity at a moisture content of 75% and higher laccase activities at a moisture content of 91% [37]. However, our study indicates that neither laccase activities nor manganese peroxidase activities showed any difference between the two sets of microcosms, although the moisture conditions in our study were 30–43% lower. Lignin peroxidase activities were low, and the highest activities were obtained on day 14, under both sets of moisture conditions.

Lignin-degrading activities, represented by the decolourisation of Remazol brilliant blue, exhibited significant increases in both set of microcosms on days 14 and 21. Activities were higher on day 14 in both sets of microcosms, but these values were not significantly different to the lower activities determined on day 21. These enzyme activities were similar even though hemicellulose decomposition was different during this period. Previous evidence indicates that Remazol brilliant blue can be degraded by laccase [38,39], manganese peroxidase [40–42] and lignin peroxidase [43]. Consequently, it is possibly indicative of overall delignifying enzyme activity, although another study reported that in studies using purified enzymes, only laccase, and not manganese peroxidase, was effective [44]. This may indicate that only certain manganese peroxidases can degrade Remazol brilliant blue. Our study appears to indicate that laccase, manganese peroxidase and lignin peroxidase were all involved in degrading Remazol brilliant blue, because all these enzymes showed activity during day 14, but none exhibited peak activities that was indicative of single-enzyme-mediated degradation.

The remaining hemicellulose contents in the less saturated microcosms showed a significant decrease after 7 days incubation, which was more noticeable in the more saturated microcosms, but, thereafter, showed only minor decreases in both sets of microcosms (Table 3). The conditions in the more saturated microcosms were more favourable for rapid fungal growth, as evidenced by the greater reduction in the remaining hemicellulose content. Although the quantity of hemicellulose in the less saturated microcosms appeared higher at the end of the experiment than in the more saturated microcosms, this was not significantly different. The cellulose contents in both sets of microcosms showed a similar increase and neither showed significant differences between each of the microcosms on the respective days. These results correlate well with another study on wheat bran, which used NMR analysis to determine the level of hemicellulose degradation by *Pleurotus ostreatus* after 62 days incubation, resulting in a total dry weight loss of 20% [45]. The fibre composition of the two sets of microcosms were highly similar when complete colonisation had occurred on day 21 even though growth patterns during the initial stages were dissimilar. Decomposition occurred more rapidly in the more saturated microcosms which was followed by a slower decomposition rate whereas decomposition was delayed in the less saturated microcosms and continued until day 21.

Table 3. Effect of moisture content of fibre degradation (white area) and percentage of total fibre degraded (light grey area) by *T. versicolor* CM13.

Moisture Content	Time (Days)	Non-Fibre	Hemicellulose	Cellulose	Lignin	Ash
43.9%, 54.6%	0	62.4 ± 1.53 [a]	22.7 ± 0.92 [a]	6.82 ± 0.31 [a]	3.23 ± 0.24 [a]	5.01 ± 0.44 [a]
43.9%	7	60.1 ± 1.23 [a]	21.7 ± 0.79 [b]	9.62 ± 0.45 [ab]	3.84 ± 0.05 [a]	3.28 ± 0.13 [b]
	14	60.1 ± 0.58 [a]	20.5 ± 0.52 [bc]	10.7 ± 0.49 [ab]	3.79 ± 0.23 [a]	3.51 ± 0.08 [b]
	21	58.0 ± 0.69 [a]	21.1 ± 0.06 [bc]	11.5 ± 0.61 [b]	4.31 ± 0.06 [a]	3.61 ± 0.51 [b]
54.6%	7	61.8 ± 0.54 [a]	19.5 ± 0.39 [c]	9.81 ± 0.24 [ab]	3.65 ± 0.13 [a]	3.65 ± 0.66 [b]
	14	62.6 ± 5.09 [a]	20.7 ± 1.38 [bc]	9.73 ± 3.43 [ab]	1.96 ± 3.04 [a]	3.53 ± 0.23 [b]
	21	58.2 ± 0.74 [a]	19.4 ± 0.51 [c]	12.5 ± 0.42 [b]	4.46 ± 0.17 [a]	4.00 ± 0.17 [ab]
43.9%	7	12.29 ± 4.39 [a]	6.63 ± 0.80 [a]	0.17 ± 0.55 [a]	0.22 ± 0.20 [a]	
	14	15.46 ± 2.79 [ab]	8.55 ± 1.40 [ab]	0.00 ± 0.80 [a]	0.47 ± 0.16 [a]	
	21	23.95 ± 0.85 [c]	10.72 ± 0.15 [c]	0.80 ± 0.24 [a]	0.64 ± 0.05 [a]	
54.6%	7	10.32 ± 1.98 [a]	7.95 ± 0.28 [a]	0.00 ± 0.37 [a]	0.26 ± 0.13 [a]	
	14	19.69 ± 1.12 [bc]	10.36 ± 0.19 [bc]	1.45 ± 2.26 [a]	0.54 ± 0.20 [a]	
	21	24.76 ± 0.87 [c]	11.96 ± 0.16 [c]	0.46 ± 0.20 [a]	0.62 ± 0.07 [a]	

Numbers with the same letters denotes similarity and those with different letters are significantly different at $p < 0.5$.

The total quantities of fibre decomposition were calculated by accounting for biomass losses, which provides a clearer indication of the types of substrates being degraded. The total non-fibre and hemicellulose which was degraded showed a significant increase on day 21 in the less saturated microcosms and earlier on day 14 in the more saturated microcosms. Hemicellulose degradation would enable lignin degradation to proceed more rapidly due to enzyme accessibility to lignin as the close association between hemicellulose and lignin is separated [46]. Consequently, there appeared to be a correlation between hemicellulose and lignin degradation. Decomposition of the non-fibre was assumed due to growth on arabinoxylans which are present in a considerable proportion in wheat bran [21] and on hemicellulose. It was apparent that no cellulose was degraded by *T. versicolor* CM13, in contrast to higher cellulose degradation by the same fungus compared with hemicellulose degradation when grown on *Fraxinus excelsior* and *Acer pseudoplatanus* separately [7]. This may be due to the higher cellulose content in both woods between 40 and 45%, compared with 7% in wheat bran, while hemicellulose contents in these lignocellulose substrates were similar at 22%.

4. Conclusions

There are a number of factors that currently limit that the potential for commercial upscaling of SSF processes, including the type and particle of the growth substrate used, heat dispersion, aeration and provision of sufficient water to support growth and metabolic activity of the organisms under investigation.

This study reported the use of wheat bran, an agricultural by-product, as a growth substrate for the cultivation of *T. versicolor* CM13, a white rot fungus which produces lignin-degrading enzymes and the influence of moisture content on growth of this organism using SSF. The growth of *T. versicolor* CM13 on wheat bran revealed that the composition in terms of hemicellulose content and biomass remaining were different during active fungal growth but became similar during the stationary phase. In general, lignin-degrading enzyme activities were unaffected by different moisture conditions and were similar throughout the active and stationary phases. These results provide an insight into how minor differences in moisture content, which may be difficult to control, may affect SSF. This study revealed that the production of lignin-degrading enzymes using SSF could be achieved without moisture content being a limiting factor affecting fungal growth during colonisation.

Author Contributions: P.W.B., A.C.: Conceptualization, P.W.B.: investigation, P.W.B.: writing—original draft preparation, A.C.: writing—review and editing, A.C.: funding acquisition. All authors have read and agreed to the published version of the manuscript.

Funding: European Union's Interreg North West Europe program, financed by the European Regional Development Fund (Biowill: project number 964) and the Welsh Government (Environmental Evidence Program) for support.

Institutional Review Board Statement: Not applicable.

Informed Consent Statement: Not applicable.

Data Availability Statement: Available on request.

Conflicts of Interest: The authors declare no conflict of interest.

References

1. de Castro, R.J.S.; Sato, H.H. Enzyme production by solid state fermentation: General aspects and an analysis of the physicochemical characteristics of substrates for agro-industrial wastes valorization. *Waste Biomass Valori.* **2015**, *6*, 1085–1093. [CrossRef]
2. Merali, Z.; Collins, S.R.; Elliston, A.; Wilson, D.R.; Käsper, A.; Waldron, K.W. Characterization of cell wall components of wheat bran following hydrothermal pretreatment and fractionation. *Biotechnol. Biofuels* **2015**, *8*, 1–13. [CrossRef]
3. Lonsane, B.K.; Ghildyal, N.P.; Budiatman, S.; Ramakrishna, S.V. Engineering aspects of solid state fermentation. *Enzym. Microb. Technol.* **1985**, *7*, 258–265. [CrossRef]
4. Teigiserova, D.A.; Bourgine, J.; Thomsen, M. Closing the loop of cereal waste and residues with sustainable technologies: An overview of enzyme production via fungal solid-state fermentation. *Sustain. Prod. Consum.* **2021**, *27*, 845–857. [CrossRef]
5. Mansour, A.A.; Arnaud, T.; Lu-Chau, T.A.; Fdz-Polanco, M.; Moreira, M.T.; Rivero, J.A.C. Review of solid state fermentation for lignocellulolytic enzyme production: Challenges for environmental applications. *Rev. Environ. Sci.* **2016**, *15*, 31–46. [CrossRef]
6. Baker, P.W.; Charlton, A.; Hale, M.D.C. Fungal pre-treatment of forestry biomass with a focus on biorefining: A comparison of biomass degradation and enzyme activities by wood rot fungi across three tree species. *Biomass Bioenergy* **2017**, *107*, 20–28. [CrossRef]
7. Ullah, M.; Liu, P.; Xie, S.; Sun, S. Recent Advancements and Challenges in Lignin Valorization: Green Routes towards Sustainable Bioproducts. *Molecules* **2022**, *27*, 6055. [CrossRef]
8. Ponnusamy, V.K.; Nguyen, D.D.; Dharmaraja, J.; Shobana, S.; Banu, J.R.; Saratale, R.G.; Chang, S.W.; Kumar, G. A review on lignin structure, pretreatments, fermentation reactions and biorefinery potential. *Bioresourc. Technol.* **2019**, *271*, 462–472. [CrossRef]
9. Ruiz-Duenas, F.J.; Lundell, T.; Floudas, D.; Nagy, L.G.; Barrasa, J.M.; Hibbett, D.S.; Martínez, A.T. Lignin-degrading peroxidases in Polyporales: An evolutionary survey based on 10 sequenced genomes. *Mycologia* **2013**, *105*, 1428–1444. [CrossRef]
10. Baker, P.W.; Charlton, A.; Hale, M.D. Increased delignification by white rot fungi after pressure refining Miscanthus. *Bioresour. Technol.* **2015**, *189*, 81–86. [CrossRef]
11. Jensen, K.A.; Bao, W.; Kawai, S.; Srebotnik, E.; Hammel, K.E. Manganese-dependent cleavage of nonphenolic lignin structures by *Ceriporiopsis subvermispora* in the absence of lignin peroxidase. *Appl. Environ. Microbiol.* **1996**, *62*, 3679–3686. [CrossRef]
12. Perez, J.; Jeffries, T.W. Mineralization of 14C-ring-labeled synthetic lignin correlates with the production of lignin peroxidase, not of manganese peroxidase or laccase. *Appl. Environ. Microbiol.* **1990**, *56*, 1806–1812. [CrossRef]

13. Arora, D.S.; Chander, M.; Gill, P.K. Involvement of lignin peroxidase, manganese peroxidase and laccase in degradation and selective ligninolysis of wheat straw. *Int. Biodeterior Biodegrad* **2002**, *50*, 115–120. [CrossRef]
14. Hofrichter, M. Review: Lignin conversion by manganese peroxidase (MnP). *Enzym. Microb. Technol.* **2002**, *30*, 454–466. [CrossRef]
15. Chang, Y.; Yang, D.; Li, R.; Wang, T.; Zhu, Y. Textile dye biodecolorization by manganese peroxidase: A review. *Molecules* **2021**, *26*, 4403. [CrossRef]
16. Chowdhary, P.; Shukla, G.; Raj, G.; Ferreira, L.F.R.; Bharagava, R.N. Microbial manganese peroxidase: A ligninolytic enzyme and its ample opportunities in research. *SN Appl. Sci.* **2019**, *1*, 1–12. [CrossRef]
17. Saikia, S.; Yadav, M.; Hoque, R.A.; Yadav, H.S. Bioremediation mediated by manganese peroxidase—An overview. *Biocatal. Biotransfor.* **2022**, *41*, 161–173. [CrossRef]
18. Gessesse, A.; Mamo, G. High-level xylanase production by an alkaliphilic *Bacillus* sp. by using solid-state fermentation. *Enzyme Microb. Technol.* **1999**, *25*, 68–72. [CrossRef]
19. Fernandes, L.; Loguercio-Leite, C.; Esposito, E.; Reis, M.M. In vitro wood decay of *Eucalyptus grandis* by the basidiomycete fungus *Phellinus flavomarginatus*. *Int. Biodeterior. Biodegrad.* **2005**, *55*, 187–193. [CrossRef]
20. Bourbonnais, R.; Paice, M.G.; Reid, D.; Lanthier, P.; Yaguchi, M. Lignin oxidation by laccase isozymes from *Trametes versicolor* and role of the mediator 2, 29-azinobis (3-ethylbenzthiazoline-6-sulfonate) in kraft lignin depolymerization. *Appl. Environ. Microbiol.* **1995**, *61*, 1876–1880. [CrossRef]
21. Eichlerová, I.; Baldrian, P. Ligninolytic enzyme production and decolorization capacity of synthetic dyes by saprotrophic white rot, brown rot, and litter decomposing Basidiomycetes. *J. Fungi* **2020**, *6*, 301. [CrossRef] [PubMed]
22. Shin, K.; Oh, I.; Kim, C. Production and purification of remazol brilliant blue R decolorizing peroxidase from the culture filtrate of *Pleurotus ostreatus*. *Appl. Environ. Microbiol.* **1997**, *63*, 1744–1748. [CrossRef] [PubMed]
23. Archibald, F.S. A new assay for lignin-type peroxidases employing the dye Azure B. *Appl. Environ. Microbiol.* **1992**, *58*, 3110–3116. [CrossRef] [PubMed]
24. Antoine, C.; Peyron, S.; Mabille, F.; Lapierre, C.; Bouchet, B.; Abecassis, J.; Rouau, X. Individual contribution of grain outer layers and their cell wall structure to the mechanical properties of wheat bran. *J. Agric. Food Chem.* **2003**, *51*, 2026–2033. [CrossRef]
25. Smits, J.P.; Rinzema, A.; Tramper, J.; Van Sonsbeek, H.M.; Knol, W. Solid-state fermentation of wheat bran by *Trichoderma reesei* QM9414: Substrate composition changes, C balance, enzyme production, growth and kinetics. *Appl. Microbiol. Biotechnol.* **1996**, *46*, 489–496. [CrossRef]
26. Vu, V.H.; Pham, T.A.; Kim, K. Improvement of fungal cellulase production by mutation and optimization of solid state fermentation. *Mycobiology* **2011**, *39*, 20–25. [CrossRef]
27. Baker, P.W.; Charlton, A.; Hale, M.D.C. Fibre degradation of wheat straw by *Pleurotus erygnii* under low moisture conditions during solid-state fermentation. *Lett. Appl. Microbiol.* **2019**, *68*, 182–187. [CrossRef]
28. Brijwani, K.; Oberoi, H.S.; Vadlani, P.V. Production of a cellulolytic enzyme system in mixed-culture solid-state fermentation of soybean hulls supplemented with wheat bran. *Process Biochem.* **2010**, *45*, 120–128. [CrossRef]
29. Sun, R.; Sun, X.F.; Tomkinson, J. Hemicelluloses and their derivatives. In *Hemicelluloses: Science and Technology*; Gatenholm, P., Tenkanen, M., Eds.; ACS Publications: Washington, DC, USA, 2003; pp. 2–22.
30. van Kuijk, S.J.; Del Río, J.C.; Rencoret, J.; Gutiérrez, A.; Sonnenberg, A.S.; Baars, J.J.; Hendriks, W.H.; Cone, J.W. Selective ligninolysis of wheat straw and wood chips by the white-rot fungus *Lentinula edodes* and its influence on in vitro rumen degradability. *J. Anim. Sci. Biotechnol.* **2016**, *7*, 1–14. [CrossRef]
31. Mao, L.; Cone, J.W.; Hendriks, W.H.; Sonnenberg, A.S. Wheat bran addition improves *Ceriporiopsis subvermispora* and *Lentinula edodes* growth on wheat straw, but not delignification. *Anim. Feed Sci. Technol.* **2020**, *259*, 114361. [CrossRef]
32. Liaud, N.; Giniés, C.; Navarro, D.; Fabre, N.; Crapart, S.; Gimbert, I.H.; Levasseur, A.; Raouche, S.; Sigoillot, J.C. Exploring fungal biodiversity: Organic acid production by 66 strains of filamentous fungi. *Fungal Biol. Biotechnol.* **2014**, *1*, 1–10. [CrossRef]
33. Bunzel, M.; Schußler, A.; Saha, G.T. Chemical characterization of Klason lignin preparations from plant-based foods. *J. Agric. Food Chem.* **2011**, *59*, 12506–12513. [CrossRef] [PubMed]
34. Salame, T.M.; Knop, D.; Levinson, D.; Yarden, O.; Hadar, Y. Redundancy among manganese peroxidases in *Pleurotus ostreatus*. *Appl. Environ. Microbiol.* **2013**, *79*, 2405–2415. [CrossRef] [PubMed]
35. Xu, X.; Lin, M.; Zang, Q.; Shi, S. Solid state bioconversion of lignocellulosic residues by *Inonotus obliquus* for production of cellulolytic enzymes and saccharification. *Bioresourc. Technol.* **2018**, *247*, 88–95. [CrossRef]
36. Osma, J.F.; Moilanen, U.; Toca-Herrera, J.L.; Rodríguez-Couto, S. Morphology and laccase production of white-rot fungi grown on wheat bran flakes under semi-solid-state fermentation conditions. *FEMS Microbiol. Lett.* **2011**, *318*, 27–34. [CrossRef] [PubMed]
37. Farani de Souza, D.; Kirst Tychanowicz, G.; Giatti Marques de Souza, C.; Peralta, R.M. Co-production of ligninolytic enzymes by *Pleurotus pulmonarius* on wheat bran solid state cultures. *J. Basic Microbiol.* **2006**, *46*, 126–134. [CrossRef]
38. Eguchi, F.; Leonowicz, A.; Higaki, M.; Fukuzumi, T. Laccase-less mutants induced by regeneration of protoplasts of *Pleurotus* basidiomycetes. *J. Jpn. Wood Res. Soc.* **1994**, *40*, 107–110.
39. Ryu, S.H.; Cho, M.K.; Kim, M.; Jung, S.M.; Seo, J.H. Enhanced lignin biodegradation by a laccase-overexpressed white-rot fungus *Polyporus brumalis* in the pretreatment of wood chips. *Appl. Biochem. Biotechnol.* **2013**, *171*, 1525–1534. [CrossRef]
40. Kumar, A.; Arora, P.K. Biotechnological applications of manganese peroxidases for sustainable management. *Front. Environ. Sci.* **2022**, *10*, 365. [CrossRef]

41. Qin, X.; Zhang, J.; Zhang, X.; Yang, Y. Induction, purification and characterization of a novel manganese peroxidase from *Irpex lacteus* CD2 and its application in the decolorization of different types of dye. *PLoS ONE* **2014**, *9*, e113282. [CrossRef]
42. Rekik, H.; Jaouadi, N.Z.; Bouacem, K.; Zenati, B.; Kourdali, S.; Badis, A.; Annane, R.; Bouanane-Darenfed, A.; Bejar, S.; Jaouadi, B. Physical and enzymatic properties of a new manganese peroxidase from the white-rot fungus *Trametes pubescens* strain i8 for lignin biodegradation and textile-dyes biodecolorization. *Int. J. Biol. Macromol.* **2019**, *125*, 514–525. [CrossRef]
43. Christian, V.; Shrivastava, R.; Shukla, D.; Modi, H.; Vyas, B.R.M. Mediator role of veratryl alcohol in the lignin peroxidase-catalyzed oxidative decolorization of Remazol Brilliant Blue R. *Enzym. Microb. Technol.* **2005**, *36*, 327–332. [CrossRef]
44. Champagne, P.P.; Ramsay, J.A. Contribution of manganese peroxidase and laccase to dye decoloration by *Trametes versicolor*. *Appl. Microbiol. Biotechnol.* **2005**, *69*, 276–285. [CrossRef] [PubMed]
45. Locci, E.; Laconi, S.; Pompei, R.; Scano, P.; Lai, A.; Marincola, F.C. Wheat bran biodegradation by *Pleurotus ostreatus*: A solid-state Carbon-13 NMR study. *Bioresourc. Technol.* **2008**, *99*, 4279–4284. [CrossRef] [PubMed]
46. Kumar, A. Biobleaching: An eco-friendly approach to reduce chemical consumption and pollutants generation. *Phys. Sci. Rev.* **2020**, *6*, 20190044.

Article

Acidogenesis of Pentose Liquor to Produce Biohydrogen and Organic Acids Integrated with 1G–2G Ethanol Production in Sugarcane Biorefineries

Guilherme Peixoto [1,2], Gustavo Mockaitis [3,4], Wojtyla Kmiecik Moreira [4], Daniel Moureira Fontes Lima [5], Marisa Aparecida de Lima [2], Filipe Vasconcelos Ferreira [4], Lucas Tadeu Fuess [4,*], Igor Polikarpov [2] and Marcelo Zaiat [4]

[1] Department of Engineering of Bioprocesses and Biotechnology, School of Pharmaceutical Sciences, São Paulo State University (UNESP), São Paulo 14800-903, SP, Brazil; guilherme.peixoto@unesp.br
[2] Molecular Biotechnology Group, São Carlos Physics Institute, University of São Paulo (IFSC/USP), São Carlos 13566-590, SP, Brazil; marisalima2000@yahoo.com.br (M.A.d.L.); ipolikarpov@ifsc.usp.br (I.P.)
[3] Interdisciplinary Research Group On Biotechnology Applied to the Agriculture and the Environment (GBMA), School of Agricultural Engineering (FEAGRI), University of Campinas (UNICAMP), Av. Candido Rondon, 501–Cidade Universitária, São Paulo 13083-875, SP, Brazil; gusmock@unicamp.br
[4] Biological Processes Laboratory, São Carlos School of Engineering, University of São Paulo (EESC/USP), São Carlos 13563-120, SP, Brazil; wojtyla.kmiecik@gmail.com (W.K.M.); filipevasconcelos@usp.br (F.V.F.); zaiat@sc.usp.br (M.Z.)
[5] Department of Civil Engineering, Federal University of Sergipe (UFS), São Cristóvão 49107-230, SE, Brazil; danielmfl@academico.ufs.br
* Correspondence: lt.fuess@usp.br; Tel.: +55-(16)-3373-8360

Citation: Peixoto, G.; Mockaitis, G.; Moreira, W.K.; Lima, D.M.F.; de Lima, M.A.; Ferreira, F.V.; Fuess, L.T.; Polikarpov, I.; Zaiat, M. Acidogenesis of Pentose Liquor to Produce Biohydrogen and Organic Acids Integrated with 1G–2G Ethanol Production in Sugarcane Biorefineries. *Waste* **2023**, *1*, 672–688. https://doi.org/10.3390/waste1030040

Academic Editors: Vassilis Athanasiadis and Dimitris P. Makris

Received: 3 July 2023
Revised: 21 July 2023
Accepted: 31 July 2023
Published: 5 August 2023

Abstract: Second-generation (2G) ethanol production has been increasingly evaluated, and the use of sugarcane bagasse as feedstock has enabled the integration of this process with first-generation (1G) ethanol production from sugarcane. The pretreatment of bagasse generates pentose liquor as a by-product, which can be anaerobically processed to recover energy and value-added chemicals. The potential to produce biohydrogen and organic acids from pentose liquor was assessed using a mesophilic (25 °C) upflow anaerobic packed-bed bioreactor in this study. An average organic loading rate of 11.1 g COD·L^{-1}·d^{-1} was applied in the reactor, resulting in a low biohydrogen production rate of 120 mL·L^{-1} d^{-1}. Meanwhile, high lactate (38.6 g·d^{-1}), acetate (31.4 g·d^{-1}), propionate (50.1 g·d^{-1}), and butyrate (50.3 g·d^{-1}) production rates were concomitantly obtained. Preliminary analyses indicated that the full-scale application of this anaerobic acidogenic technology for hydrogen production in a medium-sized 2G ethanol distillery would have the potential to completely fuel 56 hydrogen-powered vehicles per day. An increase of 24.3% was estimated over the economic potential by means of chemical production, whereas an 8.1% increase was calculated if organic acids were converted into methane for cogeneration (806.73 MWh). In addition, 62.7 and 74.7% of excess organic matter from the 2G ethanol waste stream could be removed with the extraction of organic acid as chemical commodities or their utilization as a substrate for biomethane generation, respectively.

Keywords: sugarcane bagasse; pentose liquor; dark fermentation; biohydrogen; organic acids; biorefinery

1. Introduction

Brazilian sugarcane mills are migrating from old concepts of sugar and ethanol production to a new concept of biorefinery. This concept aims to produce not only biofuels but also electricity, food, and other products that use renewable sources from sugarcane biomass [1–3]. One of the main actions to consolidate the biorefinery concept is the implementation of second-generation (2G) ethanol production integrated with first-generation (1G) plants. This industrial concept intends to use surplus bagasse that is currently used as fuel in cogeneration systems, to produce lignocellulosic ethanol (2G) instead [4]. The utilization of this remaining biomass plays a fundamental role in the biorefinery yield because 1 ton of sugarcane generates 280 kg of bagasse [5], which is very significant considering the production of 610 million tons of sugarcane [6] in the 2022/2023 harvest.

The implementation of 2G ethanol production can improve process sustainability and also ethanol production [7,8]. In this process, the bagasse is pretreated to obtain a solid fraction that is rich in cellulose, while hemicellulose is mainly hydrolyzed to pentoses (pentose liquor). The cellulose fraction obtained from this pretreatment can be enzymatically hydrolyzed to feed the 2G ethanol-producing process, and pentose liquor remains a by-product of the process. Although ethanol production is the natural destination of the hexoses (C6) fraction, the processing of pentose liquor is still not well defined, and alternatives must be studied for the application of this stream. For instance, opportunities for fermenting pentoses to obtain ethanol have been massively studied [9,10]; however, selecting efficient pentose fermenters is still challenging.

The xylose-rich nature of pentose liquor characterizes it as a highly suitable substrate for biohydrogen and organic acid production in acidogenic bioreactors because this process depends primarily on carbohydrate-rich substrates [11,12]. Hydrogen is a renewable clean energy carrier when biomass and its by-products are used as raw materials in fermentation. The biological production of hydrogen from residues and wastewater is less energy intensive and is less expensive than methane steam reforming and electrolysis [11–14]. Moreover, various organic acids and other compounds with different applications (foods, pharmaceuticals, and chemicals) can be generated during the biological production of hydrogen. Most studies have reported that biohydrogen production generates intermediates such as acetic acid, propionic acid, butyric acid, succinic acid, formic acid, butanediol, and acetone [15–20]. These compounds are of special interest because of their high market value [21,22], especially butyric and lactic acid, which are used as precursors of industrial thermoplastics [23] and biodegradable polymers [24,25]. Although only a few studies have addressed hydrogen and organic acid production using sugarcane bagasse (SCB) hydrolysate [26–30], several studies have used pure xylose as a substrate for the production of biohydrogen [31–39], indicating a gap for exploitation within the context of dark fermentation studies. Furthermore, most researchers have used batch reactors in contrast with the few studies conducted in continuous bioreactors. Neither xylose nor SCB hydrolysate have been used in upflow anaerobic packed-bed reactors, which have generally yielded high biohydrogen productivity [40–44].

This study assessed the potential application of sugarcane bagasse-derived pentose liquor as a source of biohydrogen and organic acids in a continuous packed-bed bioreactor, characterizing an alternative to add value to a voluminous by-product which will be inevitably available in integrated 1G-2G sugarcane biorefineries in coming years. A dual approach based on an experimental assessment followed by a scenarization-based exercise through a simple process application analysis in a medium-sized distillery (milling capacity of 500 ton·h^{-1}) was carried out, providing the bases to understand the potential of pentose liquor as a substrate for biotechnological applications.

2. Materials and Methods

2.1. Bioreactor and Support Material

The upflow anaerobic packed-bed reactor (UAPBR) was built in tubular acrylic with a length-to-diameter (L/D) ratio of 9.4, considering a total height of 750 mm and a working volume of 2.370 L. The bioreactor's bed region was randomly filled with recycled low-density polyethylene (LDPE) pellets to support cell attachment, similar to previous experiments with real wastewater [43–45]. Cylinder-shaped particles with a mean diameter of 12.7 mm and ca. 30 mm in height were used with specific gravity and a surface area of 0.96 g·cm^{-3} and 7.94 m^2·g^{-1}, respectively. A total of 295 g LDPE·L^{-1} was used to build the bed zone, resulting in a bed porosity of 55%. The packed bed occupied 66% of the total volume. A schematic diagram of the reactor is shown in Figure 1.

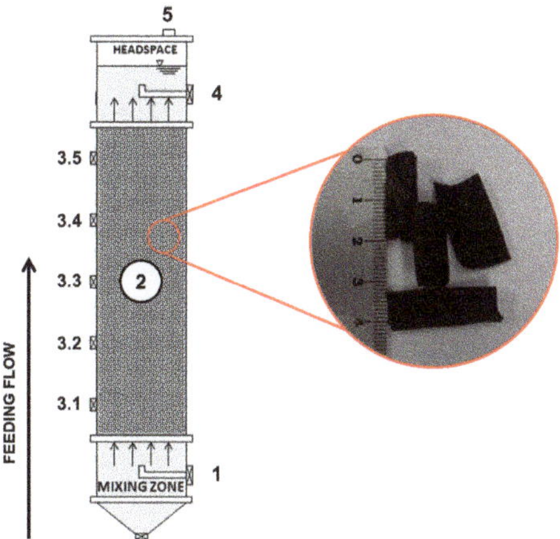

Figure 1. Sketch of the upflow anaerobic packed-bed reactor and details of the LDPE pellets. Legend: 1—inlet port, 2—packed bed, 3.1–3.5—sampling ports, 4—outlet port (liquid phase), 5—outlet port (biogas).

2.2. Substrates

The UAPBR was fed with two different substrates in distinct operating periods, namely, a xylose-based lab-made wastewater (experimental condition I) and pentose liquor derived from 2G ethanol production (experimental condition II), as indicated in Figure 2. The former was prepared using 99.9% pure D-(+)-Xylose (Sigma-Aldrich®, San Luis, MO, USA), whilst pentose liquor was obtained from the processing of residual SCB collected from a Brazilian sugarcane mill. Bagasse was subjected to physicochemical pretreatment (121 °C and 1.1 atm) utilizing a mass ratio of solid (g dry weight) to liquor (g) at 1:10 with 2% sulfuric acid (H_2SO_4) (v/v) for 60 min [46]. This pretreatment resulted in a hydrolysate with an organic matter concentration in terms of COD (chemical oxygen demand) of 14 $g \cdot L^{-1}$. Prior to feeding the bioreactor, the hydrolysate was neutralized with sodium hydroxide (NaOH) and diluted with tap water to obtain a COD of approximately 2 $g \cdot L^{-1}$. The compositional characteristics of the wastewater utilized in both experimental conditions are shown in Table 1.

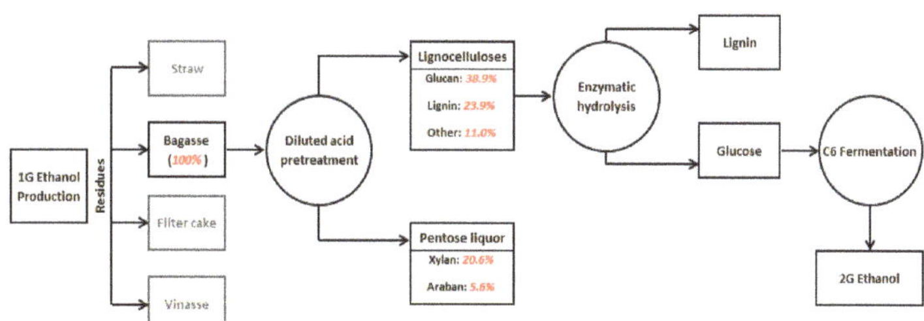

Figure 2. Processing of sugarcane bagasse to produce second generation ethanol. Sugarcane bagasse constituents by Aguilar et al. [46].

Table 1. Compositional characterization of the wastewater used in reactor feeding.

Parameters	Xylose-Based Wastewater	Pentose Liquor
Total carbohydrates ($mg \cdot L^{-1}$)	1495 ± 733	1459 ± 440
Xylose (%)	99.9	82.4
Arabinose (%)	–	7.12
Rhamnose (%)	–	6.14
Glucose (%)	–	1.58
pH	6.7 ± 0.1	5.9 ± 0.1

2.3. Experimental Procedure

Bioreactor inoculation was carried out prior to experimental conditions I and II and consisted of the natural fermentation of xylose and pentose liquor, respectively. The natural fermentation was performed by exposing 30 L of each medium for 3 days to the atmosphere at ca. 25 °C. In previous studies, such as those published elsewhere [16,40,41], this procedure yielded mainly microorganisms similar to *Clostridium* sp. (91%), *Klebsiella* sp. (97%), and *Enterobacter* sp. (93%), which are directly related to hydrogen and organic acid production [41]. Naturally fermented substrates were recirculated into the UAPBR for 7 days [16] as a strategy to promote the attachment of microorganisms to the packed bed, after which a continuous operation was started. The volumetric flow rate of the bioreactor feed was maintained at ca. 1.19 $L \cdot h^{-1}$ using a positive displacement pump (Concept Plus, ProMinent Brasil Ltd.a., São Bernardo do Campo, Brazil), resulting in a hydraulic retention time (HRT) of 2 h. The operating temperature was controlled at 25 ± 1 °C in a thermostatic chamber (410-DRE, Nova Ética, Vargem Grande Paulista, Brazil). In condition I, xylose was used as the only substrate for 36 days, whilst pentose liquor was evaluated as the substrate for 39 days in condition II. The pH of both substrates (lab-made and real wastewater) was maintained at ca. 6.0 by dosing with NaOH (500 $mg \cdot L^{-1}$) or hydrochloric acid (HCl) (10 $mol \cdot L^{-1}$) solutions. Additionally, both media were supplemented with a macro and micronutrient solution containing (in $mg \cdot L^{-1}$): CH_4N_2O (7.7), $NiSO_4 \cdot 6H_2O$ (0.15), $FeSO_4 \cdot 7H_2O$ (2.5), $FeCl_3 \cdot 6H_2O$ (0.25), $CoCl_2 \cdot 2H_2O$ (0.04), $CaCl_2 \cdot 6H_2O$ (2.06), SeO_2 (0.036), KH_2PO_4 (1.3), $KHPO_4$ (5.36), and $Na_2HPO_4 \cdot 2H_2O$ (2.76).

2.4. Monitoring Procedure and Analytical Methods

Liquid phase monitoring was based on periodic measurements of the pH, COD, and concentrations of the total carbohydrates, organic acids, and solvents. pH and COD were determined according to the Standard Methods for the Examination of Water and Wastewater [47]. Potentiometric measurements (pH) were carried out using a pH meter with a standard glass electrode (Digimed Instrumentação Analítica, São Paulo, Brazil). Total carbohydrates were measured as proposed by Dubois et al. [48], whilst organic acids and solvents were analyzed with a high-performance liquid chromatography (HPLC) system (Shimadzu Scientific Instruments, Kyoto, Japan) using the same configuration and protocols described elsewhere [49].

Gas phase monitoring was carried out by measuring both the biogas flow rate (BFR) and composition. BFR was obtained by directly coupling a gas meter (MilliGascounter-1 V30, Dr.-Ing. Ritter Apparatebau GMBH & Co. KG, Bochum, Germany) to the headspace of the reactor. The biogas composition in terms of hydrogen (H_2), nitrogen (N_2), methane (CH_4), and carbon dioxide (CO_2) was carried out using a gas chromatography set (GC-2010) equipped with a thermal conductivity detector (GC/TCD; Shimadzu Scientific Instruments, Kyoto, Japan) and argon as the carrier gas, as described elsewhere [50]. Gas samples were collected from UAPBR's headspace with 1000 μL-insulin syringes equipped with Teflon body two-way valves (Supelco® Analytical—Sigma-Aldrich, Bellefonte, PA, USA).

2.5. Scenario Assessment

Four different scenarios, namely, 1 to 4, were considered for a medium-sized 2G ethanol plant (milling capacity of 500 $ton \cdot h^{-1}$), providing a preliminary analysis of pentose liquor as raw material for producing bioenergy and biochemicals. Scenarios 1 and 2 did not include pentose as raw material, whilst scenarios 3 and 4 considered the utilization of pentoses for the energy recovery and/or generation of value-added products. In Scenario 1, 50% of SCB was considered to be used for 2G ethanol production, with the remaining fraction used in the cogeneration of electricity and steam. In Scenario 2, 100% of SCB was directed to 2G ethanol production. In Scenario 3, 2G ethanol was produced with the utilization of all available SCB. Additionally, hydrogen and volatile organic acids (VOA) were derived from pentose liquor. In Scenario 4, 2G ethanol, hydrogen, and organic acids were produced. In this last scenario, organic acid consumption in biomethane production for cogeneration was considered despite the use of chemicals, as reported in Scenario 3.

3. Results and Discussion

3.1. Biohydrogen Production

Biogas generation profiles obtained with xylose-based wastewater and pentose liquor are depicted in Figure 3, showing distinct patterns according to the type of substrate. In experimental condition I (xylose as the substrate), biogas production reached 193.8 mL·h^{-1} (4.65 L·d^{-1}) on day 10. This production peak was followed by continuous decay until the end of the operation (Figure 3A). This drop in biogas generation has also been observed in previous studies regarding hydrogen production from carbohydrates [40–42] and complex wastewaters [44,51]. A coherent explanation can be found in the major growth of homoacetogenic organisms, which are capable of using the Wood–Ljungdahl pathway because the applied specific organic load decreased with the increase in the biomass in the packed-bed reactor [52,53]. The homoacetogenic pathway explains the biogas production decrease because both H$_2$ and CO$_2$ produced by hydrogen-producing bacteria (HPB) were converted into acetate, as demonstrated in Reaction (1) [54]. Numerous studies dealing with the dark fermentation of sugarcane-derived substrates, namely, vinasse, molasses, and juice, have reported the occurrence of homoacetogenesis when applying mesophilic temperature conditions [19,55–58], which supports this hypothesis. Homoacetogenic bacteria belonging to genera *Moorella*, *Oxobacter*, and *Sporomusa*, as well as to the family *Lachnospiraceae*, were identified in vinasse-fed reactors [19,59,60]. Some clostridial groups [61], as well as some sulfate-reducing bacteria belonging to the genus *Desulfotomaculum* [62], are also capable of utilizing H$_2$ and CO$_2$.

$$2CO_2 + 4H_2 \rightarrow \text{Acetate} + 2H_2O \ (\Delta G^\circ = -104 \ \text{kJ·mol}^{-1}) \tag{1}$$

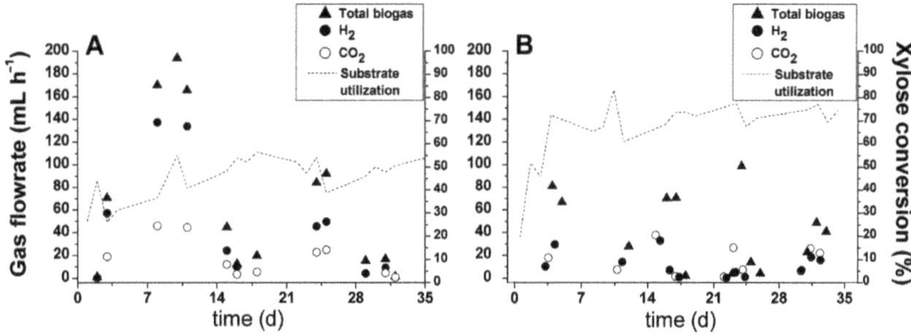

Figure 3. Biogas flowrate and composition and substrate utilization in experimental conditions (**A**) I (xylose-based wastewater) and (**B**) II (pentose liquor).

The mean biogas production rates obtained in conditions I and II were 0.026 L·L^{-1}·h^{-1} (624 mL·L^{-1}·d^{-1}) and 0.017 L·L^{-1}·h^{-1} (408 mL·L^{-1}·d^{-1}), respectively. The volumetric hydrogen production rate (VHPR) followed the same trend, with values of 0.016 (384 mL·L^{-1}·d^{-1}) and 0.005 L·h^{-1}·L^{-1} (120 mL·L^{-1}·d^{-1}) observed in experimental conditions I and II, respectively. The main reason for a higher hydrogen production rate in the bioreactor processing xylose was the production peak reported at the beginning of the operation (days 8–11). This behavior could be explained considering that the substrate in condition I consisted only of readily available xylose, which could potentially contribute to a faster and more efficient biomass adaptation [63].

By contrast, experimental condition II contained a mixture of pentoses (xylose, arabinose, and rhamnose; Table 1). These different carbon sources and the presence of furfural and acetic acid could be responsible for lower hydrogen production rate at the beginning of the bioreactor operation. In the study by Aguilar et al. [46], the sugarcane bagasse subjected to the same diluted acid pretreatment released approximately 1.2 g·L^{-1} of furfural and 4.5 g·L^{-1} of acetic acid, which are considered toxic to fermentation in these ranges [46,64]. However, the hydrolysate generated in this study was diluted by a factor of 7 (COD decreased from 13,989 to 1989 mg·L^{-1}) before its utilization in the bioreactor, which consequently resulted in lower concentrations of both constituents. Thus, the best explanation for the worst performance in biohydrogen production in condition II was the lower rate of biomass adaptation. Nevertheless, after 11 days of operation, the highest peak of substrate utilization (83.2%) was noticed in this experimental condition (pentose-containing hydrolysate). According to the biogas profile in Figure 3B, xylose was not utilized for hydrogen production. Rather, this substrate was diverted for the production of soluble metabolites. Overall, in both conditions, the reported efficiency in xylose conversion (Figure 3A,B) was considered high because UAPBR was operated at a short hydraulic retention time (2 h).

Because the pentose liquor derives from sugarcane, it is pertinent to compare the obtained results with those found in the dark fermentation of other sugarcane-derived substrates, focusing on fixed-film reactors. Overall, the VHPR observed in condition II (120 mL·L^{-1}·d^{-1}) exceeded only the value reported by Ferraz Jr. et al. [51], who achieved a mean VHPR of 84 mL·L^{-1}·d^{-1} while processing vinasse in an LDPE-filled UAPBR at mesophilic conditions (25 °C). Using the same type of substrate and reactor but at thermophilic conditions (55 °C), much better results were reported, namely, 526.8 mL·L^{-1}·d^{-1} [65], 761.7 mL·L^{-1}·d^{-1} [43] and 1604 mL·L^{-1}·d^{-1} [44]. Changing the bed conformation, i.e., replacing the random packing by orderly placing the support media further produced even better results, with VHPR reaching 2074 mL·L^{-1}·d^{-1} [66] and 3477 mL·L^{-1}·d^{-1} [67] while still considering vinasse as the substrate and temperature conditions. Comparing the results obtained herein with the utilization of vinasse is pertinent because both the pentose liquor and vinasse are highly complex materials. The use of molasses, a much simpler carbohydrate-rich sugarcane-derived by-product, led to much higher VHPR values, reaching 4504 mL·L^{-1}·d^{-1} [68] and 8479 mL·L^{-1}·d^{-1} [69]. Apart from the temperature and support material arrangement differences, in most of the comparative studies, much higher organic loading rate (OLR) levels were used (>50 g COD·L^{-1}·d^{-1}) compared to the one applied in the UAPBR (11.1 g COD·L^{-1}·d^{-1}), which could explain the high discrepancy among the VHPR values. Future studies with pentose liquor should focus on optimizing the OLR, as carried out for vinasse [65,66] and molasses [68].

3.2. Organic Acid Production

The distribution of soluble metabolites in experimental conditions I and II showed a higher amount of organic acids and solvents in the latter (Figure 4D,E). It is likely that hydrolysate compounds, especially the macro and micronutrients released during the hydrolysis of SCB, stimulated the metabolism of immobilized bacteria toward soluble metabolite production. Lactate was produced in large quantities in both experimental conditions. One plausible explanation for its high concentration is the short HRT (2 h) utilized in the bioreactor operation. The reduced HRT could have led to the accumulation of lactic acid because the reaction time was not sufficient to convert lactate and acetate to butyrate and hydrogen. This hypothesis is supported by the hydrogen profiles (Figure 3A,B) and the metabolic pathway suggested elsewhere [70], as demonstrated in Reactions (2)–(4). The results observed during the dark fermentation of other sugarcane-derived substrates, such as vinasse [20,66,67,71] and molasses [68,69,72], also corroborate this hypothesis because in all cases, lactate was identified as the primary precursor of biohydrogen and butyrate.

$$\text{Lactate} + 0.4\text{Acetate} + 0.7\text{H}^+ \rightarrow 0.7\text{Butyrate} + 0.6\text{H}_2 + \text{CO}_2 + 0.4\text{H}_2\text{O} \ (\Delta G^\circ = -183.9 \text{ kJ·mol}^{-1}) \quad (2)$$

$$\text{Lactate} + \text{Acetate} + \text{H}^+ \rightarrow \text{Butyrate} + 0.8\text{H}_2 + 1.4\text{CO}_2 + 0.6\text{H}_2\text{O} \ (\Delta G^\circ = -59.4 \text{ kJ·mol}^{-1}) \quad (3)$$

$$2\text{Lactate} + \text{H}^+ \rightarrow \text{Butyrate} + 2\text{H}_2 + 2\text{CO}_2 \ (\Delta G^\circ = -64.1 \text{ kJ·mol}^{-1}) \quad (4)$$

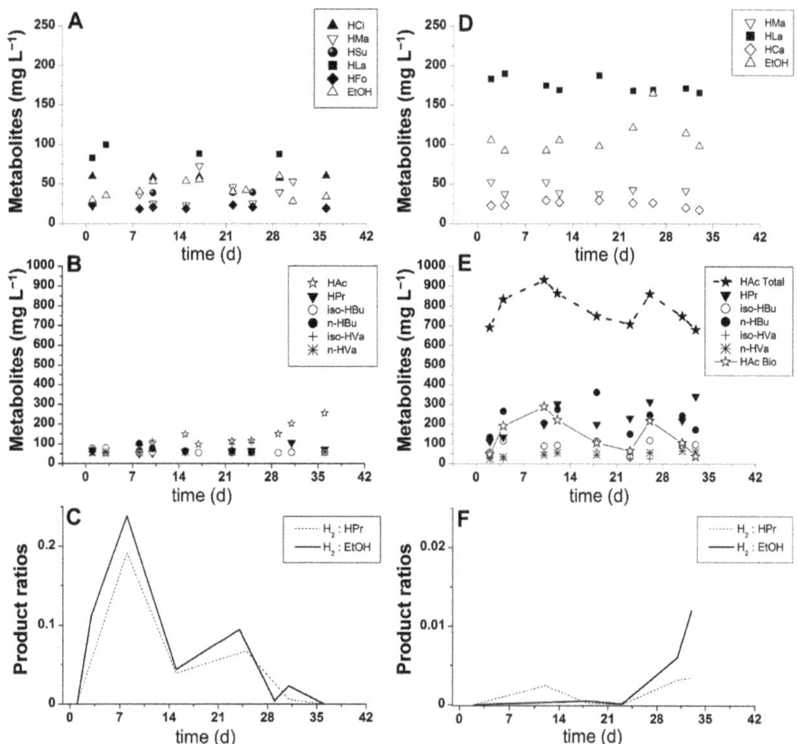

Figure 4. Condition I (xylose-based wastewater): (**A,B**) soluble phase products and (**C**) hydrogen-to-propionate (H_2:HPr) and hydrogen-to-ethanol (H_2:EtOH) ratios. Condition II (pentose liquor): (**D,E**) soluble phase products and (**F**) hydrogen-to-propionate (H_2:HPr) and hydrogen-to-ethanol (H_2:EtOH) ratios. Nomenclature: HCi—citric acid, HMa—malic acid, HSu—succinic acid, HLa—lactic acid, HFo—formic acid, HAc—acetic acid (resulting from fermentation in condition I), HAc Total—total acetic acid (resulting from both bagasse hydrolysate and substrate fermentation in condition II); HAc Bio—acetic acid (resulting from fermentation in condition II), HPr—propionic acid, HBu—butyric acid, HVa—valeric acid, HCa—caproic acid, EtOH—ethanol. Note: HAc Total calculated according to Aguilar et al. [46].

The higher concentration of malate detected in experimental condition II could possibly be linked to the release of malic acid, which is part of the natural composition of plants, including sugarcane and citrus fruits. Malic acid has been previously detected in sugarcane-derived substrates, such as vinasse [73,74]. According to Figure 4D, malate presented a marked decay profile after day 10 because it could be readily metabolized by the acclimatized bacterial consortium. Malate was most likely converted to pyruvic acid, which was integrated into the ethanol fermentation reactions involving decarboxylation to acetaldehyde with a subsequent reduction in alcohol [75]. The increasing profile of ethanol production, as shown in Figure 4D, supports this hypothesis. Malate may also have been converted to lactate following malolactic fermentation [76], which has been previously suggested to occur during vinasse fermentation [77].

Caproate was another relevant soluble metabolite produced in the fermentation of pentose liquor. According to the metabolite profiles shown in Figures 3B and 4D, caproic acid production is likely related to a pathway described elsewhere [78], in which simultaneous hydrogen and caproic acid production was feasible but resulted in low amounts of biohydrogen. These authors reported that caproic acid could also be formed through the secondary fermentation of ethanol and acetate or ethanol and butyrate. Because acetate was one of the major products found in experimental condition II (Figure 4E), it is likely that acidogenic bacteria used acetate as an alternative terminal acceptor for the reducing equivalents during caproic acid synthesis [79]. Moreover, during the fermentation of pentoses, elevated production of ethanol (Figure 4D) potentially resulted from the above-mentioned metabolic pathway. This observation is consistent with the assumption that some caproate-producing bacteria formed a syntrophic relationship with ethanol-producing bacteria [80]. Another factor supporting the existence of caproate-producing bacteria in the mixed culture is the occurrence of spore-forming *Clostridium kluyveri*, which is known to be resistant to the harsh environments and treatments (e.g., highly acidic conditions) used for selecting HPB [78].

Regarding acetate production, the fermentative pathway of xylose leading to this product can occur spontaneously [27], as shown in Reaction (5). In experimental condition I, acetic acid was the major metabolite produced (Figure 4B), thus implying improved hydrogen yields due to coupled reactions in the metabolic pathway, as demonstrated by Reaction (5). However, stable or increasing hydrogen production was not observed (Figure 3A). On the contrary, the biohydrogen generation profile demonstrated a peak followed by continuous decay, which was possibly caused by the increasing dominance of homoacetogenic bacteria (Reaction 1). A marked decrease in the volumetric biogas production indicated that carbon dioxide and hydrogen were diverted to acetate production, as stoichiometrically demonstrated in Reaction (1). In experimental condition II, a mean concentration of acetate equal to 784.55 mg·L^{-1} (HAc Total; Figure 4E) was produced; however, a major part of this production was derived from the hydrolysis of acetyl groups bound to hemicellulosic monomers of SCB. Because the protocol described by Aguilar et al. [46] was followed in this study, the calculations pointed to the physicochemical release of 642 mg·L^{-1} of acetate and enabled biologically produced acetic acid to be estimated (HAc Bio; Figure 4E), which was 142.55 mg·L^{-1} on average. Whereas in this condition, the acetate production was unstable (HAc Bio; Figure 4E), the generation of this metabolite in condition I showed an increasing profile (Figure 4B). This result indicates the occurrence of homoacetogenic reactions in condition I and the limitations of this biochemical route in experimental condition II. The high concentration of acetate derived from pretreatment could have inhibited the homoacetogenic pathway in the reactor fed with the SCB hydrolysate (condition II) because the accumulation of non-dissociated organic acids is deleterious to acidogenic bacteria [81]. However, a systemic inhibition, i.e., affecting all fermentative groups (including HPB), should be expected in this case. Hence, although continuous, the biogas production in condition II was most likely not high enough to stimulate the development of homoacetogens.

$$\text{Xylose} + 1.67\text{H}_2\text{O} \rightarrow \text{Acetate} + 1.67\text{CO}_2 + 3.33\text{H}_2 \ (\Delta G^\circ = -195.5 \ \text{kJ·mol}^{-1}) \tag{5}$$

Butyrate production was detected at concentrations ranging from 60 to 100 mg·L^{-1} in condition I and from 133 to 361 mg·L^{-1} in condition II. The reasons for these differences are likely related to the increasing utilization of the substrate in condition II (Figure 3B) and the occurrence of different terminal electron acceptors in condition I, as demonstrated by the production of acetate (Figure 4B) and metabolic trends in the formation of propionate and ethanol (Figure 4C). Similar to acetate production, the generation of butyric acid occurred simultaneously with the production of hydrogen; however, the yield was lower [82], as shown in Reaction (6). In experimental condition II, the lower biohydrogen generation was possibly determined by the biochemical yield demonstrated in Reaction (6), which implied half of the biohydrogen yield compared to that obtained when acetic acid was the by-product (Reaction 5). In both experimental conditions, the predominance of acetate and butyrate as major metabolites suggested the initial occurrence of a butyrate-acetate metabolic pathway, in which *Clostridium* sp. is the specific dominant strain. The same observation was made in previous studies utilizing UAPBR systems and natural fermentation for inoculation procedures [41,42,52]. In the study by Peixoto et al. [41], 16S rRNA sequencing analysis showed a 91% similarity to *Clostridium* sp, which comprised the butyrate producers *C. acetobutylicum*, *C. tyrobutyricum*, and *C. beijerinckii*.

$$\text{Xylose} \rightarrow 0.83\text{Butyrate} + 1.67CO_2 + 1.67H_2 \ (\Delta G° = -239.9 \ kJ \cdot mol^{-1}) \tag{6}$$

The propionate concentration in condition I increased from 50 to 105 mg·L^{-1}, with a mean production of 66.3 mg·L^{-1}. In condition II, propionate was also detected at an increment of 230 mg·L^{-1}, which resulted in an increase from 112 to 342 mg·L^{-1}. Although this increase is interesting due to the market value of propionic acid [21], the generation of this metabolite in biohydrogen-producing reactors results in a lower yield. This hydrogen-consuming pathway [13,27,34] is demonstrated by Reaction (7), usually taking place when high hydrogen partial pressures are established in the reactors. One of the factors influencing the instability of biohydrogen production in condition II probably included the peaks of propionate concentration on days 12, 23, and 33 (Figure 4E). This hypothesis was valid because the hydrogen concentration had a simultaneous decrease when these propionate peaks were detected. On the other hand, the decrease in biohydrogen production in condition I could more likely be explained by its conversion into acetate in the homoacetogenic pathway because the average production of propionate was quite low (66 mg·L^{-1}) compared to that observed in experimental condition II (228 mg·L^{-1}). In spite of the elevated production of propionic acid in condition II, the H$_2$/Propionate ratio (Figure 4F) showed an increase in its trend because hydrogen production did not cease. On the contrary, in condition I, the H$_2$/Propionate ratio was close to 0 by the end of the reactor operation due to the interruption of hydrogen production despite the production of acetate (Figure 4B).

$$\text{Xylose} + 1.67H_2 \rightarrow 1.67\text{Propionate} + 1.67H_2O \ (\Delta G° = -315.1 \ kJ \cdot mol^{-1}) \tag{7}$$

In addition to the production of hydrogen and volatile organic acids, a relatively low amount of ethanol was generated from the xylose media and SCB-derived pentose liquor. The results in Figure 4A,B show the mean ethanol productions of 43 and 110 mg·L^{-1} in conditions I and II, respectively. This decrease in the trend of the H$_2$/Ethanol ratio in condition I occurred due to the progressive termination of biohydrogen production (Figure 3A); however, in experimental condition II, the profile of the H$_2$/Ethanol ratio (Figure 4F) was different due to the opposite dynamics in biohydrogen production.

According to the mass flow rate shown in Table 2, the average missing equivalents in experimental condition I was lower than 2%, implying a 98% correspondence between the calculations and experimental data. One explanation for this unbalance could be the presence of the bacteria *Klebsiella* sp., which was previously identified in acidogenic inocula obtained from carbohydrate natural fermentation [41,83]. This microbial group was involved in the production of ethanol and 2,3-butanediol: a metabolite that was not monitored by the analytical methods used in this study. This metabolite was produced under limited oxygen and low pH conditions [84], similar to those utilized in the experiment presented herein. In experimental condition II, the sum of organic metabolites exceeded the mass balance in 207.24 mg·h^{-1}, causing an approximate lack of correspondence at 11%. In this case, it is not likely that the formation of undetected metabolites or the presence of alternative electron sinks accounted for the unbalance. Rather, the other carbon sources that were not monitored, such as glucose (1.58%), galactose (2.81%), rhamnose (6.14%), and arabinose (7.12%), were probably fermented and led to the formation of their own metabolites besides those produced with xylose (82.4%).

Table 2. Mass flowrate in experimental conditions I and II. Values in parentheses correspond to the day in which each maximum value was obtained.

Mass Flow Rate (mg·h^{-1})	Condition I (Xylose-Based Wastewater)		Condition II (Pentose Liquor)	
	Maximum	Mean	Maximum	Mean
Influent xylose	4171.12 (10)	2096.68 ± 811.4	2256.85 (33)	1886.92 ± 141.3
Effluent xylose	2591.34 (4)	1148.13 ± 491.2	1312.13 (2)	610.28 ± 173.7
Citric acid	72.28 (36)	70.71 ± 1.07	ND	ND
Malic acid	87.19 (17)	45.89 ± 16.8	69.31 (2)	56.82 ± 6.6
Succinic acid	47.22 (25)	46.88 ± 0.3	ND	ND
Lactic acid	119.36 (3)	107.37 ± 7.0	248.73 (4)	229.87 ± 9.1
Formic acid	28.11 (22)	24.86 ± 2.0	ND	ND
Acetic acid	305.76 (36)	150.66 ± 61.2	379.20 [1] (10)	186.72 [1] ± 89.8
Propionic acid	125.92 (31)	79.63 ± 17.0	447.80 (33)	298.52 ± 78.6
Isobutyric acid	94.07 (3)	72.33 ± 8.7	156.48 (26)	119.34 ± 28.0
n-Butyric acid	120.80 (8)	85.53 ± 15.2	473.39 (18)	299.23 ± 71.1
Isovaleric acid	140.55 (10)	109.02 ± 22.4	88.20 (18)	53.24 ± 18.8
n-Valeric acid	72.65 (36)	69.61 ± 1.6	85.62 (31)	64.43 ± 13.3
Caproic acid	ND	ND	32.40 (18)	17.47 ± 4.0
Ethanol	72.05 (29)	51.46 ± 11.0	215.66 (26)	144.47 ± 22.5
Hydrogen	11.24 (8)	1.84 ± 0.003	2.70 (16)	0.55 ± 0.0008
Carbon dioxide	82.43 (8)	15.70 ± 0.02	68.04 (16)	13.22 ± 0.02
Sum (soluble + gas metabolites)	NC	2079.62	NC	2094.16
Correspondence (%) [2]	NC	99.2	NC	111.0

[1] Resulting from substrate fermentation. [2] Corresponds to the ratio between the sum of metabolites and the influent xylose. Legend: ND—not detected, NC—not calculated (the sum of metabolites and their correspondence were not calculated for maximum values because they were observed in different days).

According to Table 3, the production of the main volatile organic acids was comparable to other studies reported in the literature, while H$_2$ production was significantly lower. In the studies by Wu et al. [36] and Lin et al. [34], 20 g·L^{-1} of xylose was used as the initial substrate concentration, which was more than ten-fold higher than the sugar concentrations used to feed the packed-bed reactor in this study. The effect of the temperature should also be considered because other studies with an equivalent substrate concentration reported hydrogen yields higher than those obtained in this study [26,32,34,36]. Lin et al. [34] demonstrated that increasing temperatures (from 30 to 50 °C) promoted higher hydrogen yields and production rates. In their study, temperatures of 30, 35, 40, 45, and 50 °C corresponded to 0.4, 0.5, 0.3, 0.8, and 1.3 mol H$_2$·mol^{-1}xylose, respectively. Thus, it is likely that the low biohydrogen production in the UAPBR was greatly influenced by the initial substrate's concentration (or applied OLR, as discussed in Section 3.1) and operation temperature. However, the production of both butyric and propionic acids (Table 3) in the UAPBR was greater than that reported by Wu et al. [36] and Pattra et al. [26].

3.3. Preliminary Analysis of Biohydrogen, VOA and Cogeneration Potential in a Pentose Liquor-Based Biorefinery

The experimental results provided the main parameters with which to develop and analyze a biorefinery performance in four independent scenarios, as detailed in Section 2.5 and depicted in Figure 5. According to Table 4, a comparison between Scenarios 1 and 2 suggested that exclusively producing 1G and 2G of ethanol (Scenario 2) was more profitable than using 50% of SCB for cogeneration and 50% for 2G ethanol production (Scenario 1) due to fuel [85] and electricity prices [86]. In the abovementioned cases, the economic potential per day of a sugarcane mill operation was 19.6% (Scenario 2) and 26.4% (Scenario 1), which is lower than that reported in Scenario 3. Null environmentally friendly potential was found in Scenarios 1 and 2 because pentose liquor derived from the SCB pretreatment was not used as a raw material for biohydrogen, VOA, or methane production and could result in the accumulation of 16,800 and 33,600 $m^3 \cdot d^{-1}$ of 2G ethanol byproducts, respectively. In Scenario 3, the greatest economic potential was achieved because the extraction of value-added VOA was considered. Among these acids, acetate is used as a precursor to industrial polymers, including mainly the vinyl acetate monomer [87], butyrate is used in industrial thermoplastics in the form of cellulose acetate butyrate [23], propionate is utilized as a food preservative [88], and lactate is involved in the production of biodegradable polymers, also known as polylactides [24]. In addition, the separation of these organic acids might have a low cost if performed by an ion exchange resin, as demonstrated elsewhere [16]. In terms of energy production, Scenario 4 provided relevant energetic potential; however, organic acids could not be directly recovered. Rather, they were considered substrates for methane production [89]. In this scenario, methane was considered the raw material for on-site cogeneration because this application did not require purification and also allowed local carbon offset projects with a cogeneration price (64.70 $USD \cdot MWh^{-1}$) that is more attractive than a commodity price of 2.77 USD per 10^3 ft^3 [90]. If the energetic potential depended on burning hydrogen alone, the maximum energy production would be 0.73 MWh, which is not significant for achieving profitability (Scenarios 3 and 4). However, hydrogen must be considered as an available surplus of this process because its production is obligatorily coupled to that of VOA (Reactions 5–6). An on-site application for biohydrogen could be the complete fueling of 56 hydrogen fuel cell vehicles per day with driving ranges of 240 miles [91], which is sufficient for the plant's complete fleet of cars. According to Table 4, the major advantage of Scenarios 3 and 4 is that in the production of organic acids or methane, the environmental impact (assessed as the COD reduction) of the 2G ethanol byproducts could be reduced by 62.7% through the separation/extraction of organic acids (acetate, butyrate, propionate, and lactate) or approximately 75% when employing a process based on sequential hydrogen and methane production [92]. Specifically in Scenario 3, organic matter removal could be further increased through the methanization of the VOA that were not extracted as raw materials for the chemical industry, namely the citric, malic, succinic, formic, valeric, and caproic acids.

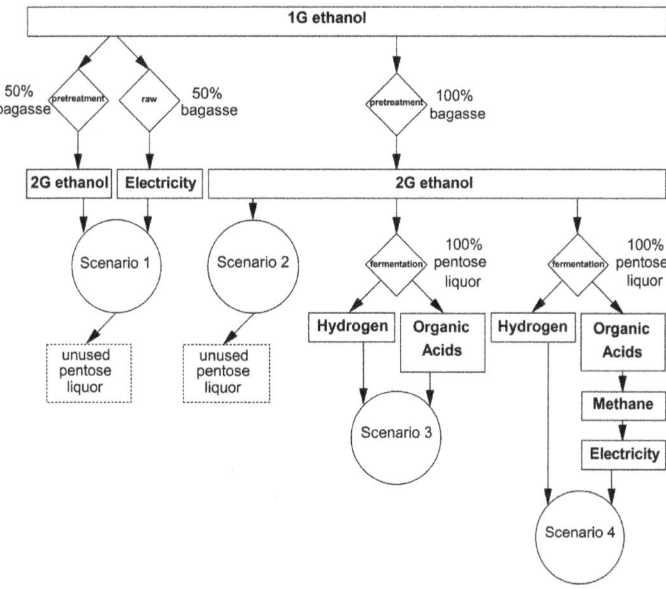

Figure 5. Second-generation sugarcane biorefinery schemes including potential biotechnological uses for pentose liquor.

Table 3. Comparative analysis of systems used for biohydrogen and organic acids production from pentoses.

Reactor	Xylose Concentration (gCOD·L^{-1})	Temp.	pH	HY	VHPR	VOA Production (mg·h^{-1})			Reference
						HBu	HPr	HAc	
CSTR	20.0	50	6.5	0.4	6600	264	1767	1601	[36]
AGSB	20.0	40	6.5	0.6	19,680	835	173	820	[36]
Batch [2]	1.5	37	5.5	1.5	-	853 [1]	4.53 [1]	149 [1]	[26]
UAPBR	1.7	25	6.7	0.08	398.4	120.8	125.9	305.76	This study
UAPBR [2]	1.7	25	5.9	0.04	120	473.4	447.8	379.20	This study

[1] Terminal soluble metabolite concentrations (measured at the end of the experimental runs). [2] Using pentose liquor. Nomenclature: Temp.—temperature (°C), HY—hydrogen yield (mol H$_2$·mol^{-1}xylose), VHPR—volumetric hydrogen production rate (mL H$_2$·L^{-1}·d^{-1}), VOA—volatile organic acids, HBu—butyric acid, HPr—propionic acid, HAc—acetic acid, CSTR—continuous stirred-tank reactor, AGSB—activated carbon-assisted agitated granular sludge bed, UAPBR—upflow anaerobic packed-bed reactor.

Table 4. Energetic, economic and environmental aspects for different second-generation sugarcane biorefinery schemes.

Scenarios/Products		1	2	3	4
		1G Ethanol 50%—2G Ethanol 50%—Cogeneration	1G Ethanol 2G Ethanol	1G Ethanol 2G Ethanol Hydrogen Organic Acids	1G Ethanol 2G Ethanol Hydrogen Methane
1G Ethanol	Production (m^3·d^{-1}) [1]	987.60	987.60	987.60	987.60
	Econ. Pot. (USD·d^{-1}) [2]	521,924.96	521,924.96	521,924.96	521,924.96
2G Ethanol	Production (m^3·d^{-1}) [3]	149.69	299.37	299.37	299.37
	Econ. Pot. (USD·d^{-1}) [2]	79,106.83	158,213.66	158,213.66	158,213.66
Hydrogen	Production (m^3·d^{-1}) [4]	NA	NA	19.06	19.06
	Econ. Pot. (USD·d^{-1}) [5]	NA	NA	2897.44	2897.44
Organic Acids (HLa, HAc, HPr, HBu)	Production (kg·d^{-1}) [6]	NA	NA	201,144.05	NA
	Econ. Pot. (USD·d^{-1}) [7]	NA	NA	162,693.67	NA
Energy (cogeneration [8] /methane [11,12])	Production (m^3·d^{-1})	NA	NA	NA	118,931.60 [11]
	Energ. Pot. (MWh)	334.32 [8]	NA	NA	806.73 [12]
	Econ. Pot. (USD·d^{-1})	23,422.82 [9]	NA	NA	52,192.02 [9]
Total Energetic potential (MWh)		334.32	NA	NA	806.73
Total Economic potential (USD·d^{-1})		622,660.68	680,138.61	845,729.72	735,228.08
Total Environmental potential (COD removal percentage)		NA	NA	62.7 [10]	74.7 [11]

[1] Medium-sized sugarcane distillery (milling capacity = 500 ton·h^{-1}) considering 1G ethanol yield [93]. [2] According to the anhydrous ethanol market price provided by the Center for Advanced Studies on Applied Economics (CEPEA/ESALQ/USP) [85]. [3] Considering the 2G ethanol yield obtained exclusively from the cellulosic fraction [94]. [4] According to the mean hydrogen yield obtained in experimental condition II. [5] According to the hydrogen market price reported by the U.S. Department of Energy [95]. [6] According to the mean organic acid yield obtained in experimental condition II. [7] According to HLa, HAc, HPr, and HBu reference prices [96]. [8] Calculated as the surplus electricity of a distillery producing only 1G ethanol [93]. [9] According to the market price defined by alternative energy auctions, as released by the Brazilian Energy Research Office (EPE/MME) [86]. [10] Calculated as the COD of all organic matter minus the COD of the extracted acids (HLa, HAc, HPr, HBu). [11] Considering 350 mL CH$_4$·g^{-1}COD removed [89] and 74.7% methanization efficiency of vinasse [92]. [12] Considering CH$_4$ properties [85] and combined cycle turbines [97]. Nomenclature: 1G— first generation, 2G—second generation, HAc—acetic acid, HBu—butyric acid, HLa—lactic acid, HPr—propionic acid, NA—not applicable.

4. Conclusions

Biohydrogen and volatile organic acids were produced using both a xylose-based media and sugarcane bagasse-derived pentose liquor. The fermentation of the pretreatment-derived pentose liquor generated a 25% higher volatile organic acid concentration compared to that obtained with the xylose-based media. The differences in the composition of both substrates triggered different fermentative metabolic pathways, leading to the stimulus of homoacetogenesis when using xylose and to the occurrence of biohydrogen production coupled to acetate and butyrate buildup when using pentose liquor. Overall, higher biohydrogen production rates could be observed when modifying the operating conditions applied in the packed-bed reactor, such as applying high organic loading rates and increasing the temperature to achieve thermophilic conditions. Nevertheless, the continuous operation of the reactor yielded both operational and performance parameters that enabled the simulation of the application of dark fermentation in a medium-sized distillery. The best estimate indicated that recovering hydrogen and organic acids from pentose liquor had the potential to enhance profits in an ethanol production plant (1G and 2G) by 24.3% with a simultaneous environmental impact reduction of 62.7%. Therefore, developing the present fermentation technology seems to be crucial for improving the sustainability of conventional 1G and 2G sugarcane biorefineries.

Author Contributions: Conceptualization, D.M.F.L.; methodology, M.A.d.L.; formal analysis, F.V.F.; investigation, W.K.M.; resources, M.A.d.L.; writing—original draft preparation, G.P. and G.M.; writing—review and editing, L.T.F.; visualization, I.P.; supervision, M.Z.; project administration, M.Z.; funding acquisition, M.Z. All authors have read and agreed to the published version of the manuscript.

Funding: This research was funded by the FUNDAÇÃO DE AMPARO À PESQUISA DO ESTADO DE SÃO PAULO, grant numbers 2008/56255-9, 2009/17539-4 and 2009/15984-0. The APC was waived by the journal.

Data Availability Statement: The data presented in this study are available on request from the corresponding author.

Conflicts of Interest: The authors declare no conflict of interest.

References

1. Lago, A.C.; Bonomi, A.; Cavalett, O.; Cunha, M.P.; Lima, M.A. Sugarcane as a carbon source: The Brazilian case. *Biomass Bioenergy* **2012**, *46*, 5–12. [CrossRef]
2. Poggi-Varaldo, H.M.; Munoz-Paez, K.M.; Escamilla-Alvarado, C.; Robledo-Narváez, P.N.; Ponce-Noyola, M.T.; Calva-Calva, G.; Ríos-Leal, E.; Galíndez-Mayer, J.; Estrada-Vázquez, C.; Ortega-Clemente, A.; et al. Biohydrogen, biomethane and bioelectricity as crucial components of biorefinery of organic wastes: A review. *Waste Manag. Res.* **2014**, *32*, 353–365. [CrossRef] [PubMed]
3. Morais, E.R.; Junqueira, T.L.; Sampaio, I.L.M.; Dias, M.O.S.; Rezende, M.C.A.F.; Jesus, C.D.F.; Klein, B.C.; Gómez, E.O.; Mantelatto, P.E.; Maciel Filho, R.; et al. Biorefinery alternatives. In *Virtual biorefinery: An Optimization Strategy for Renewable Carbon Valorization*; Bonomi, A., Cavalett, O., Cunha, M.P., Lima, M.A.P., Eds.; Springer: London, UK, 2016; pp. 53–132.
4. Dias, M.O.S.; Junqueira, T.L.; Sampaio, I.L.M.; Chagas, M.F.; Watanabe, M.D.B.; Morais, E.R.; Gouveia, V.L.R.; Klein, B.C.; Rezende, M.C.A.F.; Cardoso, T.F.; et al. Use of the VSB to assess biorefinery strategies. In *Virtual biorefinery: An Optimization Strategy for Renewable Carbon Valorization*; Bonomi, A., Cavalett, O., Cunha, M.P., Lima, M.A.P., Eds.; Springer: London, UK, 2016; pp. 189–256.
5. Rabelo, S.C.; Carrere, H.; Maciel Filho, R.; Costa, A.C. Production of bioethanol, methane and heat from sugarcane bagasse in a biorefinery concept. *Bioresour. Technol.* **2011**, *102*, 7887–7895. [CrossRef] [PubMed]
6. CONAB. *Acompanhamento da Safra Brasileira de Cana-de-Açúcar*; Safra 2023/24—Primeiro Levantamento; CONAB: Brasília, Brasil, 2023. Available online: https://www.conab.gov.br/info-agro/safras/cana/boletim-da-safra-de-cana-de-acucar/item/download/47193_22dfa86f63aa61cd8e405419b30d9262 (accessed on 27 June 2023). (In Portuguese)
7. Dias, M.O.S.; Junqueira, T.L.; Cavalett, O.; Pavanello, L.G.; Cunha, M.P.; Jesus, C.D.F.; Maciel Filho, R.; Bonomi, A. Biorefineries for the production of first and second generation ethanol and electricity from sugarcane. *Appl. Energy* **2013**, *109*, 72–78. [CrossRef]
8. Dias, M.O.S.; Junqueira, T.L.; Rossel, C.E.; Maciel Filho, R.; Bonomi, A. Evaluation of process configurations for second generation integrated with first generation bioethanol production from sugarcane. *Fuel Process. Technol.* **2013**, *109*, 84–89. [CrossRef]
9. Zhou, M.; Lü, X. Strategies on simultaneous fermentation of pentose and hexose to bioethanol. In *Advances in 2nd Generation of Bioethanol Production*; Lü, X., Ed.; Woodhead Publishing: Duxford, UK, 2021; pp. 161–211.
10. Sriariyanun, M.; Gundupalli, M.P.; Phakeenuya, V.; Phusamtisampan, T.; Cheng, Y.S.; Venkatachalam, P. Biorefinery Approaches For Production Of Cellulosic Ethanol Fuel Using Recombinant Engineered Microorganisms. *J. Appl. Sci. Eng.* **2023**, *27*, 1985–2005.
11. Ghimire, A.; Frunzo, L.; Pirozzi, F.; Trably, E.; Escudie, R.; Lens, P.N.L.; Esposito, G. A review on dark fermentative biohydrogen production from organic biomass: Process parameters and use of by-products. *Appl. Energy* **2015**, *144*, 73–95. [CrossRef]
12. Elbeshbishy, E.; Dhar, B.R.; Nakhla, G.; Lee, H.S. A critical review on inhibition of dark biohydrogen fermentation. *Renew. Sustain. Energy Rev.* **2017**, *79*, 656–668. [CrossRef]
13. Hawkes, F.R.; Dinsdale, R.; Hawkes, D.L.; Hussy, I. Sustainable fermentative hydrogen production: Challenges for process optimization. *Int. J. Hydrogen Energy* **2002**, *27*, 1339–1347. [CrossRef]

14. Show, K.Y.; Lee, D.J.; Tay, J.H.; Lin, C.Y.; Chang, J.S. Biohydrogen production: Current perspectives and the way forward. *Int. J. Hydrogen Energy* **2012**, *37*, 15616–15631. [CrossRef]
15. Das, D.; Veziroglu, T. Hydrogen production by biological processes: A survey of literature. *Int. J. Hydrogen Energy* **2001**, *26*, 13–28. [CrossRef]
16. Leite, J.A.C.; Fenandes, B.S.; Pozzi, E.; Barboza, M.; Zaiat, M. Application of an anaerobic packed-bed biobioreactor for the production of hydrogen and organic acids. *Int. J. Hydrogen Energy* **2008**, *33*, 579–586. [CrossRef]
17. Ramos, L.R.; Silva, E.L. Thermophilic hydrogen and methane production from sugarcane stillage in two-stage anaerobic fluidized bed reactors. *Int. J. Hydrogen Energy* **2020**, *45*, 5239–5251. [CrossRef]
18. Ferreira, T.B.; Rego, G.C.; Ramos, L.R.; Menezes, C.A.; Silva, E.L. Improved dark fermentation of cane molasses in mesophilic and thermophilic anaerobic fluidized bed reactors by selecting operational conditions. *Int. J. Energy Res.* **2020**, *44*, 10442–10452. [CrossRef]
19. Fuess, L.T.; Braga, A.F.M.; Eng, F.; Gregoracci, G.; Saia, F.T.; Zaiat, M.; Lens, P.N.L. Solving the bottlenecks of sugarcane vinasse biodigestion: Impacts of temperature and substrate exchange on sulfate removal during dark fermentation. *Chem. Eng. J.* **2023**, *455*, 140965. [CrossRef]
20. Menezes, C.A.; Almeida, P.S.; Delforno, T.P.; Oliveira, V.M.; Sakamoto, I.K.; Varesche, M.B.A.; Silva, E.L. Relating biomass composition and the distribution of metabolic functions in the co-fermentation of sugarcane vinasse and glycerol. *Int. J. Hydrogen Energy* **2023**, *48*, 8837–8853. [CrossRef]
21. Kim, H.; Jeon, B.S.; Sang, B.I. An efficient new process for the selective production of odd-chain carboxylic acids by simple carbon elongation using *Megasphaera hexanoica*. *Sci. Rep.* **2019**, *9*, 11999. [CrossRef]
22. Bastidas-Oyanedel, J.R.; Schmidt, J.E. Increasing profits in food waste biorefinery—A techno-economic analysis. *Energies* **2018**, *11*, 1551. [CrossRef]
23. Brady, G.S.; Clauser, H.R.; Vaccari, J.A. *Materials Handbook*, 15th ed.; McGraw-Hill Professional: New York, NY, USA, 2002.
24. Napoothiri, K.M.; Nair, N.R.; John, R.P. An overview of the recent developments in polylactide (PLA) research. *Bioresour Technol.* **2010**, *101*, 6493–8501.
25. Oliveira, G.H.D.; Niz, M.Y.K.; Zaiat, M.; Rodrigues, J.A.D. Effects of organic loading rate on polyhydroxyalkanoate production from sugarcane stillage by mixed microbial cultures. *Appl. Biochem. Biotechnol.* **2019**, *189*, 1039–1055. [CrossRef]
26. Pattra, S.; Sangyoka, S.; Boonmee, M.; Reungsang, A. Bio-hydrogen Production from the Fermentation of Sugarcane Bagasse Hydrolysate by *Clostridium butyricum*. *Int. J. Hydrogen Energy* **2008**, *33*, 5256–5265. [CrossRef]
27. Kongjan, P.; Min, B.; Angelidaki, I. Biohydrogen production from xylose at extreme thermophilic temperatures (70 °C) by mixed culture fermentation. *Water Res.* **2009**, *43*, 1414–1424. [CrossRef] [PubMed]
28. Baêta, B.E.L.; Lima, D.R.S.; Balena Filho, J.G.; Adarme, O.F.H.; Gurgel, L.V.A.; Aquino, S.F. Evaluation of hydrogen and methane production from sugarcane bagasse hemicellulose hydrolysates by two-stage anaerobic digestion process. *Bioresour Technol.* **2016**, *218*, 436–446. [CrossRef] [PubMed]
29. Thungklin, P.; Sittijunda, S.; Reungsang, A. Sequential fermentation of hydrogen and methane from steam-exploded sugarcane bagasse hydrolysate. *Int. J. Hydrogen Energy* **2018**, *43*, 9924–9934. [CrossRef]
30. Chatterjee, S.; Mohan, S.V. Simultaneous production of green hydrogen and bioethanol from segregated sugarcane bagasse hydrolysate streams with circular biorefinery design. *Chem. Eng. J.* **2021**, *425*, 130386. [CrossRef]
31. Fangkum, A.; Reungsang, A. Biohydrogen production from mixed xylose/arabinose at thermophilic temperature by anaerobic mixed cultures in elephant dung. *Int. J. Hydrogen Energy* **2011**, *36*, 13928–13938. [CrossRef]
32. Maintinguer, S.I.; Fernandes, B.S.; Duarte, I.C.S.; Saavedra, N.K.; Adorno, M.A.T.; Varesche, M.B.A. Fermentative hydrogen production with xylose by *Clostridium* and *Klebsiella* species in anaerobic batch bioreactors. *Int. J. Hydrogen Energy* **2011**, *36*, 13508–13517. [CrossRef]
33. Lin, C.; Cheng, C. Fermentative hydrogen production from xylose using anaerobic mixed microflora. *Int. J. Hydrogen Energy* **2006**, *31*, 832–840. [CrossRef]
34. Lin, C.; Wu, C.; Hung, C. Temperature effects on fermentative hydrogen production from xylose using mixed anaerobic cultures. *Int. J. Hydrogen Energy* **2008**, *33*, 43–50. [CrossRef]
35. Zhao, C.; Karakashev, D.; Lu, W.; Wang, H.; Angelidaki, I. Xylose fermentation to biofuels (hydrogen and ethanol) by extreme thermophilic (70 °C) mixed culture. *Int. J. Hydrogen Energy* **2010**, *5*, 3415–3422. [CrossRef]
36. Wu, S.Y.; Lin, C.Y.; Lee, K.S.; Hung, C.H.; Chang, J.S.; Lin, P.J.; Chang, F.Y. Dark Fermentative Hydrogen Production from Xylose in Different Bioreactors Using Sewage Sludge Microflora. *Energy Fuels* **2008**, *22*, 113–119. [CrossRef]
37. Dessi, P.; Lakaniemi, A.M.; Lens, P.N.L. Biohydrogen production from xylose by fresh and digested activated sludge at 37, 55 and 70 °C. *Water Res.* **2017**, *115*, 120–129. [CrossRef]
38. Baik, J.H.; Jung, J.H.; Sim, Y.B.; Park, J.H.; Kim, S.M.; Yang, J.; Kim, S.H. High-rate biohydrogen production from xylose using a dynamic membrane bioreactor. *Bioresour. Technol.* **2022**, *344*, 126205. [CrossRef]
39. Silva, V.; Rabelo, C.A.B.S.; Camargo, F.P.; Sakamoto, I.K.; Silva, E.L.; Varesche, M.B.A. Optimization of Key Factors Affecting Hydrogen and Ethanol Production from Xylose by *Thermoanaerobacterium calidifontis* VCS1 Isolated from Vinasse Treatment Sludge. *Waste Biomass Valorization* **2022**, *13*, 1897–1912. [CrossRef]
40. Lima, D.M.F.; Zaiat, M. The influence of the degree of back-mixing on hydrogen production in an anaerobic fixed-bed reactor. *Int. J. Hydrogen Energy* **2012**, *37*, 9630–9635. [CrossRef]

41. Peixoto, G.; Saavedra, N.K.; Varesche, M.B.A.; Zaiat, M. Hydrogen production from soft-drink wastewater in an upflow anaerobic packed-bed bioreactor. *Int. J. Hydrogen Energy* **2011**, *36*, 8953–8966. [CrossRef]
42. Penteado, E.D.; Lazaro, C.Z.; Sakamoto, I.K.; Zaiat, M. Influence of seed sludge and pretreatment method on hydrogen production in packed-bed anaerobic reactors. *Int. J.Hydrogen Energy* **2013**, *38*, 6137–6145. [CrossRef]
43. Ferraz Jr, A.D.N.; Etchebehere, C.; Zaiat, M. High organic loading rate on thermophilic hydrogen production and metagenomic study at an anaerobic packed-bed reactor treating a residual liquid stream of a Brazilian biorefinery. *Bioresour Technol.* **2015**, *186*, 81–88. [CrossRef]
44. Fuess, L.T.; Kiyuna, L.S.M.; Garcia, M.L.; Zaiat, M. Operational strategies for long-term biohydrogen production from sugarcane stillage in continuous acidogenic packed-bed reactor. *Int. J. Hydrogen Energy* **2016**, *41*, 8132–8145. [CrossRef]
45. Corbari, S.D.M.L.; Andreani, C.L.; Torres, D.G.B.; Eng, F.; Gomes, S.D. Strategies to improve the biohydrogen production from cassava wastewater in fixed-bed reactors. *Int. J. Hydrogen Energy* **2019**, *44*, 17214–17223. [CrossRef]
46. Aguilar, R.; Ramirez, J.A.; Garrote, G.; Vazquez, M. Kinetic study of the acid hydrolysis of sugarcane bagasse. *J. Food Eng.* **2002**, *55*, 309–318. [CrossRef]
47. APHA; AWWA; WEF. *Standard Methods for the Examination of Water and Wastewater*, 22nd ed.; APHA: Washington, DC, USA, 2012.
48. Dubois, M.; Gilles, K.A.; Hamilton, J.K.; Rebers, P.A.; Smith, F. Colorimetric methods for determination of sugar and related substance. *Anal. Chem.* **1956**, *28*, 350–356. [CrossRef]
49. Mockaitis, G.; Rodrigues, J.A.D.; Foresi, E.; Zaiat, M. Toxic effects of cadmium (Cd^{2+}) on anaerobic biomass: Kinetic and metabolic implications. *J. Environ. Manag.* **2012**, *106*, 75–84. [CrossRef] [PubMed]
50. Perna, V.; Castelló, E.; Wenzel, J.; Zampol, C.; Lima, D.M.F.; Borzacconi, L.; Varesche, M.B.; Zaiat, M.; Etchebehere, C. Hydrogen production in an upflow anaerobic packed bed reactor used to treat cheese whey. *Int. J. Hydrogen Energy* **2013**, *38*, 54–62. [CrossRef]
51. Ferraz, A.D.N., Jr.; Etchebehere, C.; Zaiat, M. Mesophilic hydrogen production in acidogenic packed-bed reactors (APBR) using raw sugarcane vinasse as substrate: Influence of support materials. *Anaerobe* **2015**, *34*, 94–105.
52. Lima, D.M.F.; Moreira, W.K.; Zaiat, M. Comparison of the use of sucrose and glucose as a substrate for hydrogen production in an upflow anaerobic fixed-bed reactor. *Int. J. Hydrogen Energy* **2013**, *38*, 15074–15083. [CrossRef]
53. Anzola-Rojas, M.P.; Fonseca, S.G.; Silva, C.C.; Oliveira, V.M.; Zaiat, M. The use of the carbon/nitrogen ratio and specific organic loading rate as tools for improving biohydrogen production in fixed-bed reactors. *Biotechnol. Rep.* **2015**, *5*, 46–54. [CrossRef]
54. Saady, N.M.C. Homoacetogenesis during hydrogen production by mixed cultures dark fermentation. *Int. J. Hydrogen Energy* **2013**, *38*, 13172–13191. [CrossRef]
55. Ferreira, T.B.; Rego, G.C.; Ramos, L.R.; Menezes, C.A.; Soares, L.A.; Sakamoto, I.K.; Varesche, M.B.A.; Silva, E.L. HRT control as a strategy to enhance continuous hydrogen production from sugarcane juice under mesophilic and thermophilic conditions in AFBRs. *Int. J. Hydrogen Energy* **2019**, *44*, 19719–19729. [CrossRef]
56. Menezes, C.A.; Silva, E.L. Hydrogen production from sugarcane juice in expanded granular sludge bed reactors under mesophilic conditions: The role of homoacetogenesis and lactic acid production. *Ind. Crop. Prod.* **2019**, *138*, 111586. [CrossRef]
57. Freitas, I.B.F.; Menezes, C.A.; Silva, E.L. An alternative for value aggregation to the sugarcane chain: Biohydrogen and volatile fatty acids production from sugarcane molasses in mesophilic expanded granular sludge bed reactors. *Fuel* **2020**, *260*, 116419. [CrossRef]
58. Bernal, A.P.; Menezes, C.A.; Silva, E.L. A new side-looking at the dark fermentation of sugarcane vinasse: Improving the carboxylates production in mesophilic EGSB by selection of the hydraulic retention time and substrate concentration. *Int. J. Hydrogen Energy* **2021**, *46*, 12758–12770. [CrossRef]
59. Niz, M.Y.K.; Etchelet, I.; Fuentes, L.; Etchebehere, C.; Zaiat, M. Extreme thermophilic condition: An alternative for long-term biohydrogen production from sugarcane vinasse. *Int. J. Hydrogen Energy* **2019**, *44*, 22876–22887. [CrossRef]
60. Piffer, M.A.; Oliveira, C.A.; Bovio-Winkler, P.; Eng, F.; Etchebehere, C.; Zaiat, M.; Nascimento, C.A.O.; Fuess, L.T. Sulfate- and pH-driven metabolic flexibility in sugarcane vinasse dark fermentation stimulates biohydrogen evolution, sulfidogenesis or homoacetogenesis. *Int. J. Hydrogen Energy* **2022**, *47*, 31202–31222. [CrossRef]
61. Singh, A.; Müller, B.; Fuxelius, H.H.; Schnürer, A. AcetoBase: A functional gene repository and database for formyltetrahydrofolate synthetase sequences. *Database* **2019**, *2019*, baz142. [CrossRef]
62. Klemps, R.; Cypionka, H.; Widdel, F.; Pfennig, N. Growth with hydrogen, and further physiological characteristics of *Desulfotomaculum* species. *Arch. Microbiol.* **1985**, *143*, 203–208. [CrossRef]
63. Okamoto, M.; Miyahara, T.; Mizuno, O.; Noike, T. Biological hydrogen potential of materials characteristic of the organic fraction of municipal solid wastes. *Water Sci. Technol.* **2000**, *41*, 25–32. [CrossRef]
64. Garay-Arroyo, A.; Covarrubias, A.A.; Clark, I.; Niño, I.; Gosset, G.; Martínez, A. Response to different environmental stress conditions of industrial an laboratory Saccharomyces cerevisiae strains. *Appl. Microbiol. Biotechnol.* **2004**, *63*, 734–741. [CrossRef]
65. Ferraz, A.D.N., Jr.; Wenzel, J.; Etchebehere, C.; Zaiat, M. Effect of organic loading rate on hydrogen production from sugarcane vinasse in thermophilic acidogenic packed bed reactors. *Int. J. Hydrogen Energy* **2014**, *39*, 16852–16862.
66. Fuess, L.T.; Zaiat, M.; Nascimento, C.A.O. Novel insights on the versatility of biohydrogen production from sugarcane vinasse via thermophilic dark fermentation: Impacts of pH-driven operating strategies on acidogenesis metabolite profiles. *Bioresour. Technol.* **2019**, *286*, 121379. [CrossRef]

67. Rogeri, R.C.; Fuess, L.T.; Eng, F.; Borges, A.V.; Araujo, M.A.; Damianovic, M.H.R.Z.; Silva, A.J. Strategies to control pH in the dark fermentation of sugarcane vinasse: Impacts on sulfate reduction, biohydrogen production and metabolite distribution. *J. Environ. Manag.* **2023**, *325*, 116495. [CrossRef] [PubMed]
68. Fuess, L.T.; Zaiat, M.; Nascimento, C.A.O. Molasses vs. juice: Maximizing biohydrogen production in sugarcane biorefineries to diversify renewable energy generation. *J. Water Process Eng.* **2020**, *37*, 101534. [CrossRef]
69. Fuess, L.T.; Fuentes, L.; Bovio-Winkler, P.; Eng, F.; Etchebehere, C.; Zaiat, M.; Nascimento, C.A.O. Full details on continuous bio-hydrogen production from sugarcane molasses are unraveled: Performance optimization, self-regulation, metabolic correlations and quanti-qualitative biomass characterization. *Chem. Eng. J.* **2021**, *414*, 128934. [CrossRef]
70. Jo, J.H.; Lee, D.S.; Park, D.; Park, J.M. Biological hydrogen production by immobilized cells of *Clostridium tyrobutyricum* JM1 isolated from a food waste treatment process. *Bioresour. Technol.* **2008**, *99*, 6666–6672. [CrossRef]
71. Piffer, M.A.; Zaiat, M.; Nascimento, C.A.O.; Fuess, L.T. Dynamics of sulfate reduction in the thermophilic dark fermentation of sugarcane vinasse: A biohydrogen-independent approach targeting enhanced bioenergy production. *J. Environ. Chem. Eng.* **2021**, *9*, 105956. [CrossRef]
72. Oliveira, C.A.; Fuess, L.T.; Soares, L.A.; Damianovic, M.H.R.Z. Thermophilic biohydrogen production from sugarcane molasses under low pH: Metabolic and microbial aspects. *Int. J. Hydrogen Energy* **2020**, *45*, 4182–4192. [CrossRef]
73. Ramos, L.R.; Silva, E.L. Continuous hydrogen production from agricultural wastewaters at thermophilic and hyperthermophilic temperatures. *Appl. Biochem. Biotechnol.* **2017**, *182*, 846–869. [CrossRef]
74. Fuess, L.T.; Garcia, M.L.; Zaiat, M. Seasonal characterization of sugarcane vinasse: Assessing environmental impacts from fertirrigation and the bioenergy recovery potential through biodigestion. *Sci. Total Environ.* **2018**, *634*, 29–40. [CrossRef]
75. Lepe, J.A.S.; Leal, B.I. *Microbiología Enológica: Fundamentos de Vinificacion*, 3rd ed.; Ediciones Mundi-Prensa: Madrid, Spain, 2004. (In Spanish)
76. Kolb, S.; Otte, H.; Nagel, B.; Schink, B. Energy conservation in malolactic fermentation by *Lactobacillus plantarum* and *Lactobacillus sake*. *Arch. Microbiol.* **1992**, *157*, 457–463. [CrossRef]
77. Fuess, L.T.; Santos, G.M.; Delforno, T.P.; Moraes, B.S.; Silva, A.J. Biochemical butyrate production via dark fermentation as an energetically efficient alternative management approach for vinasse in sugarcane biorefineries. *Renew. Energy* **2020**, *158*, 3–12. [CrossRef]
78. Ding, H.B.; Amy, T.G.; Wang, J.Y. Caproate formation in mixed-culture fermentative hydrogen production. *Bioresour Technol.* **2010**, *101*, 9550–9559. [CrossRef]
79. Cavalcante, W.A.; Leitão, R.C.; Gehring, T.A.; Angenent, L.T.; Santaella, S.T. Anaerobic fermentation for n-caproic acid production: A review. *Process Biochem.* **2017**, *54*, 106–119. [CrossRef]
80. Chen, W.M.; Tseng, Z.J.; Lee, K.S.; Chang, J.S. Fermentative hydrogen production with Clostridium butyricum CGS5 isolated from anaerobic sewage sludge. *Int. J. Hydrogen Energy* **2005**, *30*, 1063–1070. [CrossRef]
81. Infantes, D.; González del Campo, A.; Villaseñor, J.; Fernández, F.J. Influence of pH, temperature and volatile fatty acids on hydrogen production by acidogenic fermentation. *Int. J. Hydrogen Energy* **2011**, *36*, 15595–15601. [CrossRef]
82. Thauer, R.K.; Jungermann, K.; Decker, K. Energy conservation in chemotrophic anaerobic bacteria. *Bacteriol. Rev.* **1977**, *41*, 100–180. [CrossRef]
83. Fernandes, B.S.; Peixoto, G.; Albrecht, F.R.; Del Aguila, N.K.S.; Zaiat, M. Potential to produce biohydrogen from various wastewaters. *Energy Sustain. Dev.* **2010**, *14*, 143–148. [CrossRef]
84. Wu, K.J.; Saratale, G.D.; Lo, Y.C.; Chen, W.M.; Tseng, Z.J.; Chang, M.C.; Tsai, B.C.; Sud, A.; Chang, J.S. Simultaneous production of 2,3-butanediol, ethanol and hydrogen with a Klebsiella sp. strain isolated from sewage sludge. *Bioresour. Technol.* **2008**, *99*, 7966–7970. [CrossRef]
85. CEPEA. CEPEA/ESALQ Hydrous Ethanol Index (Fuel)—São Paulo State. Available online: https://cepea.esalq.usp.br/en/indicator/ethanol.aspx (accessed on 27 June 2023).
86. EPE. Leilões. Available online: https://www.epe.gov.br/pt/leiloes-de-energia/leiloes (accessed on 28 June 2023). (In Portuguese)
87. Harahap, B.M.; Ahring, B.K. Acetate Production from Syngas Produced from Lignocellulosic Biomass Materials along with Gaseous Fermentation of the Syngas: A Review. *Microorganisms* **2023**, *11*, 995. [CrossRef]
88. Sauer, M.; Porro, D.; Mattanovich, D.; Branduardi, P. Microbial production of organic acids: Expanding the markets. *Trends Biotechnol.* **2008**, *26*, 100–108. [CrossRef]
89. Angelidaki, I.; Sanders, W. Assessment of the anaerobic biodegradability of macropollutants. *Rev. Environ. Sci. Biotechnol.* **2004**, *3*, 117–129. [CrossRef]
90. EIA. Open Data: Natural Gas. Available online: https://www.eia.gov/opendata/browser/natural-gas/pri (accessed on 28 June 2023).
91. Honda. Honda FCEV Concept Makes World Debut at Los Angeles International Auto Show. November 2013. Available online: https://hondanews.com/en-US/releases/release-47803fbebd894dddb275faf0fb0c0a28-honda-fcev-concept-makes-world-debut-at-los-angeles-international-auto-show (accessed on 27 June 2023).
92. Peixoto, G.; Pantoja-Filho, J.L.R.; Agnelli, J.A.B.; Barboza, M.; Zaiat, M. Hydrogen and methane production, energy recovery, and organic matter removal from effluents in a two-stage fermentative process. *Appl. Biochem. Biotechnol.* **2012**, *168*, 651–671. [CrossRef] [PubMed]

93. Dias, M.O.S.; Junqueira, T.L.; Cavalett, O.; Cunha, M.P.; Jesus, C.D.F.; Mantelatto, P.E.; Rosell, C.E.V.; Maciel-Filho, R.; Bonomi, A. Cogeneration in integrated first and second generation ethanol from sugarcane. *Chem. Eng. Res. Des.* **2013**, *91*, 1411–1417. [CrossRef]

94. Dantas, G.A.; Legey, L.F.L.; Mazzone, A. Energy from sugarcane bagasse in Brazil: An assessment of the productivity and cost of different technological routes. *Renew. Sustain. Energy Rev.* **2013**, *21*, 356–364. [CrossRef]

95. Clean Cities. *Alternative Fuel Price Report*; U.S. Department of Energy: Washington, DC, USA, 2016. Available online: http://www.afdc.energy.gov/uploads/publication/alternative_fuel_price_report_april_2016.pdf (accessed on 27 June 2023).

96. MOLBASE. Available online: https://www.molbase.com/ (accessed on 28 June 2023).

97. Chiesa, P.; Macchi, E. A Thermodynamic Analysis of Different Options to Break 60% Electric Efficiency in Combined Cycle Power Plants. *J. Eng. Gas Turbines Power* **2004**, *126*, 770–785. [CrossRef]

 *waste*

Article

Rosaceae Nut-Shells as Sustainable Aggregate for Potential Use in Non-Structural Lightweight Concrete

Veronica D'Eusanio [1,*], Lucia Bertacchini [2], Andrea Marchetti [1,3,4], Mattia Mariani [1], Stefano Pastorelli [2], Michele Silvestri [2] and Lorenzo Tassi [1,3,4]

[1] Department of Chemical and Geological Sciences, University of Modena and Reggio Emilia, 41125 Modena, Italy; andrea.marchetti@unimore.it (A.M.); 239950@studenti.unimore.it (M.M.)
[2] Litokol S.p.A., 42048 Rubiera, Italy; lucia.bertacchini@litokol.it (L.B.); stefano.pastorelli@litokol.it (S.P.); michele.silvestri@litokol.it (M.S.)
[3] Consorzio Interuniversitario Nazionale per la Scienza e Tecnologia dei Materiali (INSTM), 50121 Firenze, Italy
[4] Interdepartmental Research Center BIOGEST-SITEIA, University of Modena and Reggio Emilia, 42124 Reggio Emilia, Italy
* Correspondence: veronica.deusanio@unimore.it

Abstract: Apricot (AS), peach (PS), and plum shells (PlS) were examined as sustainable aggregates for non-structural lightweight concrete. The extraction of natural resources has a significant environmental impact and is not in line with the Sustainable Development Goals (SDGs) of Agenda 2030. Recycling agri-food waste, such as fruit shells, fully respects circular economy principles and SDGs. The chemical and physical properties of the shells were investigated using scanning electron microscopy (SEM) for microstructure analysis and TG-MS-EGA for thermal stress behavior. Two binding mixtures were used to prepare the concrete samples, one containing lime only (mixture "a") and one containing both lime and cement (mixture "b"). Lime is a more sustainable building material but it compromises mechanical strength and durability. The performance of lightweight concrete was determined based on the type of aggregate used. PS had a high-water absorption capacity due to numerous micropores, resulting in lower density (1000–1200 kg/m^3), compressive strength (1–4 MPa), and thermal conductivity (0.15–0.20 W/mK) of PS concrete. AS concrete showed the opposite trend (1120–1260 kg/m^3; 2.8–7.0 MPa; 0.2–0.4 W/mK) due to AS microporosity-free and denser structure. PlS has intermediate characteristics in terms of porosity, density, and water absorption, resulting in concrete with intermediate characteristics (1050–1240 kg/m^3; 1.9–5.2 MPa; 0.15–0.3 W/mK).

Keywords: sustainability; green building; recycle; food waste; lightweight concrete; lime concrete; fruit shells; coarse aggregate replacement

Citation: D'Eusanio, V.; Bertacchini, L.; Marchetti, A.; Mariani, M.; Pastorelli, S.; Silvestri, M.; Tassi, L. Rosaceae Nut-Shells as Sustainable Aggregate for Potential Use in Non-Structural Lightweight Concrete. *Waste* **2023**, *1*, 549–568. https://doi.org/10.3390/waste1020033

Academic Editors: Vassilis Athanasiadis, Dimitris P. Makris and Catherine N. Mulligan

Received: 16 March 2023
Revised: 23 April 2023
Accepted: 25 May 2023
Published: 6 June 2023

1. Introduction

The climate emergency confronts us with the need to minimize the exploitation of our planet's non-renewable resources. The extraction of raw materials has a dramatic impact on the environment, degrading landscapes, polluting soils and waters, irreparably damaging biodiversity, and inefficiently consuming a huge amount of energy [1]. Furthermore, the extraction of natural resources has accelerated exponentially since the 21st century and has grown significantly globally [2]. Indeed, global population growth, unbridled industrialization, and increased consumption have led to an increase in their demand [3]. The indiscriminate exploitation of non-renewable raw materials leads to their depletion: their cost is expected to increase significantly, and many of them may no longer be available in the near future [4,5]. Instead, renewable materials can be produced.

Ah indefinitely with strong environmental benefits, especially if they are waste by-products from other supply chains. This is the fundamental principle of the circular economy: someone's waste becomes a valuable resource for someone else. The transition to a circular system provides the opportunity to address this problem by reducing the use of raw materials, protecting material resources, and reducing the carbon footprint [6,7]. It is also expected to bring economic benefits such as an increase in gross domestic product, net savings in raw materials, growth in employment, and reduced volatility in material and supply prices [8,9].

The building materials sector is a major contributor to environmental deterioration as it is one of the largest exploiters of resources, half of which are non-renewable [10,11]. Thus, global government policies are driving the need to use low-energy and renewable building materials for construction, with the aim of combating climate change and minimizing its effects [12]. The UN Agenda 2030 sets 17 Sustainable Development Goals (SDGs) to end poverty, protect the planet, and improve the lives and prospects of everyone and everywhere [13]. The construction sector is closely associated with several SDGs including clean water and sanitation (SDG 6), affordable clean energy (SDG 7), industry, innovation, and infrastructure (SDG 9), and sustainable cities and communities (SDG 11). In addition, also SGD 12 (ensuring sustainable consumption and production patterns) provides interesting insights. This concerns the substantial reduction of waste through prevention, reduction, recycling, and reuse, which are core principles of the circular economy. In fact, the use of waste or by-products as a substitute for fossil raw materials represents an important opportunity, as it allows for plugging the problems related to waste disposal while reducing the exploitation of the resources of our planet. The construction sector can only be considered truly sustainable when it starts using renewable materials or materials recycled from construction waste and demolition residues [14]. The conservation of natural resources must be maximized and the environmental impact during the entire life cycle of the building project must be reduced. The development of environmentally friendly building materials is an inevitable path to sustainable construction and the SDGs of Agenda 2030.

Cement concrete is undoubtedly the most widely used building material [14,15], with a worldwide production of more than 10 billion tons. The enormous demand for concrete has led to the exploitation of a massive quantity of aggregates, causing their depletion or exhaustion in natural basins, as well as significant environmental issues [14,15]. The utilization of recycled or bio-based materials as a substitute for natural aggregates has been identified as a highly effective strategy to enhance sustainability in the construction industry. Several studies have explored the potential of industrial byproducts, agri-food wastes, or demolition wastes as lightweight aggregates (LWAs). LWAs have a lower average density and higher porosity, providing concrete with lower density and thermal and acoustic insulation properties. Lightweight concrete has a dry density of up to 2000 kg/m^3 and a thermal conductivity usually lower than 1 W/(m·°C). Therefore, it is used when low weight and insulating properties are relevant. The thermal conductivity of concrete refers to its ability to transfer heat. High thermal conductivity can lead to unwanted energy loss, which can increase energy costs and reduce the comfort levels of indoor spaces. In contrast, low thermal conductivity in concrete can promote energy efficiency and thermal comfort. One way to improve the thermal conductivity of concrete is by adding lightweight aggregates, which typically have good insulation properties. For example, Real et al. [16] reported that the use of lightweight concrete in buildings can reduce heating energy by 15% compared to normal-weight concrete.

Although lime concrete is an old material used in civil engineering [17–19], only a few studies have investigated its potential as an alternative to Portland cement concrete for structural components [20,21]. Lime has some environmental advantages over concrete-based materials: it requires less energy for its production, since limestone, the basic raw material, can be burned at lower temperatures (900–1000 °C), whereas silicate rocks for concrete require at least 1300 °C. Furthermore, part of the CO_2 generated during the production process is reabsorbed by hardened lime [22]. Therefore, the main aim of this study was to develop lime-based non-structural lightweight concrete using waste materials from the agri-food chain as coarse aggregates. In particular, *Rosaceae* fruit shells are a widely available waste, as a significant part of the harvested fruit is processed, resulting in a huge amount of waste kernels [23].

Peach (*Prunus persica* L.), apricot (*Prunus armeniaca* L.), and plum (*Prunus domestica* L.), belonging to the same *Prunus* family, are widely cultivated fruits [24] above all for their relevance to human health [25–29] as an important source of phenolic compounds, cyanogenic glucosides, vitamins, mineral salts, and phytoestrogens. According to the "Food and Agriculture Organization (FAO)", in 2020, global peach production was approximately 24 million tons, apricot one 3.7 million tons, and plum one 12 million tons. Their pulp is still the only part that is most appreciated and used by agro-industries, whereas pits are considered low-value agro-industrial residues. In addition to being directly consumed as fresh fruit, most Rosaceae fruits are processed into juices, canned fruits, jams, and sweet snacks. All these productive sectors give rise to a large quantity of kernels as waste, estimated at around 10% of the total mass [30]. Thanks to their high calorific value, the current alternative to landfilling fruit shells is incineration in biomass heating systems. However, this seasonal activity requires temporary kernel storage in large open-air stacks. This leads to some issues such as space availability, environmental hygiene problems, and the development of odorous exhalations due to uncontrolled fermentation of pulp residues and decomposition of organic material [31]. A second and more important factor concerns the serious environmental effects caused by the incineration of these materials. It is estimated that the combustion of agricultural residues, such as wood, leaves, trees, and grass, generates approximately 40% of CO_2 emissions, 32% of CO emissions, 20% of particulate matter, and 50% of polycyclic aromatic hydrocarbons (HAPs) [32]. Fruit shells exhibit several characteristics that make them an interesting alternative to common coarse aggregates. For example, their degradation under natural conditions is difficult and slow [33,34], unlike other food waste by-products. Moreover, it is a widely available and low-cost waste material, and its reuse and valorization are perfectly aligned with Agenda 2030 for Sustainable Development [13] and, in particular, with SGD 12.

Several studies have reported the use of agricultural waste materials, such as oil palm shells, palm oil clinkers, wood, mussel shells, date seeds, and coconut shells [35–38], for the production of environmentally friendly concrete. Some studies have used fruit shells to prepare cement-based concrete [39–41], but there is little evidence of lime-based materials. The effects induced on the physical and mechanical properties of lime-based concrete of peach, apricot, and plum shells were investigated, considering density, compressive strength, and thermal conductivity. In this preliminary study, the goal was to evaluate the potential of these materials as LWAs and identify any differences between the different shell types. The compositional and morphological characteristics of the aggregates were evaluated and correlated with the performance of lightweight concrete. Scanning electron microscopy (SEM) was used for the morphological study of the LWAs, and TGA-MS-EGA was used to obtain compositional information and study their behavior under thermal stress.

2. Materials and Methods

2.1. Raw Materials Properties and Specimens Preparation

2.1.1. Binder Mixture

The main binder used was hydrated lime (Litokol S.p.A., Rubiera, Italy). To improve the mechanical properties, a few sets of specimens were prepared with the addition of Type I 52.5 grade Portland cement (Litokol S.p.A., Rubiera, Italy). The physical and chemical properties of the binders are listed in Table 1.

Table 1. Physical and chemical properties of the binders.

	Hydrated Lime	Cement 52.5
Chemical analysis (wt %)		
SiO_2	-	19.8
CaO	75.68	63.89
Al_2O_3	-	4.43

Table 1. *Cont.*

	Hydrated Lime	**Cement 52.5**
Fe_2O_3	-	3.08
SO_3	-	3.77
MgO	-	1.02
Na_2O	-	0.09
K_2O	-	0.67
TiO_2	-	0.18
Physical Properties		
Bulk density (kg/m^3)	450	770
Specific gravity (g/cm^3)	2.24	2.75
Compressive strength 7 days (N/mm^2)	-	30
Compressive strength 28 days (N/mm^2)	-	52.5

2.1.2. Coarse and Fine Aggregates

Crushed shells were used as alternative coarse aggregates (Figure 1). Peaches, apricots, and plums were obtained from a local orchard in Modena, Italy. The pulp was separated from the pits, cleaned before use, and the residual dried pulp and dust on their surfaces were removed. The pits were preliminarily air dried for 30 days to remove residual moisture. Preliminary coarse grinding allowed the separation of the internal kernel from the external shell. The dried pits were crushed with a crushing machine and sieved with 4.5 and 9.5 mm sieves. Natural alluvial silica sand was used as the fine aggregate (Litokol S.p.A., Rubiera, Italy). The physical properties of the aggregates are listed in Table 2, and the proximate chemical compositions of the fruit shells are listed in Table 3.

Figure 1. Crushed peach shells (PS, sx), crushed apricot shells (AS, centre), and crushed plum shells (PlS, sx).

Table 2. Properties of coarse and fine aggregates were used in this study.

Physical Property	**Coarse Aggregate**			**Fine Aggregate**
	PS	**AS**	**PlS**	**Sand**
Particle size (mm)	4.5–9.5	4.5–9.5	4.5–9.5	1
Specific gravity (kg/dm^3)	1.28	1.44	1.37	1.5
Bulk density (kg/m^3)	556	630	591	1560
Water absorption (24 h) (%)	15.2	10.9	12.6	1.1
Shape	Flaky	Flaky	Flaky	Tout-venant

PS = Peach Shells; AS = Apricot Shells; PlS = Plum Shells.

Table 3. Proximal chemical composition of nutshells from different *Rosaceae* fruits.

	PS	**AS**	**PlS**
Moisture content (%)	4.2 ± 0.7	2.8 ± 0.4	3.8 ± 0.5
Ash (%)	0.99 ± 0.2	1.12 ± 0.6	1.09 ± 0.6
* Lignin (%)	41.7	51.4	49.5
* Hemicellulose (%)	21.8	20.8	20.2
* Cellulose (%)	23.8	22.4	23.2
C (%)	47.7 ± 0.5	47.0 ± 0.5	47.3 ± 0.5
H (%)	5.73 ± 0.10	6.13 ± 0.09	5.73 ± 0.09
N (%)	0.19 ± 0.05	0.17 ± 0.04	0.36 ± 0.04
O (%) (from difference)	45.39	45.58	45.52
Protein content (%)	1.19 ± 0.10	1.06 ± 0.11	2.25 ± 0.11
Fat (%)	0.09 ± 0.01	0.12 ± 0.02	0.11 ± 0.01

* Data taken from the literature [42–44]; PS = Peach Shells; AS = Apricot Shells; PlS = Plum Shells.

The methods recommended by the Association of Official Analytical Chemists [45] were used to determine the levels of moisture, ash, crude protein, and residual oil. Moisture content was determined by drying the samples at 105 °C to a constant weight. The ash content was determined using a laboratory furnace at 550 °C and the temperature was gradually increased. Nitrogen content was determined using the Dumas method and converted to protein content by multiplying by a factor of 6.25. The residual fat fraction was recovered using the Soxhlet method, exhaustively extracting 10 g of each sample using petroleum ether (boiling point range 40–60 °C) as the extractant solvent. Each measurement was performed in triplicate, and the results were averaged.

Finally, we emphasize that the chemical composition and physical properties of vegetable matrices are significantly influenced by certain factors, including the geographical origin, degree of ripeness, and cultivar to which they belong [46].

2.1.3. Lime-Concrete Design and Specimen Preparation

Normal tap water was used in this study. The mix proportions of all specimens are listed in Table 4. For each shell, the mix proportion of the related concrete was kept constant (PSC = Peach Shell Concrete; ASC = Apricot Shell Concrete; PlSC = Plum Shell Concrete). Specimens were removed from the mold after 24 h. They were stored in a laboratory room with a relative humidity of 95 ± 5% and a temperature of 20 ± 2 °C until the test age. Binder mixture a (PSC_a, ASC_a, PlSC_a) only includes lime, while mixture b (PSC_b, ASC_b, PlSC_b) involves the addition of cement. Three sets of specimens were prepared: one for the compressive strength test, one for the demolded, air-dry, and oven-dry density evaluation, and one for the thermal conductivity test. Each set contained three cubic specimens ($100 \times 100 \times 100$ mm^3), and the average values were obtained for each test result.

Table 4. Mix proportions of concrete (kg/m^3).

Sample	Lime	Cement	Sand	Lightweight Aggregate	w/b Ratio *
PSC_a	585	-	625	350	0.45
ASC_a	585	-	625	350	0.45
PlSC_a	585	-	625	350	0.45
PSC_b	390	195	625	350	0.40
ASC_b	390	195	625	350	0.40
PlSC_b	390	195	625	350	0.40

* water-binder ratio.

The specimens were prepared as follows: river sand, lime, and cement were poured into a blender and dry-mixed for 1 min. Water was added and the mixture was mixed for 3 min. The lightweight aggregates were finally added to and mixed for 5 min. After mixing, fresh mixtures were then poured into the mold and compacted. The specimens were placed in the laboratory room and were removed from the molds after approximately 24 h.

2.2. Experimental Methods

2.2.1. Morphological Analysis of the Aggregates

The field emission scanning electron microscope (SEM) instrument (Nova NanoSEM 450, FEI, Hillsboro, OR, USA) was used to evaluate the microscopic morphology of coarse lightweight aggregates.

2.2.2. TGA-MSEGA

A Seiko SSC 5200 thermal analyzer (Seiko Instruments Inc., Chiba, Japan) was used to perform the thermogravimetric analysis (TGA) in an inert atmosphere. A coupled quadrupole mass spectrometer (ESS, GeneSys Quadstar 422) was used to analyze the gases released during the thermal reactions (MS-EGA) (ESS Ltd., Cheshire, UK). Sampling was performed using an inert and fused silicon capillary system, which was heated to prevent condensation. The intensity of the signal of selected target gases was collected in multiple ion detection mode (MID); a secondary electron multiplier operating at 900 V was collected in multiple ion detection mode (MID), the intensity of the signal of selected target gases. The signal intensities of m/z ratios of 18 for H_2O, 44 for CO_2, 60 for $C_2H_4O_2$ (acetic acid), 94 for C_6H_6O (phenol), 39 for $C_3H_3^+$ (furfural fragment), 96 for $C_5H_4O_2$ (furfural), and 151 for $C_8H_8O_3$ (vanillin) were measured, where m/z is the ratio between the mass number and charge of the ion. The heating conditions were 20 °C/min in the thermal range of 25–1000 °C using ultrapure He at a flow rate of 100 μL/min as the purging gas.

2.2.3. Demolded, Air-Dry, and Oven-Dry Densities

Demodeld, air-dry, and oven-dry densities were determined following ASTM C567 [47]. The demolded mass was measured after demolding (after 24 h of curing), and the air-dry mass was measured after 28 days of curing. The test method for oven-dry density is more complex. The specimens were immersed in water (at about 20 °C) for 48 h, then the surface water was removed, and the saturated surface-dry mass was measured. Then, it was suspended in water with a wire, and the apparent mass of the suspended-immersed specimens was determined. The samples were then oven-dried at 110 °C for 72 h. The oven-dry density was calculated from Equation (1):

$$O_m = \frac{D \times 997}{F - G} \tag{1}$$

where O_m is the measured oven-dry density (kg/m^3); D is the specimen mass (kg); F is the mass of saturated surface-dry specimen (kg); G is the apparent mass of suspended-immersed specimen (kg).

2.2.4. Mechanical Test

The compressive strength test was performed after 28 and 56 days using a Technotest compression test machine (Technotest, Modena, Italy). The average value of at least three specimens was used as the test result. It was performed in conformity with the European standard for structural concrete (EN 12390-3:2009), although our concrete had no structural purpose. Lime mortar (EN 1015-11:1999) would be more suitable for the intended use, but the presence of coarse aggregates prevents its application.

2.2.5. Thermal Conductivity of Lime-Concrete Specimens

A KD2 Pro thermal properties analyzer (Decagon Inc., Pullman, WA 99163, USA) was used for thermal conductivity measurements. It is a portable device fully compliant with ASTM D5334-08 and is used to measure the thermal properties of materials based on probe/sensor methods (transient line heat source), as confirmed by Decagon Devices Inc. Operator Manual version 11. It consists of a portable controller and sensors probe to be inserted into the medium to be measured. The measurement consists of heating the probe for a certain time and monitoring the temperature during heating and cooling. The influence of the ambient temperature on the samples must be minimized to obtain more accurate values. The measurement range of thermal conductivity of KD2 Pro is 0.02 to 2.00 W/(mK). In this study, three cubic specimens for each sample ($100 \times 100 \times 100$ mm^3), at 28 days of curing were selected to measure thermal conductivity at dry conditions. The samples were oven-dried for 24 h at 100 °C prior to testing.

3. Results and Discussion

3.1. Lightweight Aggregates

LWAs, their nature, and compositional characteristics critically define the physical and chemical properties of concrete [48–50]. Depending on the type of aggregate used, the uses and functions of the final product change drastically [51]. Therefore, defining the compositional characteristics of our fruit shells is fundamental to understanding their potential as LWAs.

Bulk density is one of the most important characteristics [17] because it significantly affects the final density of concrete, which, in turn, determines its mechanical and insulation properties. This depends on the size and shape of the aggregates, the moisture content, and the porosity. PS had the lowest bulk density, whereas AS had the highest density. The SEM analysis (Section 3.2) highlights the marked morphological differences, which fully explains these density trends.

The water absorption of lightweight aggregates is generally significantly higher than that of conventional coarse aggregates. This is certainly due to the greater porosity but also to the different chemical compositions. In particular, fruit shells, lignocellulosic biomass, are composed mainly of the three main natural polymers, cellulose, lignin, and hemicellulose [52,53], and the content of these components in the examined matrices are collected in Table 2. Unlike lignin, which is highly hydrophobic, cellulose has a marked ability to absorb water. The latter acts as the glue that connects cellulose and hemicellulose [54]. In fact, in the cell wall of certain biomasses, especially wood species, lignin functions to cement cellulose fiber [55]. It is a three-dimensional strongly cross-linked macromolecule. Cellulose, on the other hand, differs markedly, as it is a linearly structured homopolymer, and in plants, it plays a fundamental role as a supporting matrix for the cell membrane. Hemicellulose is a heterogeneous, completely amorphous, weak polymer. Hemicellulose is decidedly more soluble and labile [53]. From Table 3, it can be seen that PS showed greater water absorption than AS and PlS. This phenomenon can certainly be attributed to the increased porosity (as will be explained later in Section 3.2). However, it is possible to draw conclusions by analyzing the chemical composition of the shells. In fact, the lignin content of PS was the lowest compared to that of AS and PlS, while that of cellulose and hemicellulose was higher. Considering the greater affinity of the latter towards the water and the hydrophobicity of lignin, the greater water absorption can be easily explained. Water absorption affects some important properties of concrete such as its strength, density, and time-dependent deformation [17,48].

3.2. Morphological Analysis of the Aggregates

The surfaces of peach shells (PS), apricot shells (AS), and plum shells (PlS) are shown in Figures 2–4, respectively. We reported only images relating to the external surface of the shells, in which we identified the most significant differences between PS, AS, and PlS. The internal one, in fact, was extremely smooth and compact for all three types of shells. Therefore, it seemed more important to pay attention only to the outer part of the shells, since the outer surface layers probably contribute more to the properties of the aggregates and, consequently, to the behavior of the specimens.

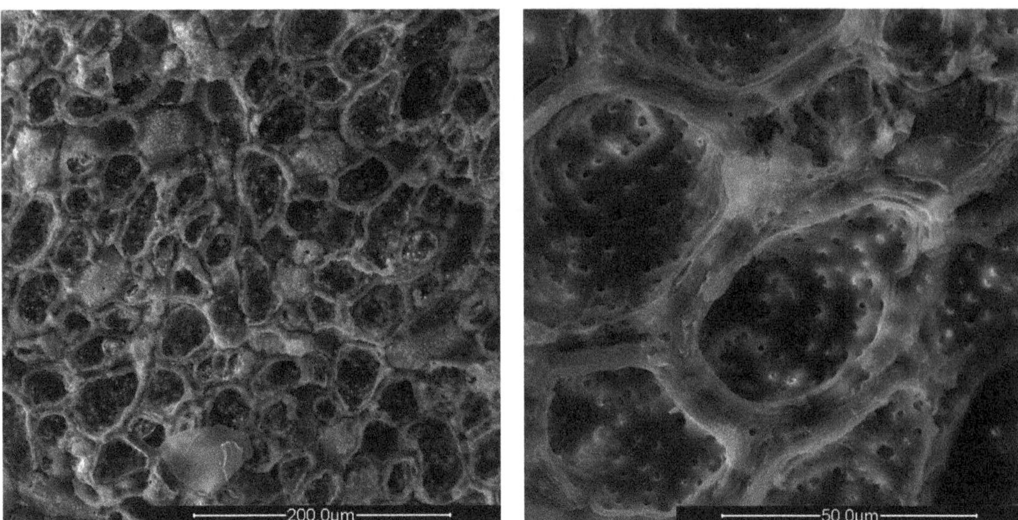

Figure 2. SEM images of crushed peach shells (PS) at different enlargements.

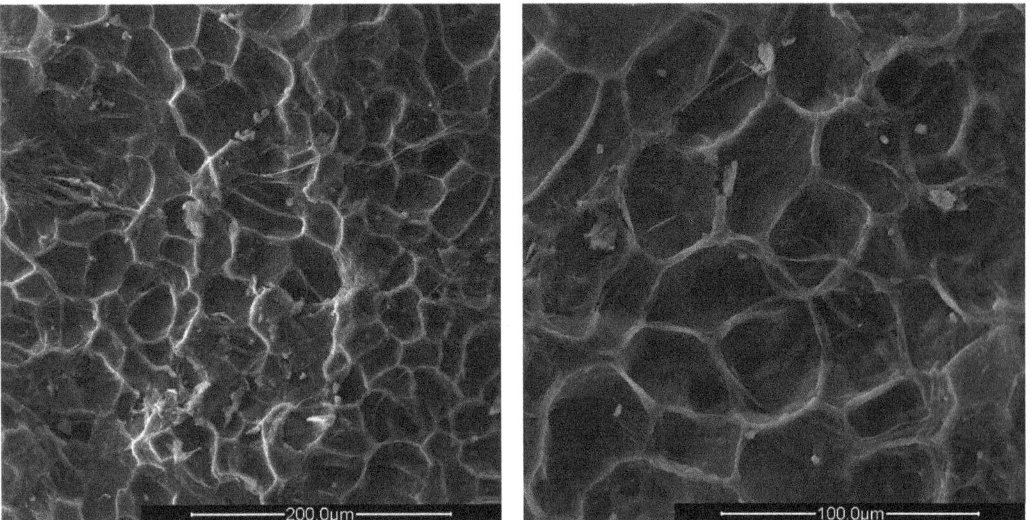

Figure 3. SEM images of crushed apricot shells (AS) at ×500 (**left**) and ×1000 magnification (**right**).

The surfaces of all shells appear rough, irregular, and have many cavities. The PS and AS cavities are ovoidal shaped, where the grater diameter is approximately 50 μm and the smaller one is about half, 25 μm. The PlS cavities, on the other hand, are much more irregular and lack a specific shape. The most evident observation concerns the presence of microporosity inside the cavities, which are only present in PS and PlS. The size of the microporosities was approximately 2.0 μm, which can be better viewed in Figure 5.

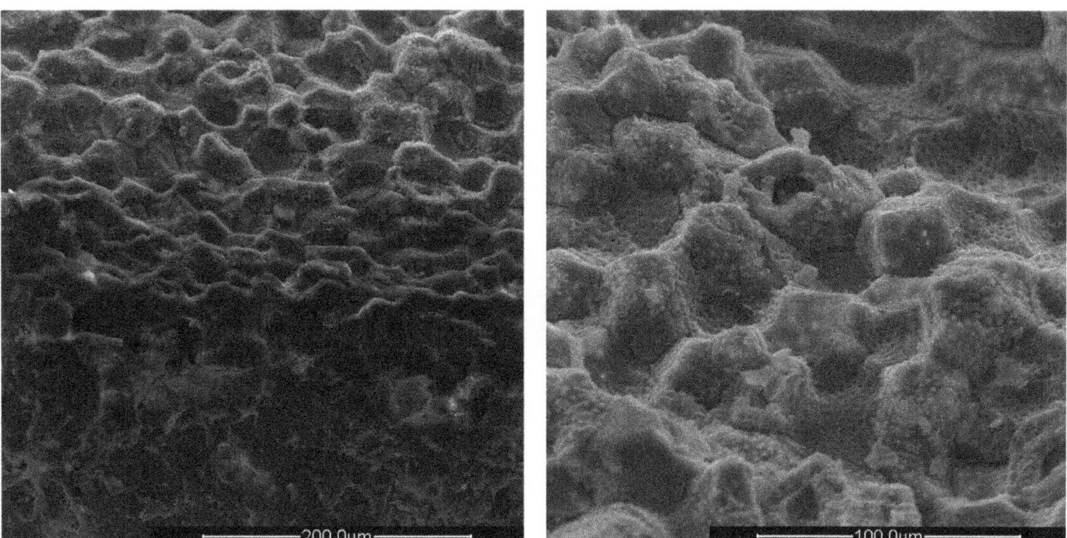

Figure 4. SEM images of crushed plum shells (PlS) at ×500 (**left**) and ×1000 magnification (**right**).

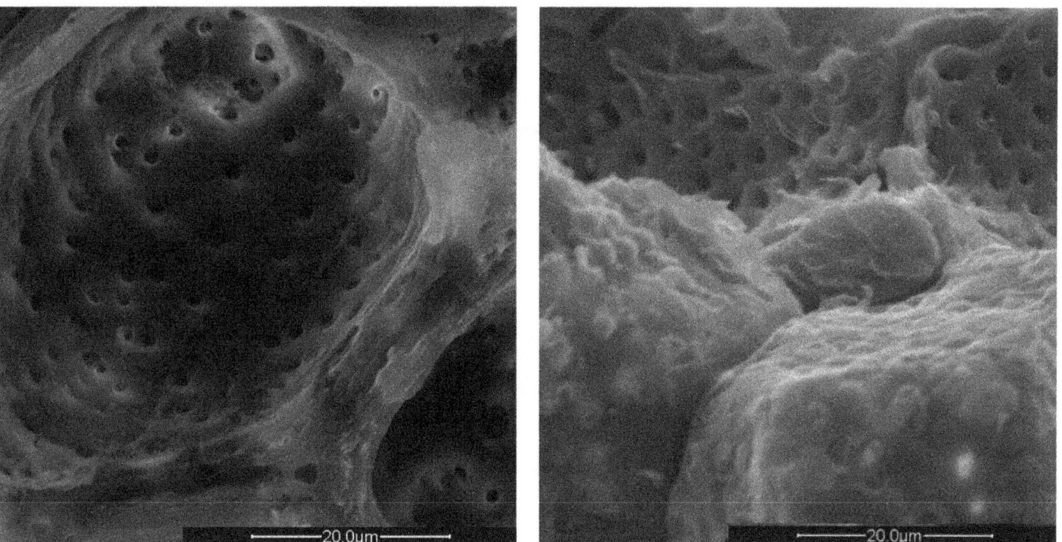

Figure 5. SEM images of the micropores of PS (**left**) and PlS (**right**), 4000× magnification.

The PlS micropores were denser, although some were not completely empty. It is likely that a small fraction of the fruit pulp remained trapped inside the micropores, which made it difficult to remove by simple mechanical separation. The SEM observations can be correlated with the data shown in Tables 2 and 3. The greater AS bulk density was justified by the absence of porosity and the consequent greater compactness. The PlS bulk density was higher than that of PS, which could be due to the more superficial and shallower porosities. The water absorption (24 h, %) data were also in line. PS has the highest value, indicating greater trapping of water inside the pores, AS has the lowest value, and PlS has an intermediate value.

3.3. TG-MS-EGA Analysis

Fruit shells are primarily composed of lignocellulosic material. TG-MS-EGA analysis allows us to obtain information about the different degradation processes involving all constituents, which occur in defined thermal ranges identifiable in the thermogram. Materials with complex compositions give rise to different degradative reactions that can occur simultaneously, and the thermogram profile is the sum of the various contributions. In these cases, deconvolution and interpretation of the signals are not particularly easy, especially if different processes lead to the formation of the same reaction products, such as H_2O and CO_2. For effective interpretation of thermograms, the entire temperature range is usually divided into thermal regions of different sizes and characteristics, as shown in Table 5. Furthermore, Table 5 shows some thermal windows or subdomains of the regions where particular deformations of the TG/DTG profiles are observed, corresponding to specific behaviors due to some degradation processes of the studied samples.

Table 5. Representative values of TG/DTG profiles of Figures 6–8 obtained in inert atmosphere (He).

Region	Thermal Window	Thermally Activated Processes
I	30–120	Removal of moisture and VOCs up to ~100 °C
II	~120–210	Removal of bound water, NH_3 from protein denaturation, low-boiling VOCs, and loss of CO and CO_2
III	~210–260	Shoulder related to protein degradation
	~260–430	Removal of reaction water, high-boiling VOCs and SVOCs, decarboxylation of acids with CO_2 loss, degradation of polysaccharides, plasticization, and pseudo-vitrification of the sample. Removal of hydrocarbons, fat degradation, water of constitution, CO and CO_2, and volatilization of other metabolites
IV	~430–490	Removal of reaction water, CO_2, and other metabolites
	~490–700	Weak reactions related to slow volatilization of CO_2, carbon residues, and other molecules. Removal of reaction water, CO and CO_2, and other metabolites
V	~700–1000	Volatilization of carbon residues, probably C20–C40 fragments
Residual ashes at 1000 °C		Inorganic compounds and carbon residues

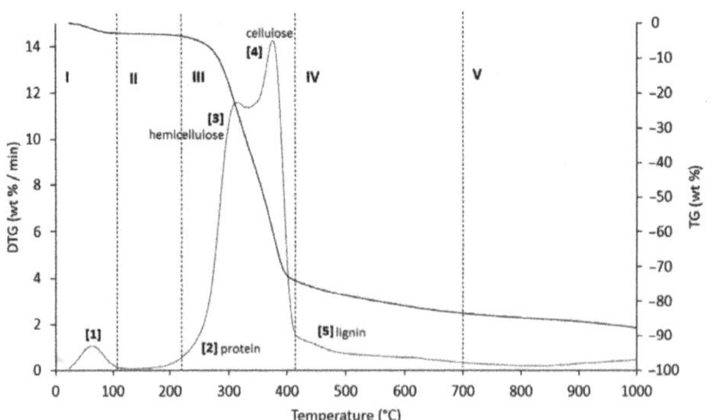

Figure 6. TG (black line) and DTG (grey line) curves of AS sample at heating rate of 20 °C/min in He atmosphere. Vertical dashed lines delimit the five thermal regions (I–V) described in the text. For the meaning of the numbers in parentheses, see Table 6.

Table 6. Representative values of TG/DTG profiles of Figure 6 (AS sample), obtained in inert atmosphere (He).

Region	Thermal Step	T_o	T_m	T_c	$\Delta m\%$
I	(1)	30	70.9	120	−2.8
II	(2)	120	-	261.7	−3.4
III	(3)	261.7	310.2	332.5	−29.2
	(4)	332.5	377.4	423.2	−39.4
IV	(5)	423.2	-	489.3	−4.0
		489.3	-	700	−5.0
V		700	-	1000	−4.7

T_o = onset temperature (beginning of thermal step processes); T_m = maximum temperature for the largest mass loss rate; T_c = conclusion temperature (end of thermal step processes).

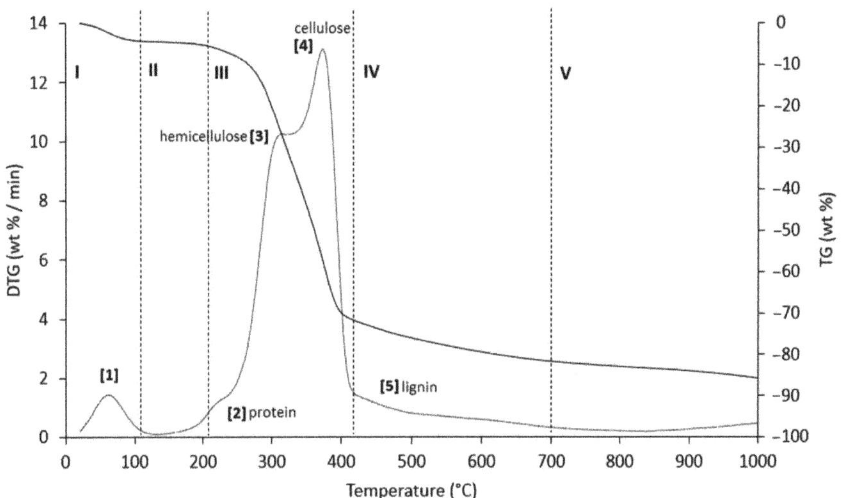

Figure 7. TG (black line) and DTG (grey line) curves of PlS sample at heating rate of 20 °C/min in He atmosphere. Vertical dashed lines delimit the five thermal regions (I–V) described in the text. For the meaning of the numbers in parentheses, see Table 7.

Table 7. Representative values of TG/DTG profiles of Figure 7 (PlS sample) obtained in inert atmosphere (He).

Region	Thermal Step	T_o	T_m	T_c	$\Delta m\%$
I	(1)	30	66.7	120	−3.8
II	(2)	120	220.4	261.7	−4.8
III	(3)	261.7	314.7	332.5	−26.4
	(4)	332.5	375.1	423.2	−36.4
IV	(5)	423.2	-	490	−3.7
		490	-	700	−6.0
V		700	-	1000	−4.0

T_o = onset temperature (beginning of thermal step processes); T_m = maximum temperature for the largest mass loss rate; T_c = conclusion temperature (end of thermal step processes).

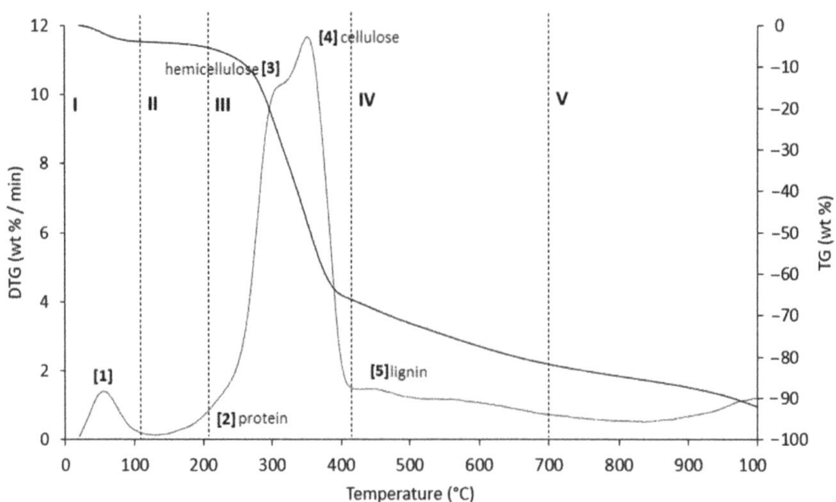

Figure 8. TG (black line) and DTG (grey line) curves of PS sample at heating rate of 20 °C/min in He atmosphere. Vertical dashed lines delimit the five thermal regions (I–V) described in the text. For the meaning of the numbers in parentheses, see Table 8.

Table 8. Representative values of TG/DTG profiles of Figure 8 (PS sample) obtained in inert atmosphere (He).

Region	Thermal Step	T_o	T_m	T_c	$\Delta m\%$
I	(1)	30	64.2	120	−4.2
II	(2)	120	-	261.7	−5.5
III	(3)	261.7	304.9	332.5	−28.1
	(4)	332.5	355.1	423.2	−29.2
IV	(5)	423.2	465	489.3	−4.4
		489.3	-	700	−10.4
V		700	-	1000	−10.2

T_o = onset temperature (beginning of thermal step processes); T_m = maximum temperature for the largest mass loss rate; T_c = conclusion temperature (end of thermal step processes).

Table 3 shows that ~90% of AS, PS, and PlS consisted of cellulose, hemicellulose, and lignin. Therefore, the TG-MS-EGA analysis showed the thermal steps leading to the degradation of these three fractions. For this reason, the evolution of some analytes characteristic of the degradation of hemicellulose and cellulose (i.e., acetic acid, m/z = 60; furfural, m/z = 96 and 39 for its fragment $C_3H_3^+$) and lignin (i.e., phenol, m/z = 98; and vanillin m/z = 151), was evaluated [56]. This analysis allows for better differentiation of the thermal processes and better identification of the process temperature range that involves every component of the matrix. Unfortunately, no significant results have been obtained regarding the evolution of the emitted gases phenol and vanillin, which will not be reported below.

The result of the TG, together with its first derivative (DTG), runs in inert atmosphere (He) as shown in Figures 6–8, and the related quantitative considerations are summarized in Table 5. For each thermogram, a summary table is provided (Tables 6–8), that collects the representative values of the TG/DTG profile.

The thermograms are divided into five regions (I, II, III, IV, and V), each representing the behavior of the samples following specific processes. Before examining the TG/DTG curves relating to the three nut-shell samples, it must be emphasized that the separation limits of the various regions are not rigidly identified with values of T/°C in an absolute way. Conversely, these are limits with a mobile index because the material fractions that undergo thermally activated processes can produce experimental evidence typical of the sample rather than of the type of matrix. Therefore, any attempt to generalize the thermal intervals of each region could lead to a speculative investigation, which is incompatible with the inspiring criteria of this study.

Region I, which covers a temperature range up to ~120 °C, is attributed to the removal of moisture and particularly volatile organic compounds (VOCs). The mass loss in this thermal step was greater for PS (Δm% = −4.2%), followed by PlS (Δm% = −3.8%) and AS (Δm% = −2.8%). Within this region, other thermally activated processes occur without mass loss, such as protein denaturation by unfolding [57,58].

Region II, which covers the temperature range from ~120 °C to ~215 °C, concerns the mass loss related to bound water, i.e., water typically retained by the inorganic fraction, such as the crystallization of mineral salt water. In this region, semi-volatile compounds with medium-low vapor pressure (SVOCs), which are present in the initial matrix or formed during the heating phase, are completely removed. The removal of structural water begins at 160 °C, following condensation reactions of the -OH groups mainly present in simple non-cellulosic carbohydrates [59]. The formation and removal of the reaction water passes through the entire thermogram up to and including region IV. Therefore, near the upper-temperature limit (~180 °C), free amino acids begin to undergo thermal degradation [60], while proteins persist up to ~200–220 °C. Thus, the processes occurring in this region suggest that the chemical structure of the biomass begins to destabilize, partly depolymerize, and plasticize.

Region III, in the temperature range from ~215 °C to ~423 °C, is the main pyrolysis window, where structural decay reactions of proteins (~240 °C), hemicellulose (~300 °C) [61,62], and cellulose (~370 °C) [57,61,63] take place. This was confirmed by the emission of acetic acid and furfural in this thermal window, as shown in Figure 9. These analytes are formed by the thermal degradation of cellulose and hemicellulose, as reported in several studies [56,64,65].

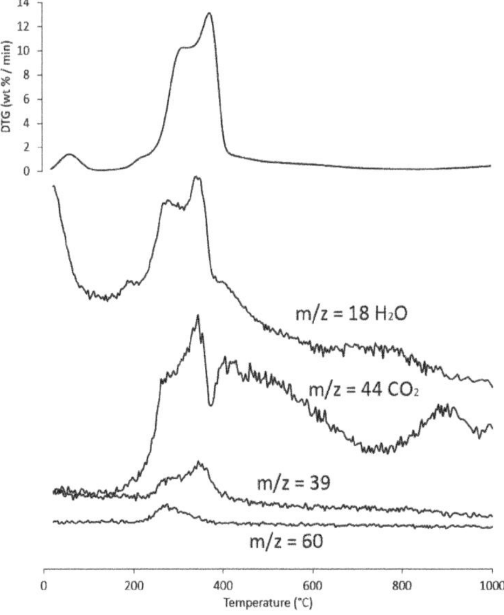

Figure 9. Evolution trend of H_2O, CO_2, furfural fragment $C_3H_3^+$ (m/z = 39), and acetic acid (m/z = 60) during the heating of PS sample; for ease of comparison the DTG curve is also shown. Intensity of m/z is in arbitrary units.

Region IV begins at ~413 °C and extends up to ~695 °C. In this thermal window, the gradual mass decrease is mainly due to the slow pyrolysis of the lignin fraction [66], which is associated with the sample vitrification and volatilization of carbon microparticles. The additional small mass loss can be attributed to the thermal decomposition of carbonaceous matter (biochar), which is mostly related to the hemicellulose fraction [67], although lignin components can also degrade [68].

In region V, above ~700 °C, up to the final temperature (1000 °C), the last residues of biomass degradation can be observed. This is the typical carbon pyrolysis window with the thermal decomposition of low volatile matter such as carbon fragments C20–C40 in the presence of mineral ash. This thermal process was confirmed by the evolution profile of CO_2 (Figure 9), where an increase in the signal occurred between 800 °C and 1000 °C. The TG/DTG profiles of PS, AS, and PlS are typical of lignocellulosic raw materials, highlighting the contents of hemicellulose, cellulose, and lignin. This observation was confirmed by the proximate composition analysis (Section 2.1.2, Table 3).

As an example, the evolution of gases is reported below (Figure 9) for only one sample (PS), as the results were almost similar for the three matrices.

3.4. Density and Compressive Strength of the Lime-Based Concrete

Lightweight concrete can be classified according to its density, which normally ranges from 320 to 1920 kg/m^3 according to the ACI Committee 213 Guide for Structural Lightweight Aggregate Concrete [69]. The classification of concrete according to density provides three groups of materials: (i) low-density concretes (300–800 kg/m^3); (ii) moderate-strength concretes (800–1350 kg/m^3); (iii) structural concretes (1350–1920 kg/m^3) [70]. These three classes are also associated with specific strength range: 0.7–2.0 MPa, 7–14 MPa, and 17–63 MPa, respectively [70]. Density is one of the most important variables in concrete design, as compressive strength depends on it [71]. The reduction of concrete density implies the increment of its porosity, which is achieved by the direct replacement of normal-weight aggregates with LWAs.

The density data obtained are collected in Table 9.

Table 9. Demolded, air-dry, and oven-dry density values of PSC, ASC, and PlSC.

	Density (kg/m^3) *		
Sample	**Demoulded (24 h)**	**Air-Dry (28 d)**	**Oven-Dry (28 d)**
PSC_a	1270.7 ± 1.5	1107.7 ± 1.4	1031.9 ± 1.4
ASC_a	1308.3 ± 0.9	1148.3 ± 1.0	1124.0 ± 0.8
PlSC_a	1288.4 ± 1.2	1124.0 ± 1.0	1053.0 ± 1.1
PSC_b	1464.5 ± 0.9	1295.2 ± 0.8	1204.8 ± 0.7
ASC_b	1513.7 ± 1.1	1342.6 ± 1.0	1251.9 ± 1.1
PlSC_b	1489.2 ± 1.0	1315.2 ± 1.2	1232.3 ± 0.9

* Data are expressed as mean ± SD.

All the samples are in the density range relating to moderate-strength lightweight concretes. As expected, the PSC has lower density values, as PS has a lower bulk density and specific gravity than AS and PlS. As observed by the SEM analysis, PS has a high porosity, AS low, and PlS intermediate between the two. The porosity of an aggregate significantly defines concrete density, as it affects both the porosity of the concrete itself and allows it to trap more air inside. Furthermore, some studies reported that the aggregate shape affects concrete density. In fact, the flaky shape easily traps air inside the concrete, increasing its porosity and consequently reducing its density. This phenomenon has been observed in concrete containing seashells [72], peach shells [38], and recycled polyolefin waste [73]. In particular, irregular shapes hinder the complete compaction of concrete, thus contributing to higher air content. In addition to this, there is also trapped air due to the high porosity of LWAs. Furthermore, in these studies, it is reported that the organic matter content is also able to increase the air inclusion in the concrete. Moreover, the extremely irregular shape of the aggregates leads to a difficult compaction of concrete, which leads inevitably to an increase in the occluded air [35]. This decrease in density, however, involves the reduction of the compressive strength [74], as explained below.

It is also important to underline that the use of lime as a binder allows it to obtain lower density values when compared with specimens prepared with similar lightweight aggregates [36–39], because of its lower specific gravity and bulk density. This is confirmed by the higher density values observed in concrete-containing specimens (Table 5).

The results of the compressive strength tests at 28 and 56 days for the concrete specimens are shown in Table 10.

Table 10. Compressive strength at 28-day and 56-day.

	Compressive Strength (MPa) *	
Sample	28-Day	56-Day
PSC_a	1.38 ± 0.20	1.99 ± 0.14
ASC_a	2.87 ± 0.12	3.35 ± 0.15
PlSC_a	1.95 ± 0.34	2.12 ± 0.17
PSC_b	4.01 ± 0.16	4.97 ± 0.21
ASC_b	6.98 ± 0.31	7.71 ± 0.13
PlSC_b	5.11 ± 0.17	6.01 ± 0.19

* Data are expressed as mean ± SD.

The compressive strength of the specimens prepared with the cement-free mixture is less than 3 MPa. This value is too low to allow the material to fall into the category of moderate-strength concretes. At the same time, the density value of these specimens is too high for them to be considered "low-densities concretes". However, not falling into a specific category does not preclude possible applications. For example, non-structural mortar beds for wooden floors have larger density allowances (1400–1600 kg/m^3) and low strength requirements [17]. Compressive strength between 1 and 2 MPa is recommended in this case. This is a clear example of how depending on the specific application to be assigned to a material, specific ranges of density, and compressive strength are required. Specimens prepared with the cement-containing mixture "b" fall perfectly into the category "moderate strength concrete", as they have a density between 800 and 1350 kg/m^3 and a compressive strength exceeding 3.4 MPa. The addition of cement, even if in a small percentage, entails a significant improvement in mechanical properties and little compromises the density value, slightly greater than lime-based concrete. Moderate strength lightweight concrete is a versatile material that can be used for various purposes in construction. One of its most useful applications is as a non-structural filler for thermal and acoustic insulation. Non-structural infills are materials that do not bear any significant weight or load in a building but provide important functions such as insulation. One benefit of lightweight moderate-strength concrete is its relative ease of installation and transport, owing to its relatively low weight. This can be particularly advantageous in situations where access to the construction site is limited or there are restrictions on the use of heavy machinery.

Several studies in the literature demonstrated that the compressive strength of concrete is mainly affected by the properties and volume content of aggregate [75,76]. LWAs are also relatively weak if compared with normal-weight coarse aggregates, and their strength is an additional limiting factor affecting concrete strength [77]. As previously mentioned, the most important characteristic of LWAs is its internal porosity, which results in a lower density and higher water absorption. These factors adversely affect the compressive strength and making concrete less compact and porous. In particular, the greater water absorption by aggregates leads to greater porosity of concrete [78]. This results in lower density and lower compressive strength. PS showed increased water absorption, as explained in Section 3.1, due to the increased cellulose content and higher porosity. AS, on the other hand, having greater lignin content, a hydrophobic polymer, and free of surface porosity showed lower water absorption. These observations are in line with the values given in

Table 6: PSC specimens have a lower strength, given the greater porosity of the aggregates. On the contrary, those ASC show the highest values, while PlSC intermediate ones.

3.5. Thermal Insulating Properties

Thermal conductivity is a fundamental parameter in the design and application of thermal-insulating lightweight concrete. These materials are becoming important in the context of the climate crisis. The development of energy efficiency strategies is increasingly being studied since the design of energy-efficient buildings is crucial for the realization of a sustainable future [79]. Room air conditioning, ventilation, and occupant comfort account for 29 of CO_2 emissions from the building sector. By increasing the energy efficiency of buildings, also using thermal insulation materials, it is possible to reduce consumption and reduce the environmental impact and CO_2 emissions significantly [80,81].

Several factors affect the thermal properties of concrete: type and content of aggregates, air voids content, pore size distribution and geometry, moisture content, w/b ratio, and types of admixtures [82]. In particular, the thermal conductivity of conventional building materials is inversely proportional to the porosity ratio. This trend is due to the relatively low thermal conduction of air (0.025 W/mK at room temperature and free of convection) and the interfaces promoted by the pores. Microstructural characteristics are thus critical factors for the consequent thermal conductivity of concrete.

The results are collected in Table 11.

Table 11. Thermal conductivity.

Sample	Thermal Conductivity Coefficient (W/mK) *
PSC_a	0.15 ± 0.01
ASC_a	0.28 ± 0.03
PlSC_a	0.19 ± 0.04
PSC_b	0.20 ± 0.01
ASC_b	0.37 ± 0.05
PlSC_b	0.28 ± 0.03

* Data are expressed as mean of three replicates ± SD.

PSC has a lower thermal conductivity, due to the highly porous structure of the lightweight aggregate and the consequent high porosity of the concrete. Generally, low-compaction concrete has better thermal insulation properties because more air bubbles are carried into the concrete during the mixing. Consequently, the thermal insulation properties improve with increasing porosity of both the lightweight aggregate and the related concrete. For the same reason, there is a significative correlation between concrete density and its thermal conductivity. Lightweight aggregates change density by forming voids, incorporating more air inside the concrete. AS, being practically porosity-free, is an aggregate that does not involve a significant inclusion of air, and therefore, does not provide a significant decrease in thermal conductivity. The addition of cement (mixture b) results in better compactness of the samples and a consequent worsening of the thermal insulation properties. This trend is in agreement with the observation reported in the literature, which suggests that the thermal conductivity of concrete decreases as its density decreases [83].

Moderate-strength lightweight concrete is known to have a thermal conductivity ranging from 0.2 to 0.6 W/mK [84]. The values in Table 8 indicate that all the concretes obtained fall within this range. Notably, some of these materials, specifically those produced using mixture "a" consisting of PS and PlS, exhibit even lower thermal conductivities, indicating enhanced thermal insulation properties. In general, all the materials obtained, having reduced thermal conductivity, have a marked potential for application as non-structural fillers to improve energy savings in buildings, thus improving environmental sustainability.

4. Conclusions

The potential of peach, plum, and apricot shells as lightweight aggregates was evaluated. The TG/DTA profile is typical of lignocellulosic material, confirming the proximal analysis. The use of lime as the main binder allowed it to obtain more eco-friendly building materials, both because it is more ecological than cement, and because it gives particularly favorable thermal insulation properties due to its greater porosity. PS, AS and PlS prepared several specimens of non-structural lightweight concrete. The specimens containing only lime as binder had an oven-dry density between 1000 and 1200 kg/m^3, a low 28-day compressive strength (<3 MPa), and low thermal conductivity values. PSC had lower conductivity and density values, and this is mainly due to the high porosity of PS highlighted by SEM analysis. ASC instead showed the highest values and is practically free of porosity. PlSC showed intermediate characteristics, which reflects the reduced porosity content of PlS. In addition, PS showed greater water absorption, probably due to the higher content of cellulose. This parameter greatly affects the chemical–physical characteristics of concrete, leading to a worsening of compactness, the reduction of density, the formation of greater voids, and consequently the lowering of thermal conductivity. The addition of cement greatly improves the mechanical properties but negatively affects thermal conductivity.

This study showed that there is a feasibility of application of these agro-industrial wastes, which can, therefore, be reused and valorized, reducing dependence on natural raw materials.

Author Contributions: Conceptualization, V.D. and A.M.; methodology, V.D. and M.M.; software, V.D., S.P. and M.S.; validation, L.T., S.P., V.D. and A.M.; formal analysis, V.D.; investigation, V.D., L.B. and M.M.; resources, L.T. and S.P.; data curation, V.D., M.M. and M.S.; writing—original draft preparation, V.D. and L.B.; writing—review and editing, A.M. and L.T.; visualization, V.D.; supervision, L.T. and S.P.; project administration, V.D.; funding acquisition, V.D. and L.T. All authors have read and agreed to the published version of the manuscript.

Funding: This research received no external funding.

Institutional Review Board Statement: Not applicable.

Informed Consent Statement: Not applicable.

Data Availability Statement: Not applicable.

Conflicts of Interest: The authors declare no conflict of interest.

References

1. Dulias, R. *The Impact of Mining on the Landscape: A Study of the Upper Silesian Coal Basin in Poland*, 1st ed.; Springer International Publishing: Cham, Switzerland, 2016.
2. Arendt, R.; Bach, V.; Finkbeiner, M. The Global Environmental Costs of Mining and Processing Abiotic Raw Materials and Their Geographic Distribution. *J. Clean. Prod.* **2022**, *361*, 132232. [CrossRef]
3. Krausmann, F.; Lauk, C.; Haas, W.; Wiedenhofer, D. From Resource Extraction to Outflows of Wastes and Emissions: The Socioeconomic Metabolism of the Global Economy, 1900–2015. *Glob. Environ. Change* **2018**, *52*, 131–140. [CrossRef]
4. Benton, D.; Hazell, J. *Resource Resilient UK: A Report from the Circular Economy Task Force*; Green Alliance: London, UK, 2013; Available online: https://green-alliance.org.uk/wp-content/uploads/2021/11/Resource-resilient-UK.pdf (accessed on 25 November 2022).
5. *Ecorys Mapping Resource Prices: The Past and the Future*; Ecorys: Rotterdam, The Netherlands, 2012.
6. MacArthur Foundation; McKinsey & Company. Towards the Circular Economy: Accelerating the Scale-Up Across Global Supply Chains. Available online: http://www3.weforum.org/docs/WEF_ENV_TowardsCircularEconomy_Report_2014.pdf (accessed on 25 November 2022).
7. Pratt, K.; Lenaghan, M. *The Carbon Impacts of the Circular Economy Summary Report*; Zero Waste Scotland: Stirling, UK, 2015.
8. EEA (European Environment Agency). *Circular Economy in Europe Developing the Knowledge Base*; Eur. Environ. Agency: Copenhagen, Denmark, 2016.
9. Morgan, J.; Mitchell, P. *Employment and the Circular Economy: Job Creation in a More Resource Efficient Britain*; Green Alliance: London, UK, 2015; Available online: http://www.green-alliance.org.uk/resources/Employment (accessed on 26 November 2022).
10. Spence, R.; Mulligan, H. Sustainable Development and the Construction Industry. *Habitat Int.* **1995**, *19*, 279–292. [CrossRef]
11. Curwell, S.; Cooper, I. The Implications of Urban Sustainability. *Build. Res. Inf.* **1998**, *26*, 17–28. [CrossRef]

12. Wieser, A.A.; Scherz, M.; Maier, S.; Passer, A.; Kreiner, H. Implementation of Sustainable Development Goals in Construction Industry—A Systemic Consideration of Synergies and Trade-Offs. *IOP Conf. Ser. Earth Environ. Sci.* **2019**, *323*, 012177. [CrossRef]
13. UN General Assembly. *Transforming Our World: The 2030 Agenda for Sustainable Development*; A/RES/70/1; United Nations: New York, NY, USA, 2015.
14. Peris Mora, E. Life Cycle, Sustainability and the Transcendent Quality of Building Materials. *Build. Environ.* **2007**, *42*, 1329–1334. [CrossRef]
15. Mehta, K.P. Reducing the Environmental Impact of Concrete. *Concr. Int.* **2001**, *23*, 61–66.
16. Real, S.; Gomes, M.G.; Moret Rodrigues, A.; Bogas, J.A. Contribution of Structural Lightweight Aggregate Concrete to the Reduction of Thermal Bridging Effect in Buildings. *Constr. Build. Mater.* **2016**, *121*, 460–470. [CrossRef]
17. Sala, E.; Zanotti, C.; Passoni, C.; Marini, A. Lightweight Natural Lime Composites for Rehabilitation of Historical Heritage. *Constr. Build. Mater.* **2016**, *125*, 81–93. [CrossRef]
18. Carran, D.; Hughes, J.; Leslie, A.; Kennedy, C. A Short History of the Use of Lime as a Building Material Beyond Europe and North America. *Int. J. Archit. Herit.* **2012**, *6*, 117–146. [CrossRef]
19. Elert, K.; Rodriguez-Navarro, C.; Pardo, E.S.; Hansen, E.; Cazalla, O. Lime Mortars for the Conservation of Historic Buildings. *Stud. Conserv.* **2002**, *47*, 62–75. [CrossRef]
20. Boynton, R.S. *Chemistry and Technology of Lime and Limestone*, 2nd ed.; John Wiley & Sons, Inc.: New York, NY, USA, 1980.
21. Saberian, M.; Jahandari, S.; Li, J.; Zivari, F. Effect of Curing, Capillary Action, and Groundwater Level Increment on Geotechnical Properties of Lime Concrete: Experimental and Prediction Studies. *J. Rock Mech. Geotech. Eng.* **2017**, *9*, 638–647. [CrossRef]
22. Bevan, R.; Woolley, T. *Hemp Lime Construction*; IHS/BRE Press: Bracknell, Berkshire, UK, 2008.
23. Pelentir, N.; Block, J.M.; Monteiro Fritz, A.R.; Reginatto, V.; Amante, E.R. Production and Chemical Characterization of Peach (Prunus Persica) Kernel Flour: Characterization of Peach Kernel Flour. *J. Food Process Eng.* **2011**, *34*, 1253–1265. [CrossRef]
24. Rodriguez, C.E.; Bustamante, C.A.; Budde, C.O.; Müller, G.L.; Drincovich, M.F.; Lara, M.V. Peach Fruit Development: A Comparative Proteomic Study Between Endocarp and Mesocarp at Very Early Stages Underpins the Main Differential Biochemical Processes Between These Tissues. *Front. Plant Sci.* **2019**, *10*, 715. [CrossRef] [PubMed]
25. Noratto, G.; Porter, W.; Byrne, D.; Cisneros-Zevallos, L. Identifying Peach and Plum Polyphenols with Chemopreventive Potential against Estrogen-Independent Breast Cancer Cells. *J. Agric. Food Chem.* **2009**, *57*, 5219–5226. [CrossRef] [PubMed]
26. Vizzotto, M.; Cisneros-Zevallos, L.; Byrne, D.H.; Ramming, D.W.; Okie, W.R. Large Variation Found in the Phytochemical and Antioxidant Activity of Peach and Plum Germplasm. *J. Am. Soc. Hortic. Sci.* **2007**, *132*, 334–340. [CrossRef]
27. Chun, O.K.; Kim, D.-O.; Moon, H.Y.; Kang, H.G.; Lee, C.Y. Contribution of Individual Polyphenolics to Total Antioxidant Capacity of Plums. *J. Agric. Food Chem.* **2003**, *51*, 7240–7245. [CrossRef]
28. Gil, M.I.; Tomás-Barberán, F.A.; Hess-Pierce, B.; Kader, A.A. Antioxidant Capacities, Phenolic Compounds, Carotenoids, and Vitamin C Contents of Nectarine, Peach, and Plum Cultivars from California. *J. Agric. Food Chem.* **2002**, *50*, 4976–4982. [CrossRef] [PubMed]
29. Radi, M.; Mahrouz, M.; Jaouad, A.; Tacchini, M.; Aubert, S.; Hugues, M.; Amiot, M.J. Phenolic Composition, Browning Susceptibility, and Carotenoid Content of Several Apricot Cultivars at Maturity. *HortScience* **1997**, *32*, 1087–1091. [CrossRef]
30. Plazzotta, S.; Ibarz, R.; Manzocco, L.; Martín-Belloso, O. Optimizing the Antioxidant Biocompound Recovery from Peach Waste Extraction Assisted by Ultrasounds or Microwaves. *Ultrason. Sonochem.* **2020**, *63*, 104954. [CrossRef] [PubMed]
31. Wechsler, A.; Molina, J.; Cayumil, R.; Núñez Decap, M.; Ballerini-Arroyo, A. Some Properties of Composite Panels Manufactured from Peach (Prunus Persica) Pits and Polypropylene. *Compos. Part B Eng.* **2019**, *175*, 107152. [CrossRef]
32. Kambis, A.D.; Levine, J.S. Biomass Burning and the Production of Carbon Dioxide: Numerical Study. In *Biomass Burning and Global Change Volume 1: Remote Sensing, Modeling and Inventory Development, and Biomass Burning in Africa*; MIT Press: Cambridge, MA, USA, 1996.
33. Duc, P.A.; Dharanipriya, P.; Velmurugan, B.K.; Shanmugavadivu, M. Groundnut Shell -a Beneficial Bio-Waste. *Biocatal. Agric. Biotechnol.* **2019**, *20*, 101206. [CrossRef]
34. Zheng, W.; Phoungthong, K.; Lü, F.; Shao, L.-M.; He, P.-J. Evaluation of a Classification Method for Biodegradable Solid Wastes Using Anaerobic Degradation Parameters. *Waste Manag.* **2013**, *33*, 2632–2640. [CrossRef] [PubMed]
35. Traore, Y.B.; Messan, A.; Hannawi, K.; Gerard, J.; Prince, W.; Tsobnang, F. Effect of Oil Palm Shell Treatment on the Physical and Mechanical Properties of Lightweight Concrete. *Constr. Build. Mater.* **2018**, *161*, 452–460. [CrossRef]
36. Wu, F.; Liu, C.; Zhang, L.; Lu, Y.; Ma, Y. Comparative Study of Carbonized Peach Shell and Carbonized Apricot Shell to Improve the Performance of Lightweight Concrete. *Constr. Build. Mater.* **2018**, *188*, 758–771. [CrossRef]
37. Adefemi, A.; Nensok, M.; Kaase, E.T.; Wuna, I.A. Exploratory Study of Date Seed as Coarse Aggregate in Concrete Production. *Civ. Environ. Res.* **2013**, *3*, 85–92.
38. Wu, F.; Liu, C.; Sun, W.; Ma, Y.; Zhang, L. Effect of Peach Shell as Lightweight Aggregate on Mechanics and Creep Properties of Concrete. *Eur. J. Environ. Civ. Eng.* **2020**, *24*, 2534–2552. [CrossRef]
39. Wu, F.; Liu, C.; Sun, W.; Zhang, L. Mechanical Properties of Bio-Based Concrete Containing Blended Peach Shell and Apricot Shell Waste. *Mater. Tehnol.* **2018**, *52*, 645–651. [CrossRef]
40. Wu, F.; Liu, C.; Sun, W.; Zhang, L.; Ma, Y. Mechanical and Creep Properties of Concrete Containing Apricot Shell Lightweight Aggregate. *KSCE J. Civ. Eng.* **2019**, *23*, 2948–2957. [CrossRef]

41. Ahmad, J.; Zaid, O.; Aslam, F.; Shahzaib, M.; Ullah, R.; Alabduljabbar, H.; Khedher, K.M. A Study on the Mechanical Characteristics of Glass and Nylon Fiber Reinforced Peach Shell Lightweight Concrete. *Materials* **2021**, *14*, 4488. [CrossRef]

42. Blasi, C.D.; Galgano, A.; Branca, C. Exothermic Events of Nut Shell and Fruit Stone Pyrolysis. *ACS Sustain. Chem. Eng.* **2019**, *7*, 9035–9049. [CrossRef]

43. Wei, L.; Xu, S.; Zhang, L.; Zhang, H.; Liu, C.; Zhu, H.; Liu, S. Characteristics of Fast Pyrolysis of Biomass in a Free Fall Reactor. *Fuel Process. Technol.* **2006**, *87*, 863–871. [CrossRef]

44. Cagnon, B.; Py, X.; Guillot, A.; Stoeckli, F.; Chambat, G. Contributions of Hemicellulose, Cellulose and Lignin to the Mass and the Porous Properties of Chars and Steam Activated Carbons from Various Lignocellulosic Precursors. *Bioresour. Technol.* **2009**, *100*, 292–298. [CrossRef]

45. Associataion of Official Analytical Chemist. *AOAC Official Methods of Analysis of the Association of Official's Analytical Chemists*, 14th ed.; Associataion of Official Analytical Chemist: Washington, DC, USA, 1990; pp. 223–225, 992–995.

46. Maletti, L.; D'Eusanio, V.; Durante, C.; Marchetti, A.; Tassi, L. VOCs Analysis of Three Different Cultivars of Watermelon (Citrullus lanatus L.) Whole Dietary Fiber. *Molecules* **2022**, *27*, 8747. [CrossRef] [PubMed]

47. *ASTM C567*; Standard Test Method for Determining Density of Structural Lightweight Concrete. ASTM International: West Conshohocken, PA, USA, 2020. [CrossRef]

48. European Committee for Concrete (CEB); International Federation for Prestressing (FIP). *Manual of Design and Technology, Lightweight Aggregate Concrete*; Construction Press: Lancaster, UK, 1977.

49. Palomar, I.; Barluenga, G.; Puentes, J. Lime–Cement Mortars for Coating with Improved Thermal and Acoustic Performance. *Constr. Build. Mater.* **2015**, *75*, 306–314. [CrossRef]

50. Silva, L.M.; Ribeiro, R.A.; Labrincha, J.A.; Ferreira, V.M. Role of Lightweight Fillers on the Properties of a Mixed-Binder Mortar. *Cem. Concr. Compos.* **2010**, *32*, 19–24. [CrossRef]

51. Lo, T.Y.; Tang, W.C.; Cui, H.Z. The Effects of Aggregate Properties on Lightweight Concrete. *Build. Environ.* **2007**, *42*, 3025–3029. [CrossRef]

52. Mohamed, A.R.; Mohammadi, M.; Darzi, G.N. Preparation of Carbon Molecular Sieve from Lignocellulosic Biomass: A Review. *Renew. Sustain. Energy Rev.* **2010**, *14*, 1591–1599. [CrossRef]

53. Bajpai, P. Wood and Fiber Fundamentals. In *Biermann's Handbook of Pulp and Paper*; Elsevier: Amsterdam, The Netherlands, 2018.

54. Watkins, D.; Nuruddin, M.; Hosur, M.; Tcherbi-Narteh, A.; Jeelani, S. Extraction and Characterization of Lignin from Different Biomass Resources. *J. Mater. Res. Technol.* **2015**, *4*, 26–32. [CrossRef]

55. Carrott, P.J.M.; Carrott, M.R. Lignin—From Natural Adsorbent to Activated Carbon: A Review. *Bioresour. Technol.* **2007**, *98*, 2301–2312. [CrossRef]

56. González Martínez, M.; Anca Couce, A.; Dupont, C.; da Silva Perez, D.; Thiéry, S.; Meyer, X.; Gourdon, C. Torrefaction of Cellulose, Hemicelluloses and Lignin Extracted from Woody and Agricultural Biomass in TGA-GC/MS: Linking Production Profiles of Volatile Species to Biomass Type and Macromolecular Composition. *Ind. Crops Prod.* **2022**, *176*, 114350. [CrossRef]

57. Johnson, C.M. Differential Scanning Calorimetry as a Tool for Protein Folding and Stability. *Arch. Biochem. Biophys.* **2013**, *531*, 100–109. [CrossRef] [PubMed]

58. Ojeda-Galván, H.J.; Hernández-Arteaga, A.C.; Rodríguez-Aranda, M.C.; Toro-Vazquez, J.F.; Cruz-González, N.; Ortíz-Chávez, S.; Comas-García, M.; Rodríguez, A.G.; Navarro-Contreras, H.R. Application of Raman Spectroscopy for the Determination of Proteins Denaturation and Amino Acids Decomposition Temperature. *Spectrochim. Acta. A Mol. Biomol. Spectrosc.* **2023**, *285*, 121941. [CrossRef] [PubMed]

59. Şen, D.; Gökmen, V. Kinetic Modeling of Maillard and Caramelization Reactions in Sucrose-Rich and Low Moisture Foods Applied for Roasted Nuts and Seeds. *Food Chem.* **2022**, *395*, 133583. [CrossRef] [PubMed]

60. Weiss, I.M.; Muth, C.; Drumm, R.; Kirchner, H.O.K. Thermal Decomposition of the Amino Acids Glycine, Cysteine, Aspartic Acid, Asparagine, Glutamic Acid, Glutamine, Arginine and Histidine. *BMC Biophys.* **2018**, *11*, 2. [CrossRef]

61. Wang, S.; Dai, G.; Yang, H.; Luo, Z. Lignocellulosic Biomass Pyrolysis Mechanism: A State-of-the-Art Review. *Prog. Energy Combust. Sci.* **2017**, *62*, 33–86. [CrossRef]

62. Salema, A.A.; Ting, R.M.W.; Shang, Y.K. Pyrolysis of Blend (Oil Palm Biomass and Sawdust) Biomass Using TG-MS. *Bioresour. Technol.* **2019**, *274*, 439–446. [CrossRef] [PubMed]

63. Ding, Y.; Huang, B.; Li, K.; Du, W.; Lu, K.; Zhang, Y. Thermal Interaction Analysis of Isolated Hemicellulose and Cellulose by Kinetic Parameters during Biomass Pyrolysis. *Energy* **2020**, *195*, 117010. [CrossRef]

64. Shen, D.K.; Gu, S. The Mechanism for Thermal Decomposition of Cellulose and Its Main Products. *Bioresour. Technol.* **2009**, *100*, 6496–6504. [CrossRef]

65. Shen, D.K.; Gu, S.; Bridgwater, A.V. Study on the Pyrolytic Behaviour of Xylan-Based Hemicellulose Using TG–FTIR and Py–GC–FTIR. *J. Anal. Appl. Pyrolysis* **2010**, *87*, 199–206. [CrossRef]

66. Yang, H.; Yan, R.; Chin, T.; Liang, D.T.; Chen, H.; Zheng, C. Thermogravimetric Analysis−Fourier Transform Infrared Analysis of Palm Oil Waste Pyrolysis. *Energy Fuels* **2004**, *18*, 1814–1821. [CrossRef]

67. Zhao, C.; Jiang, E.; Chen, A. Volatile Production from Pyrolysis of Cellulose, Hemicellulose and Lignin. *J. Energy Inst.* **2017**, *90*, 902–913. [CrossRef]

68. Yeo, J.Y.; Chin, B.L.F.; Tan, J.K.; Loh, Y.S. Comparative Studies on the Pyrolysis of Cellulose, Hemicellulose, and Lignin Based on Combined Kinetics. *J. Energy Inst.* **2019**, *92*, 27–37. [CrossRef]

69. ACI Committee 213. Guide for Structural Lightwieght-Aggregate Concrete. *J. Am. Concr. Inst. Proc.* **1967**, *64*, 433–469.
70. Chaipanich, A.; Chindaprasirt, P. The Properties and Durablity of Autoclaved Aerated Concrete Masonry Blocks. In *Eco-Efficient Masonry Bricks and Blocks*; Elsevier: Amsterdam, The Netherlands, 2015; pp. 215–230.
71. Alengaram, U.J.; Muhit, B.A.A.; Jumaat, M.Z. bin Utilization of Oil Palm Kernel Shell as Lightweight Aggregate in Concrete—A Review. *Constr. Build. Mater.* **2013**, *38*, 161–172. [CrossRef]
72. Martínez-García, C.; González-Fonteboa, B.; Martínez-Abella, F.; Carro- López, D. Performance of Mussel Shell as Aggregate in Plain Concrete. *Constr. Build. Mater.* **2017**, *139*, 570–583. [CrossRef]
73. Colangelo, F.; Cioffi, R.; Liguori, B.; Iucolano, F. Recycled Polyolefins Waste as Aggregates for Lightweight Concrete. *Compos. Part B Eng.* **2016**, *106*, 234–241. [CrossRef]
74. Pacheco Menor, M.C.; Serna Ros, P.; Macías García, A.; Arévalo Caballero, M.J. Granulated Cork with Bark Characterised as Environment-Friendly Lightweight Aggregate for Cement Based Materials. *J. Clean. Prod.* **2019**, *229*, 358–373. [CrossRef]
75. Ünal, O.; Uygunoğlu, T.; Yildiz, A. Investigation of Properties of Low-Strength Lightweight Concrete for Thermal Insulation. *Build. Environ.* **2007**, *42*, 584–590. [CrossRef]
76. Yang, C.-C.; Huang, R. A Two-Phase Model for Predicting the Compressive Strength of Concrete. *Cem. Concr. Res.* **1996**, *26*, 1567–1577. [CrossRef]
77. Elastic Compatibility and the Behavior of Concrete. *ACI J. Proc.* **1986**, *83*, 10422. [CrossRef]
78. Lo, T.Y.; Cui, H.Z.; Tang, W.C.; Leung, W.M. The Effect of Aggregate Absorption on Pore Area at Interfacial Zone of Lightweight Concrete. *Constr. Build. Mater.* **2008**, *22*, 623–628. [CrossRef]
79. Raj, B.P.; Meena, C.S.; Agarwal, N.; Saini, L.; Hussain Khahro, S.; Subramaniam, U.; Ghosh, A. A Review on Numerical Approach to Achieve Building Energy Efficiency for Energy, Economy and Environment (3E) Benefit. *Energies* **2021**, *14*, 4487. [CrossRef]
80. Kircher, K.; Shi, X.; Patil, S.; Zhang, K.M. Cleanroom Energy Efficiency Strategies: Modeling and Simulation. *Energy Build.* **2010**, *42*, 282–289. [CrossRef]
81. Arkar, C.; Vidrih, B.; Medved, S. Efficiency of Free Cooling Using Latent Heat Storage Integrated into the Ventilation System of a Low Energy Building. *Int. J. Refrig.* **2007**, *30*, 134–143. [CrossRef]
82. Asadi, I.; Shafigh, P.; Abu Hassan, Z.F.B.; Mahyuddin, N.B. Thermal Conductivity of ConcreteA Review. *J. Build. Eng.* **2018**, *20*, 81–93. [CrossRef]
83. Cheboub, T.; Senhadji, Y.; Khelafi, H.; Escadeillas, G. Investigation of the Engineering Properties of Environmentally-Friendly Self-Compacting Lightweight Mortar Containing Olive Kernel Shells as Aggregate. *J. Clean. Prod.* **2020**, *249*, 119406. [CrossRef]
84. Samson, G.; Phelipot-Mardelé, A.; Lanos, C. A Review of Thermomechanical Properties of Lightweight Concrete. *Mag. Concr. Res.* **2017**, *69*, 201–216. [CrossRef]

waste

Article

Promoting Sustainable Fruit and Vegetable Biowaste Management and Industrial Symbiosis through an Innovative Web Platform

Ioannis Varvaringos [1], Eva Skourtanioti [1,*], Georgios Letsos [1], Evgenia Rizoudi [1], Ektoras Makras [1], Margarita Panagiotopoulou [2], Sofia Papadaki [2] and Katerina Valta [1]

[1] DRAXIS ENVIRONMENTAL SA, 317 Mesogeion Avenue & Lokridos, Halandri, 15231 Athens, Greece; ivarvaringos@draxis.gr (I.V.); gletsos@draxis.gr (G.L.); erizoudi@draxis.gr (E.R.); emakras@draxis.gr (E.M.); katvalta@draxis.gr (K.V.)

[2] Laboratory of Process Analysis and Design, School of Chemical Engineering, National Technical University of Athens, Zografou Campus, 9 Iroon Polytechneiou, 15780 Athens, Greece; panagiot.marg@gmail.com (M.P.); sofiachemeng@hotmail.com (S.P.)

* Correspondence: eskourtanioti@draxis.gr; Tel.: +30-210-9247134

Abstract: Sustainable bioeconomy is a promising pathway towards the transition to a circular and climate-neutral economy. The valorization of biowaste is a key player in this direction. This paper presents the design and development of the AgriPLaCE Platform, which aims to promote synergies that enable the utilization of biowaste from the fruit and vegetable supply chain. The platform consists of the AgriPLaCE Waste Management Database, which provides users with an extended list of potential utilization methods for various types of fruit and vegetable biowaste streams, and the AgriPLaCE Synergies Tool, which facilitates synergies between different actors involved in the biowaste-to-resource value chain from agricultural waste production to waste treatment and new valuable products' exploitation. Initially, the conceptual design of both tools took place based on analysis of user needs and services alongside the system architecture. Following this, the AgriPLaCE Platform was developed with the implementation of all the necessary subsystems. The results of the platform's implementation demonstrated its potential to generate multiple collaborations and synergies while users can also deepen their knowledge about alternative and emerging treatment technologies and valuable products from a wide range of fruit and vegetable biowaste streams.

Keywords: fruit and vegetable; biowaste; synergies; matchmaking; platform; bioeconomy; industrial symbiosis; waste management

Citation: Varvaringos, I.; Skourtanioti, E.; Letsos, G.; Rizoudi, E.; Makras, E.; Panagiotopoulou, M.; Papadaki, S.; Valta, K. Promoting Sustainable Fruit and Vegetable Biowaste Management and Industrial Symbiosis through an Innovative Web Platform. *Waste* **2023**, *1*, 532–548. https://doi.org/10.3390/waste1020032

Academic Editors: Catherine N. Mulligan, Dimitris P. Makris and Vassilis Athanasiadis

Received: 6 April 2023
Revised: 29 May 2023
Accepted: 1 June 2023
Published: 5 June 2023

1. Introduction

In December 2019, the European Union (EU) announced its commitment to achieving climate neutrality by 2050. This commitment was outlined in the European Green Deal, in alignment with the goals and targets set under the Paris Agreement [1,2]. According to the position of the European Parliament [3], to protect, preserve, and improve the quality of the environment and public health, waste management should be improved and transformed into sustainable material management, promoting the principles of circular economy, enhancing the use of renewable energy, increasing energy efficiency, providing new economic opportunities and contributing to long-term competitiveness. To achieve a truly circular economy, additional measures on sustainable production must be taken, focusing on the entire life cycle of products in a way that preserves resources and closes the loops [4]. In this context, highlighting waste as a valuable material can provide greater independence in terms of raw material import needs and can create important opportunities for local economies and stakeholders to strengthen the circular economy's alignment with various policy objectives, and deliver environmental benefits through reductions in greenhouse gas emissions. This is in line with the "industrial symbiosis" concept, which encourages collaboration and resource sharing among traditionally separate entities [5]. Originally, the concept of industrial symbiosis focused on the physical transfer of materials (such as waste, stock, or by-products), energy, and/or water between geographically proximate companies. This exchange entails one company utilizing another company's residual flow as an input, resulting in mutual benefits and reduced environmental impact. Currently, industrial symbiosis encompasses various forms of intercompany cooperation, including resource exchange or sharing, underutilized resource utilization, utility sharing, infrastructure collaboration, and service cooperation [6,7].

Connecting industrial symbiosis and the bioeconomy can offer synergistic opportunities to enhance sustainability and resource efficiency. The European Parliament emphasizes the need to encourage investments in the development of a sustainable bioeconomy where fossil-intensive materials are replaced by renewable and bio-based ones since a switch to sustainable materials can play a significant role in the transition to a climate-neutral economy [8]. One of the most promising routes to a resource-efficient circular economy is toward bio-products and bio-based value chains. Industrial symbiosis can facilitate the exchange of bio-based resources, by-products, or waste streams between different companies within the bioeconomy. Several research studies have been carried out to explore the valorization of diverse biowaste streams, including agricultural waste [9] and industrial biowaste [10–14]. Additionally, there has been increasing attention towards plant-origin biowaste, specifically fruit and vegetable biowaste [12], as an area of interest for valorization efforts. Fruit and vegetable biowaste from the fruit and vegetable supply chain can originate from agriculture production, the food and beverage processing industry, retail, and the hotel, restaurant, and catering (HoReCa) sector.

Overall, while industrial symbiosis offers numerous benefits, several barriers can hinder its widespread adoption, depending on the specific context and stakeholders involved. Such barriers can be the lack of awareness and knowledge, regulatory and legal constraints, financial constraints, infrastructure limitations, and a lack of trust and data exchange between stakeholders [5–7]. Several tools have been created in recent years to overcome some of these obstacles. Existing tools have been categorized into different types depending on their facilitation approach in a recent literature review of information systems aiding the promotion of industrial symbiosis [15]. There are five of them: (1) open online waste markets, (2) facilitated synergy identification systems, (3) industry sector synergy identification, (4) social network platforms and communities, and (5) tools to identify potential waste-to-resource exchanges at the industry sector level [16]. They represent digital environments for businesses to find and establish partnerships for waste-to-resource exchanges.

Several recent studies [6,7,17–19] have explored the development of web platforms that address the utilization of various waste types as valuable resources; however, none of these platforms specifically emphasize the matchmaking of synergies for biowaste utilization. Additionally, the Sustainable Agriculture Initiative Platform (SAI Platform) [20] was established in 2002. The SAI Platform is a global membership organization dedicated to promoting sustainable agriculture on a global scale. It achieves this objective through capacity building and communication in the field of sustainable agriculture. The platform offers tools like the Sustainability Performance Assessment tool, which enables self-assessment or external assessment of sustainable agricultural practices. Nevertheless, a platform with a specialized focus on supporting the efficient utilization of fruit and vegetable biowaste through web-based matchmaking algorithms, eliminating intermediaries, is currently absent. Considering the substantial volume of fruit and vegetable biowaste generated, it is imperative to develop mechanisms for reutilizing them and fostering synergies among suppliers and demanders.

To cover this gap, an electronic platform, the AgriPLaCE Platform, focusing on fruit and vegetable biowaste valorization, was developed. In particular, this study presents the development of the AgriPLaCE Platform, which aims to promote synergies for the utilization of fruit and vegetable biowaste in the production of secondary materials. The platform consists of two (2) tools, i.e., the AgriPLaCE Waste Management Database and the AgriPLaCE Synergies Tool. Overall, the process involved formulating the concept of AgriPLaCE, conducting a literature review on fruit and vegetable biowaste valorization methods, implementing the platform, and validating its functionality and effectiveness.

2. Materials and Methods

2.1. The Concept behind the AgriPLaCE Platform

The *general objective* of this study was to facilitate the efficient and sustainable utilization of fruit and vegetable biowaste for resource recovery through a digital platform, the *AgriPLaCE Platform*. To this end, the *first specific objective (SO1)* was to provide a comprehensive knowledge base of fruit and vegetable biowaste management practices to various stakeholders, including agricultural producers, food processors, waste management companies, and potential end-users, and facilitate informed decision-making in this field. To achieve SO1, the *AgriPLaCE Waste Management Database* was designed and developed. This database serves as a centralized repository for fruit and vegetable biowaste management information. Moreover, the *second specific objective (SO2)* was to foster collaboration among various stakeholders from the supply chain of fruit and vegetable biowaste generation; the waste management industry, and the new valuable products industries. With this objective in mind, the *AgriPLaCE Synergies Tool* was created and implemented. This tool facilitates synergies between different actors involved in the biowaste-to-resource value chain. Figure 1 presents the AgriPLaCE concept and interconnections between the objectives and the AgriPLaCE tools.

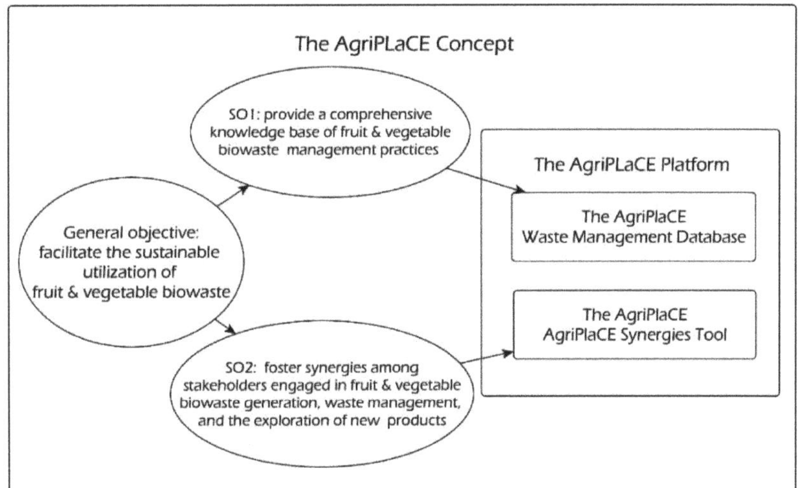

Figure 1. The AgriPLaCE concept and interconnections of objectives and AgriPLaCE tools.

Both the AgriPLaCE Waste Management Database and the AgriPLaCE Synergies Tool cater to specific user categories, i.e., waste suppliers (WS) and waste demanders (WD). The waste suppliers (WS) category encompasses a range of entities involved in the generation of fruit and vegetable biowaste, including farmers, the food and beverage industry, retailers (grocery stores, supermarkets, farmers' markets), and HoReCa representatives. On the other hand, the waste demanders (WD) category includes owners of waste treatment technologies and industrial partners that are interested in developing technologies to produce products from biowaste.

The conceptual design was based on the assumption that there is a mutual demand and that each user category (WS or WD) requires information from the other user category to cover their needs. In specific, the WS requires information from the WD to understand how their (WS) waste could be exploited, and the WD need information regarding which WS could provide them with waste that could be exploitable by them (WD). Consequently, the need for a common *waste stream categorization vocabulary* arose. Additionally, it was essential to define the *processes generating fruit and vegetable biowaste streams* as well as the *processes to treat those waste streams* both through conventional waste treatment methods and through alternative and emerging ones. Moreover, it was crucial to define *the sectors that would act as end-users of tio-based products*. Finally, establishing the interconnections between these elements was also essential. The interconnections are depicted in Figure 2, while further details for each element are given afterwards.

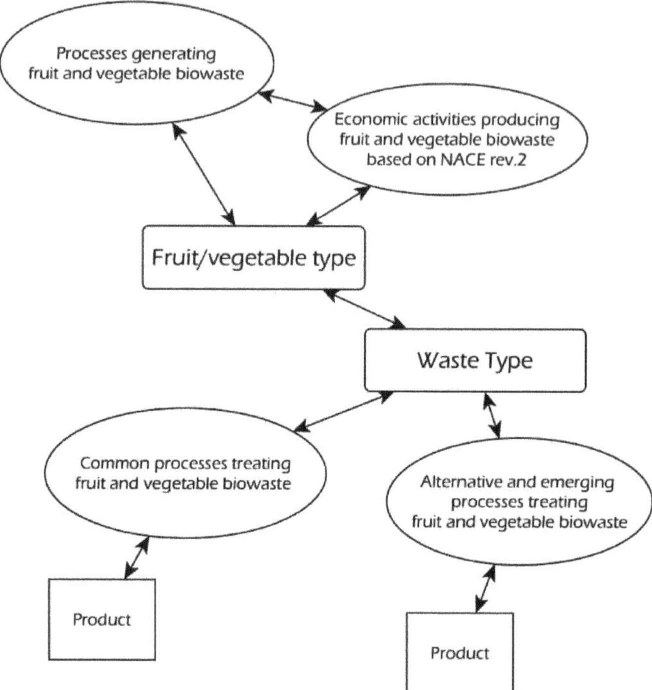

Figure 2. Interconnections between the elements of the AgriPLaCE concept.

2.1.1. Waste Streams Categorization Vocabulary

To address the mutual demand between WS and WD, it was necessary to investigate the fruit and vegetable biowaste streams as generated by the WSs and also, to investigate the fruit and vegetable biowaste streams specifications as they are required by the WDs for being able to receive the waste. The findings showed that waste generated in agricultural fields could be categorized per type of fruit/vegetable and then further sub-categorized into leaves, stems, twigs, and fruits/vegetables unsuitable for human consumption due to factors, such as quality standards or aesthetic preferences. Regarding the fruit and vegetable biowaste generated by the food industry and the HoReCa sector, it encompasses various components of fruits and vegetables, including pits, kernels, seeds, and peels. Additionally, unsold fruits and vegetables are also included in this category of biowaste. The reasons for the generation of such biowaste from those sectors can be attributed to factors, such as quality standards, aesthetic preferences, shelf-life limitations, overstocking or excess inventory, fluctuations in market demand, and evolving consumer preferences, among others.

The waste streams that can be exploitable by the WD were the following:

- *Waste stream category "a"*: This category comprises waste streams that consist of **only one** part of **one** specific type of fruit/vegetable. Examples include kernels from apricots, peels from tomatoes, or seeds from lemons.
- *Waste stream category "b"*: In this category, waste streams are formed by **one part** of **more than one or all** types of fruit/vegetable plant/s. Examples include twigs from all types of fruit and vegetable plants, or fruits/vegetables unsuitable for human consumption derived from all categories of fruit/vegetable plants.
- *Waste stream category "c"*: This category includes waste streams that consist of **different parts** from **more than one or all** registered types of fruit/vegetable plants. Examples can include leaves, twigs, and seeds from all kinds of fruit/vegetable plants.

To ensure that the needs of both WS and WD would be met, it was essential to account for all possible waste streams. To achieve this, the adopted waste stream classification was defined as presented in Table 1.

Table 1. Waste classification in AgriPLaCE Platform.

Type of Fruit/Vegetable	Waste Type (Part of Fruit/Vegetable Plant)
Type 1	peels
	pits/kernels/seeds
	twigs
	leaves
	stems
	fruits/vegetables unsuitable for human consumption
Type 2	. . .
. . .	. . .

2.1.2. Processes Generating Fruit and Vegetable Biowaste Streams

As was already mentioned, the economic activity, the agro-industrial sector, and the processes generating different fruit and vegetable biowaste streams were investigated. As a result, information regarding the waste stream origin was provided per waste type (part of fruit/vegetable) for each type of fruit and vegetable. This was achieved by both indicating the relevant code(s) of the economic activity(ies) generating the different waste streams based on the Statistical Classification of Economic Activities in the European Community (NACE Rev. 2) and also by providing information about the process generating each waste stream (e.g., food processing).

2.1.3. Processes Treating Fruit and Vegetable Biowaste Streams for Bio-Based Products

Regarding the fruit and vegetable biowaste treatment methods for the generation of valuable bio-based products, two categories were taken into consideration:

- The established and frequently applied waste management methods;
- The alternative and emerging biowaste management methods.

While the project aimed to promote alternative biowaste management solutions and emerging management techniques, conventional waste management methods, such as processes producing biogas, pellets, animal food, etc., were also considered per waste type (part of a fruit/vegetable plant) for each type of fruit or vegetable, because a lot of biowaste streams are not exploited even with conventional methods and are just discarded; however, the main part of the study was focused on defining alternative biowaste management solutions and emerging management techniques. To this end, an extended literature review was conducted. As a result, a plethora of information was extracted regarding innovative existing and emerging technologies for the treatment of the fruit and vegetable biowaste stream categories "a", "b" and "c".

2.1.4. Sectors That Can Act as End-Users of the Bio-Based Products

The products derived from the methods described above have a wide range of applications in different industrial sectors, including the food, cosmetic and pharmaceutical industries, polymer production, energy, etc. To this end, those sectors were identified and matched with each specific bio-based product.

2.2. AgriPLaCE Platform Development

The AgriPLaCE Platform was implemented with the latest technologies in web development, through the use of open-source platforms, offering flexible access through mobile devices and personal computers to allow easy access and use by any user from any device connected to the internet. The AgriPLaCE Platform [21], available in the Greek language, was the output of the AgriPLaCE project: *'Valorisation of Agricultural waste for the production of innovative Plastic materials in line with the Circular Economy'* co-financed by the ERDF of EU and Greek funds (project code: Τ6ΥΒΠ-00220, MIS 5048495).

During the implementation of the AgriPLaCE Platform, the subsequent methodological phases were followed:

1. Preliminary conceptualization of the actual functionalities of the two tools.
2. Preparation of draft documents with expected visuals of the platform.
3. Enrichment of the draft documents and development of wireframes.
4. Drafting of mock-ups.
5. Mock-up revision and finalization.
6. Development of the AgriPLaCE framework. The completion of the mock-ups was followed by the actual development of the AgriPLaCE Platform.
7. Deployment of the AgriPLaCE Platform release candidate version. A fully functional version was deployed and provided for testing and further contributions.
8. Validation exercise: A validation exercise with the technical partners was organized. All comments were incorporated into the final release.
9. Final release of the AgriPLaCE Platform.

2.2.1. AgriPLaCE Platform System Architecture

The subsystems that form the AgriPLaCE platform stack followed a simple and intelligent architecture consisting of three layers, i.e., (i) the Presentation layer, (ii) the Application layer, and (iii) the Database layer [22], as depicted in Figure 3.

The user interacts with the platform by way of the Presentation layer, which provides all necessary user interfaces for performing the tools' functionalities. Every action in the system triggers an operation on the Application layer. Conversely, when the user requests information by performing the relevant actions in the Presentation layer, the Application layer retrieves any relevant data from the Database layer, transforms it to the appropriate format, and then sends it to the Presentation layer to present the data to the user.

The Presentation layer uses JavaScript with VueJS [23] as the primary framework and additionally Vuelidate [24] for model-based validation and Vuetify [25] framework for the components. Vuetify is a complete user interface framework built on top of VueJS. The goal of this approach was to provide the AgriPLaCE platform with all the necessary features needed to build a rich and engaging user experience. Additionally, Vuetify takes a "mobile first" approach to design, which means that the platform just works out of the box—whether it's on a phone, tablet, or desktop computer.

The Application layer incorporates the Laravel Framework [26], a free and open-source PHP web framework intended for the development of web applications. In that manner, the web application dynamically interacts with the user providing faster and smoother transitions making the website feel more responsive [27].

The Database layer encompasses a PostgreSQL database [28] for computational operations [29]. PostgreSQL is a free and open-source relational database management system that operates as a data repository for the user input data, the waste management catalogue data, and the synergies tool data.

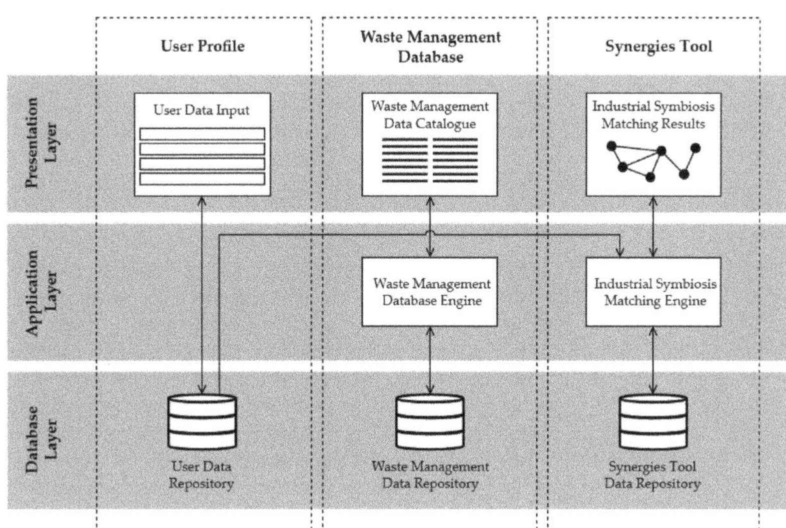

Figure 3. The AgriPLaCE System Architecture.

The server is the physical container that hosts the AgriPLaCE Platform and makes all its functionality available to the public. NGINX [30] and the Apache HTTP web server software [31] resides on the server hosting all the platform's layers. NGINX was used as a reverse proxy server for handling static content requests and Apache as the backend to serve dynamic content, offering enhanced server performance for the web platform. Each web server software has Docker containers installed to provide all the software infrastructure to host the platform. The AgriPLaCE docker contained a standard unit of software that packages up the AgriPLaCE code and all its dependencies for the application to run quickly and reliably from one computing environment to another [32].

2.2.2. AgriPLaCE User Profile

To access the full functionality of the AgriPLaCE Platform, users are prompted to register on the website by submitting relevant information regarding their waste management needs. Apart from login details, users are requested to provide the enterprise type and geographical location, as well as information regarding their waste supply or needs. These data are stored in the User Data Repository and used in the Synergies Tool for calculation of the industrial symbiosis matching. Based on the information stored in their profile, the users can view a ranked catalogue of potential waste management partners and contact them through the Synergies Tool of the AgriPLaCE Platform.

2.2.3. AgriPLaCE Waste Management Database

The AgriPLaCE Waste Management Database provides an extended list of potential utilization methods for a variety of fruit and vegetable biowaste types and acts as a catalogue for waste management applications for its users. The AgriPLaCE Database is open to the public and users do not have to be registered to the platform to access it. The list provides detailed information related to the waste, specifically:

1. Type of fruit or vegetable;
2. Waste type;
3. Processes generating waste;
4. Economic activity from which the waste originates based on NACE Rev. 2;
5. Conventional methods of waste management;
6. Emerging technologies of waste management;

7. Products derived from conventional and emerging waste management methods;
8. Economic activity and industrial sectors valorize products derived from waste.

The database users can use filters and perform an advanced search for their waste type of interest by typing keywords in the search field. The algorithm then recategorizes the results based on the specific filtering provided. Finally, users can download their search results in .xls file format for further editing.

2.2.4. AgriPLaCE Synergies Tool

The successful implementation of industrial symbiosis relies heavily on stakeholders' willingness to collaborate and share information. Consequently, stakeholders must have trust in the platform provider regarding matters of confidentiality [7]. Thus, the AgriPLaCE Synergies Tool is accessible only to registered users. This approach ensures that any sensitive information shared on the tool remains confidential and secure. In particular, upon registration, users are requested to provide information related to their waste supply or demand needs. This process occurs in the user profile subsystem, which stores the user data input in the User Data Repository. Once the registered user enters the AgriPLaCE Synergies Tool page, the Industrial Symbiosis Matching Engine calculates the priority of each user profile stored in the user data repository and sorts all users based on their assigned priority.

Upon calculation, the user is presented with a sorted list of potential industrial symbiosis matches. Results (other platform users) with the same priority value are shown alphabetically, and the user can further categorize them by performing an advanced search of the results with the use of filters and searching by keywords. The engine then recategorizes the results based on the specific filtering provided and presents the updated list to the user. It is also possible to download the industrial symbiosis matching list in .xls file format for further editing.

Finally, the AgriPLaCE Synergies Tool incorporates a communication feature within its contact module, enabling registered platform users to connect. Specifically, following the matchmaking process, the user can send a message through the platform that reaches the receiver via e-mail. Importantly, the receiver's e-mail is shared only if a reply is initiated, ensuring compliance with the General Data Protection Regulation (GDPR) principles.

2.3. Industrial Symbiosis Matching Algorithm

The Industrial Symbiosis Matching Engine of the Synergies Tool encompasses the profile matching algorithm, which was based on weighted variables in a weighted sum model [33,34]. Specifically, there were five indexes used for ranking the potential matching of two users. Each index was assigned one variable, which receives the value 1 in case two users have a common index value and 0 in case the users have different index values.

The priority calculation formula is:

$$a \cdot K + b \cdot W + c \cdot R + d \cdot U + e \cdot M = P, \tag{1}$$

where K, W, R, U, and M are the index variables, and a, b, c, d, e are the weights of each variable, and P is the priority value.

For each index variable, a unique weight was assigned based on the importance of each index to the matchmaking process. The five indexes and the respective variables and weights are presented in Table 2.

The most important indexes (core indexes) were the "Type of fruit or vegetable" (K, a = 0.35) and the "Waste type" (W, b = 0.35) since those were the necessary indexes to achieve an agreement for the material to create synergy. For this reason, each one of these core indexes ("Type of fruit or vegetable", "Waste type") receives a higher weight than the weight of indexes that are related to the location ("Region", "Regional unit", "Municipality"). Moreover, the total weight of the location indexes (0.1 + 0.1 + 0.1) is lower than the individual weight of each of the core indexes ("Type of fruit or vegetable", "Waste type") which each is equal to 0.35.

Table 2. Synergies Tool indexes with respective variables and weights.

Index	Variable	Weight
Type of fruit or vegetable	K	a = 0.35
Waste type	W	b = 0.35
Region	R	c = 0.10
Regional unit	U	d = 0.10
Municipality	M	e = 0.10

The priority calculation matrix, based on the index variable values, is presented in Table 3.

Table 3. User matching priority matrix based on index variable value.

Priority	K	W	R	U	M
1	1	1	1	1	1
0.90	1	1	1	1	0
0.80	1	1	1	0	0
0.70	1	1	0	0	0
0.65	0	1	1	1	1
0.55	0	1	1	1	0
0.45	0	1	1	0	0
0.35	0	1	0	0	0
0.30	0	0	1	1	1
0.20	0	0	1	1	0
0.10	0	0	1	0	0
0	0	0	0	0	0

Users receiving a higher priority value are then classified as more suitable candidates for waste transactions and appear higher on the AgriPLaCE Synergies Tool matching list. Users receiving the same priority value are then listed alphabetically.

At this point, it should be noted that when K = 0 and W = 1, the type of fruit or vegetable (K) is determined by only one of the matched users (WS or WD), while the waste type (W) is defined by both of them. For instance, this applies in situations where a WS indicates that they offer "seeds" (waste type) from "tomato", and the WD expresses the desire to receive "seeds" without specifying the fruit's specific type.

Users can be waste suppliers, demanders, or both based on the information stored on their profile. The AgriPLaCE Synergies tool provides two independent lists:

- Suppliers list, presenting waste suppliers matching the user's waste demands;
- Demanders list, presenting waste demanders matching the user's waste supplies.

In case the user falls under only one category, e.g., is a waste supplier, the waste supplier list performs the same algorithm showcasing the user's competition based on the same criteria.

3. Results

3.1. Results Related to AgriPLaCE Waste Management Database

The extended literature review on processes generating fruit and vegetable biowaste resulted in the identification of a diverse range of alternative and emerging biowaste treatment technologies with the potential to produce valuable materials that have applications across a wide spectrum of industrial sectors. These emerging biowaste treatment technologies descriptions are provided to the users through the AgriPLaCE Waste Management

Database. The user can access information about waste treatment technologies for waste streams of interest (waste stream categories "a", "b" and "c") or obtain insights into which stream categories ("a", "b" and "c") can be effectively treated using the waste treatment technologies that align with their interests. This inclusive approach ensures that both user categories (WS and WD) can obtain specific and pertinent information.

Following is a summary of the identified fruit and vegetable biowaste treatment methods applicable to different waste stream categories ("a", "b" and "c"):

- *Valuable components' extraction processes* (relevant to waste stream category "a"). For instance, these processes may include the extraction of various types of oils from the peels and kernels of fruits such as lemons, oranges, or from vegetables. Additionally, extraction methods have been applied to obtain lycopene from tomato peels, extract pectin from watermelon peels, and isolate valuable compounds, such as chlorogenic acid, vanillic acid, and ferulic acid, from pear peels. Numerous other valuable ingredients have also been extracted from various waste streams using similar processes [12,35–66].
- *Volatile fatty acid (VFA)* production (relevant to waste stream category "b"). Fruit and vegetable waste (flesh) have been used as raw materials for the production of VFAs. VFAs are chemical building blocks, which are globally demanded by the chemical industry. Due to their functional groups, they are suitable precursors necessary for the production of chemicals, such as biopolymers, polyhydroxyalkanoates (PHAs), polylactic acids (PLAs), ketones, esters, alcohols, aldehydes, alkanes and biofuels, such as CH_4 and H_2 [67–71].
- *PHA and PLA production* (relevant to waste stream category "b"). Traditionally, bioplastics have been primarily produced from food crops, such as corn starch, tapioca roots, or sugar cane; however, in recent years, there has been a growing focus on utilizing bio-waste, including the flesh of fruit and vegetable waste as a raw material for bioplastic production [72].
- *Insect protein production for animal feed* (relevant to waste stream category "b"). Many varieties of insects are used for animal feed, so mass production of edible insects for animal feed appears to be a viable method to meet the growing demand for animal protein. Regarding the feed provided to farmed animals, restrictions have been established at EU level. In this context, insects intended for farmed animals' feed must be fed exclusively with materials of plant origin. Therefore, fruit and vegetable waste (flesh) has been used as a suitable raw material for insect farming [73,74].
- *Production of "green" building materials* (relevant to waste stream category "c"). Certain types of biowaste, such as leaves, stems, and branches, can be effectively used as a valuable source of fibers to reinforce building materials, such as panels or for the production of biocomposites.
- *2,3-butanediol production* (relevant to waste stream category "c"). 2,3-butanediol is a valuable chemical building block with a wide variety of applications in areas, such as chemical, energy, food, and polymer production. The industrial production of this compound is carried out by chemical methods from fossil sources and requires high energy intensity and the use of expensive catalysts. 2,3-butanediol is an alcohol that has been produced by fermentation of sugars derived from a wide range of plant raw materials: garden waste, vegetable and fruit waste (fruits, stems, leaves, and branches) [75].
- Microalgae production for animal feed (relevant to waste stream category "c"). The use of fruit and vegetable waste (flesh) could be applied to the production of microalgae. Fruit and vegetable flesh is a rich source of nutrients necessary for the growth of algae [76–79].
- Biochar production (relevant to "c" waste stream). Biochar is produced by a process called carbonization. This process involves pyrolysis in the absence of oxygen. The pyrolysis process has been employed on both the small and industrial scale to generate

biochar from residual forest and agricultural biomass; hence, fruit and vegetable waste can be utilized for biochar production [80,81].

Figure 4 presents a summary representation of fruit and vegetable biowaste treatment methods for all waste stream categories.

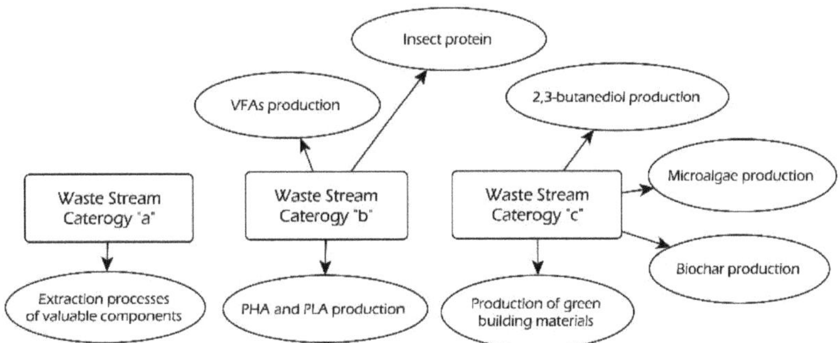

Figure 4. Fruit and vegetable biowaste treatment methods for all waste stream categories.

3.2. Results Related to the AgriPLaCE Synergies Tool

To illustrate the functionality of the AgriPLaCE Synergies Tool, a hypothetical use case of five potential users (one waste supplier and four waste demanders) was presumed, where the main user is a waste supplier (WS) seeking potential partners (waste demanders) in need of the supplied waste (WD 1, 2, 3 and 4). Each user has provided information on their profile upon registration, where they have selected their needs from a given list of items. For the Synergies Tool, the functionality was based on five indexes, where the users selected the values presented in Table 4.

Table 4. Use case of a waste supplier (WS) and four waste demanders (WD) index values.

Index	WS	WD 1	WD 2	WD 3	WD 4
Type of fruit or vegetable	Tomato	Tomato	-	Tomato	Melon
Waste type	Seeds	Seeds	Seeds	Seeds	Rind
Region	Attica	Attica	Attica	Attica	Attica
Regional unit	Piraeus	Athens Central	Piraeus	Piraeus	Piraeus
Municipality	Perama	Athens	Piraeus	Perama	Perama

These data were extracted from the User Data Repository and fed to the Industrial Symbiosis Matching Engine as inputs to perform calculations. Based on the common index values between the waste supplier and each waste demander, each potential collaboration pair received a binary value for each index variable. The index variable matrix for each potential pair is presented in Table 5.

Table 5. Use Case of a waste supplier (WS) and four waste demanders (WD) index variable values.

Variable	WS—WD 1	WS—WD 2	WS—WD 3	WS—WD 4
K	1	0	1	0
W	1	1	1	0
R	1	1	1	1
U	0	1	1	1
M	0	0	1	1

Then, the engine calculated each demander's priority for the given supplier profile and sorted the synergies tool list based on the users with the highest priority ranking. The flow diagram of the use cases for the user priority calculation and Synergies Tool list sorting is depicted in Figure 5.

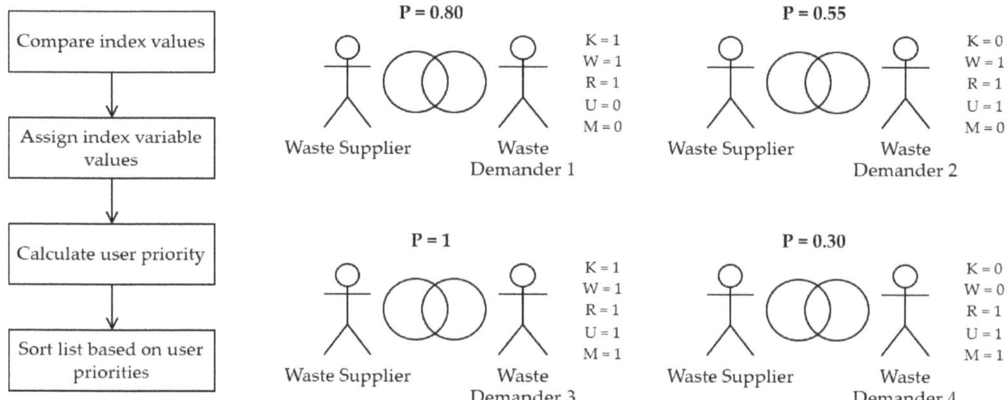

Figure 5. Synergies Tool functionality flow for the use case scenario.

On the left side of Figure 5, the flow diagram of the Industrial Symbiosis Matching Engine algorithm is depicted, while on the right side, the user profile data are compared, and index variables take the value 1 when the respective index values are the same and the value 0 when the respective index values are different. In these cases, the priority equations are calculated as follows:

$$0.35 \cdot 1 + 0.35 \cdot 1 + 0.10 \cdot 1 + 0.10 \cdot 0 + 0.10 \cdot 0 = 0.80 \tag{2}$$

$$0.35 \cdot 0 + 0.35 \cdot 1 + 0.10 \cdot 1 + 0.10 \cdot 1 + 0.10 \cdot 0 = 0.55 \tag{3}$$

$$0.35 \cdot 1 + 0.35 \cdot 1 + 0.10 \cdot 1 + 0.10 \cdot 1 + 0.10 \cdot 1 = 1 \tag{4}$$

$$0.35 \cdot 0 + 0.35 \cdot 0 + 0.10 \cdot 1 + 0.10 \cdot 1 + 0.10 \cdot 1 = 0.30 \tag{5}$$

Based on these calculations, the system presented the list to the waste supplier, sorted by the highest priority:

1. Waste Demander 3
2. Waste Demander 1
3. Waste Demander 2
4. Waste Demander 4

Apart from the use case presented above, the Synergies Tool was tested with over 20 different user profiles (both waste suppliers and demanders,) showcasing multiple potential users and successful collaborations that arise through the use of the AgriPLaCE Platform.

4. Discussion

Numerous efforts have been undertaken to foster industrial symbiosis synergies via a digital platform, targeting the valorization of diverse waste and wastewaters. In the subsequent paragraphs, the most pertinent initiatives will be presented and compared with the AgriPLaCE Platform for comparative analysis.

Akrivou et al. [6] presented a conceptual framework for industrial symbiosis matchmaking, specifically designed for a waste valorization marketplace. This framework aimed to provide users with a comprehensive evaluation that incorporates both quantitative and qualitative information, thereby contributing to the decision-making process. The total relevance score was calculated taking into consideration the sub-scores derived from three different level scoring systems: (i) level A: stakeholders' compatibility regarding industrial symbiosis compliance level assessed via a questionnaire, expressed as sub-score A; (ii) level B: flow analysis, expressed as sub-score B; and (iii) level C: environmental impact of a potential synergy through Life Cycle Assessment; however, it is important to note that the research did not encompass the development of a matchmaking algorithm, similar to the case of the AgriPLaCE platform. Nevertheless, it is worth mentioning that, akin to the AgriPLaCE Platform, the study conducted by Akrivou et al. [6] also considered compatibility based on flow characteristics (both quality and quantity) as a fundamental aspect of their matching concept. This research also proposes quantifying the impact assessments for the synergies created, aiming to provide potential users with information regarding the most environmentally friendly synergies. This aspect, which is not currently included in the AgriPLaCE matching algorithm, could be seen as a potential area for future improvement of the AgriPLaCE Platform.

The e-Symbiosis platform [19] was a web-based platform that enabled users to participate in industrial symbiosis activities to improve resource efficiency across the economy. The creation of synergies on the e-Symbiosis platform between users was achieved through a matchmaking algorithm. The three different user categories were waste providers, solution providers, and waste demanders. The users were matched based on the resources' categorization, the availability and quantity of the resources, the location, and the supply pattern. While there are certain similarities between the AgriPLaCE Synergies Tool and the e-Symbiosis Synergies Tool, a key difference lies in the scope of their coverage. The e-Symbiosis platform aimed to encompass all types of waste that may arise within a given region. In contrast, AgriPLaCE specifically focused on biowaste originating from the fruit and vegetable value chain. Moreover, in the e-Symbiosis tool, biowaste types were classified into broader categories, whereas AgriPLaCE offered a precise classification specifically for fruit and vegetable biowaste. This meticulous classification system in AgriPLaCE facilitated the identification of highly specific utilization opportunities, enabling the application of a wide range of valorization techniques.

On the other hand, the main objective of the first Industrial Symbiosis Platform in Italy [18] was to launch industrial symbiosis through a geo-referenced information system. It served as a tool to support businesses and the local community. It incorporated a taxonomy designed for capturing input–output data from companies, encompassing resources, such as "materials, energy, services, and skills," while utilizing official national codes for numerous inventories; however, it should be noted that the waste classification in this platform is not as comprehensive as in AgriPLaCE, as it primarily focuses on broader categories.

The Online Brine Platform (OBP) [17], which is focused on industrial saline wastewater, aimed to create direct synergies between its users. While both the AgriPLaCE Synergies Tool and the Online Brine Platform considered the quantity, quality characteristics, and location of flows in the matchmaking process between users, there is a notable distinction. In the case of the OBP, the roles are not limited to waste receivers and waste providers but also include waste providers (and heat providers), technology providers, recovered material demanders and recovered material providers. It is important to highlight that the OBP's primary focus of interest lies in saline wastewater, which is entirely different from the focus of the AgriPLaCE Platform.

Finally, the SAI Platform [20] is a tool designed to assess outcomes and track progress in four impact areas of agricultural activities related to soil, biodiversity, water, and climate. The SAI Platform is a portal where registered users can use the Farm Sustainability Assessment (FSA) and Sustainable Dairy Partnership (SDP) online training, technical documents, and guidance materials exclusively for members. The FSA enables food and beverage businesses to assess, improve, and validate on-farm sustainability in their supply chains. In addition, this platform provides a space for knowledge-sharing, networking, and collaboration among stakeholders in the agro-industrial sector. It facilitates discussions on sustainable practices, promotes research findings and technological developments, and creates opportunities for partnerships and projects related to the valorization of agro-industrial waste. While the platform indirectly contributes to addressing the issue of waste management and utilization of biological origin, its primary focus is on broader aspects of sustainable agriculture rather than directly connecting waste producers with waste demanders from the fruit and vegetable supply chain.

The implementation of the AgriPLaCE Digital Platform has demonstrated promising results in terms of its potential contribution to the sustainable bioeconomy and circular economy, specifically through the effective utilization of fruit and vegetable biowaste. By facilitating the exchange of resources, the platform plays a crucial role in waste reduction. Additionally, it can enhance economic performance by lowering costs and fostering the emergence of novel business opportunities and circular business models.

Furthermore, an important characteristic of the AgriPLaCE Platform is its scalability, which allows for potential expansion to encompass additional industries and resources. This scalability offers the opportunity for even greater environmental and economic benefits. As the platform continues to grow and evolve, new synergies and opportunities for resource exchange are likely to arise, further bolstering sustainability efforts and diminishing the environmental impact within the fruit and vegetable supply chain and the biowaste-to-resource value chain.

While the results of the evaluation were positive, there were certain challenges and limitations to the implementation of the AgriPLaCE Platform. One notable challenge is ensuring the active participation and engagement of all relevant industries within the region. While some users may readily embrace the platform and recognize its potential benefits, others may exhibit reluctance in terms of sharing information or engaging in resource exchange. This emphasizes the significance of ongoing outreach and education efforts to effectively communicate the advantages of industrial symbiosis and the digital platform.

Another limitation pertains to the necessity for continuous maintenance and updates of the AgriPLaCE Platform. As new industries and resources are integrated, it becomes crucial to ensure that the database remains up-to-date and accurate. Furthermore, continuous user feedback and evaluation are essential in identifying areas for improvement and ensuring that the platform effectively caters to the needs of its users. By addressing these challenges and limitations, the platform can further enhance its functionality and maximize its potential impact.

Author Contributions: Conceptualization, E.S. and K.V.; methodology, I.V. and E.S.; software, I.V., G.L., E.R. and E.M.; validation, I.V., G.L., E.R. and E.M.; formal analysis, I.V., E.S., M.P. and S.P.; investigation, I.V., E.S., M.P. and S.P.; resources, I.V., E.S., M.P. and S.P.; data curation, I.V.; writing—original draft preparation, I.V., E.S., M.P. and S.P.; writing—review and editing, I.V., E.S., K.V., M.P. and S.P.; visualization I.V.; supervision, K.V.; project administration, K.V. and E.S. All authors have read and agreed to the published version of the manuscript.

Funding: This research has been co-financed by the ERDF of EU and Greek national funds through the Operational Program Competitiveness, Entrepreneurship, and Innovation, under the special actions "Aquaculture"—"Industrial Materials"—"Open Innovation in Culture" (project code: T6ΥΒΠ-00220, MIS 5048495).

Data Availability Statement: Not applicable.

Conflicts of Interest: The authors declare no conflict of interest. The funders had no role in the design of the study; in the collection, analyses, or interpretation of data; in the writing of the manuscript, or in the decision to publish the results.

References

1. Dimitrov, R.S. The Paris Agreement on Climate Change: Behind Closed Doors. *Glob. Environ. Politics* **2016**, *16*, 1–11. [CrossRef]
2. European Commission. *The European Green Deal. Communication from the Commission. COM (2019) 640 Final*; European Commission: Brussels, Belgium, 2019.

3. European Parliament and Council (2018b) Directive (EU) 2018/851 of the European Parliament and of the Council of 30 May 2018 Amending Directive 2008/98/EC on Waste. Available online: https://eur-lex.europa.eu/legal-content/EN/TXT/?uri=celex%3A32018L0851 (accessed on 28 March 2023).

4. European Commission. Communication from the Commission to the European Parliament, the Council, the European Economic and Social Committee and the Committee of the Regions: A new Circular Economy Action Plan For a cleaner and more competitive Europe. COM (2020) 98 Final. 2020. Available online: https://eur-lex.europa.eu/legal-content/EN/TXT/HTML/?uri=CELEX:52020DC0098 (accessed on 20 May 2023).

5. Papathanasoglou, A.; Panagiotidou, M.; Valta, K.; Loizidou, M. Institutional Barriers and Opportunities for the Implementation of Industrial Symbiosis in Greece. *Environ. Pract.* **2016**, *18*, 253–259. [CrossRef]

6. Akrivou, C.; Lekawska-Andrinopoulou, L.; Manousiadis, C.; Tsimiklis, G.; Papadaki, S.; Oikonomopoulou, V.; Krokida, M.; Amditis, A. Industrial symbiosis marketplace concept for waste valorization pathways. *E3S Web Conf.* **2022**, *349*, 11005. [CrossRef]

7. Akrivou, C.; Lekawska-Andrinopoulou, L.; Tsimiklis, G.; Amditis, A. Industrial symbiosis platforms for synergy identification and their most important data points: A systematic review. *Open Res. Eur.* **2021**, *1*, 101. [CrossRef]

8. Gaffey, J.; McMahon, H.; Marsh, E.; Vos, J. Switching to Biobased Products—The Brand Owner Perspective. *Ind. Biotechnol.* **2021**, *17*, 109–116. [CrossRef]

9. Toop, T.A.; Ward, S.; Oldfield, T.; Hull, M.; Kirby, M.E.; Theodorou, M.K. AgroCycle—Developing a circular economy in agriculture. *Energy Procedia* **2017**, *123*, 76–80. [CrossRef]

10. Sadh, P.K.; Duhan, S.; Duhan, J.S. Agro-industrial wastes and their utilization using solid state fermentation: A review. *Bioresour. Bioprocess.* **2018**, *5*, 1. [CrossRef]

11. Valta, K.; Damala, P.; Orli, E.; Papadaskalopoulou, C.; Moustakas, K.; Malamis, D.; Loizidou, M. Valorisation Opportunities Related to Wastewater and Animal By-Products Exploitation by the Greek Slaughtering Industry: Current Status and Future Potentials. *Waste Biomass Valorization* **2015**, *6*, 927–945. [CrossRef]

12. Valta, K.; Damala, P.; Panaretou, V.; Orli, E.; Moustakas, K.; Loizidou, M. Review and assessment of waste and wastewater treatment from fruits and vegetables processing industries in Greece. *Waste Biomass Valorization* **2017**, *8*, 1629–1648. [CrossRef]

13. Valta, K.; Aggeli, E.; Papadaskalopoulou, C.; Panaretou, V.; Sotiropoulos, A.; Malamis, D.; Moustakas, K.; Haralambous, K.J. Adding value to olive oil production through waste and wastewater treatment and valorisation: The case of Greece. *Waste Biomass Valorization* **2015**, *6*, 913–925. [CrossRef]

14. Valta, K.; Damala, P.; Angeli, E.; Antonopoulou, G.; Malamis, D.; Haralambous, K.J. Current treatment technologies of cheese whey and wastewater by Greek cheese manufacturing units and potential valorisation opportunities. *Waste Biomass Valorization* **2017**, *8*, 1649–1663. [CrossRef]

15. van Capelleveen, G.; Amrit, C.; Yazan, D.M. A Literature Survey of Information Systems Facilitating the Identification of Industrial Symbiosis. In *From Science to Society*; Otjacques, B., Hitzelberger, P., Naumann, S., Wohlgemuth, V., Eds.; Progress in IS; Springer: Cham, Switzerland, 2017; pp. 155–169.

16. Low, J.S.C.; Tjandra, T.B.; Yunus, F.; Chung, S.Y.; Tan, D.Z.L.; Raabe, B.; Ting, N.Y.; Yeo, Z.; Bressan, S.; Ramakrishna, S.; et al. A Collaboration Platform for Enabling Industrial Symbiosis: Application of the Database Engine for Waste-to-Resource Matching. *Procedia CIRP* **2018**, *69*, 849–854. [CrossRef]

17. Bakogianni, D.; Skourtanioti, E.; Meimaris, D.; Xevgenos, D.; Loizidou, M. Online Brine Platform: A Tool for Enabling Industrial Symbiosis in Saline Wastewater Management Domain. In Proceedings of the 15th International Conference on Distributed Computing in Sensor Systems (DCOSS), Santorini, Greece, 29–31 May 2019.

18. Cutaia, L.; Luciano, A.; Barberio, G.; Sbaffoni, S.; Mancuso, E.; Scagliarino, C.; La Monica, M. The experience of the first industrial symbiosis platform in Italy. *Environ. Eng. Manag. J.* **2015**, *14*, 1521–1533. [CrossRef]

19. Cecelja, F.; Raafat, T.; Trokanas, N.; Innes, S.; Smith, M.; Yang, A.; Zorgios, Y.; Korkofygas, A.; Kokossis, A. e-Symbiosis: Technology-enabled support for Industrial Symbiosis targeting Small and Medium Enterprises and innovation. *J. Clean. Prod.* **2015**, *98*, 336–352. [CrossRef]

20. SAI Platform. Available online: https://saiplatform.org/ (accessed on 20 May 2023).

21. AgriPLaCE Platform. Available online: https://www.agriplace.gr/ (accessed on 20 May 2023).

22. Liu, X.; Heo, J.; Sha, L. Modeling 3-tiered Web applications. In Proceedings of the 13th IEEE International Symposium on Modeling, Analysis, and Simulation of Computer and Telecommunication Systems, Atlanta, GA, USA, 27–29 September 2005.

23. Vue.js—The Progressive JavaScript Framework | Vue.js. Available online: https://vuejs.org/ (accessed on 20 May 2023).

24. Vuelidate | A Vue.js Model Validation Library. Available online: https://vuelidate.js.org/ (accessed on 22 May 2023).

25. Vuetify—A Vue Component Framework. Available online: https://vuetifyjs.com/en/ (accessed on 22 May 2023).

26. Laravel—The PHP Framework for Web Artisans. Available online: https://laravel.com/ (accessed on 22 May 2023).

27. Tang, L. *Building Websites with Laravel and VueJS*; Centria University of Applied Sciences: Kokkola, Finland, 2019; Available online: https://www.theseus.fi/bitstream/handle/10024/261248/linh_tang.pdf?sequence=2&isAllowed=y (accessed on 22 May 2023).

28. PostgreSQL: The World's Most Advanced Open-Source Database. Available online: https://www.postgresql.org/ (accessed on 22 May 2023).

29. Worsley, J.C.; Drake, J.D. *Practical PostgreSQL*, 1st ed.; O'Reilly & Associates, Inc.: Sebastopol, CA, USA, 2001.

30. NGINX: Advanced Load Balancer, Web Server, & Reverse Proxy. Available online: https://www.nginx.com/ (accessed on 22 May 2023).

31. Apache HTTP Server—The Apache Software Foundation! Available online: https://httpd.apache.org/ (accessed on 22 May 2023).

32. Xu, Q.; Awasthi, M.; Malladi, K.; Bhimani, J.; Yang, J.; Annavaram, M.; Hsieh, M. Performance Analysis of Containerized Applications on Local and Remote Storage. In Proceedings of the 33rd International Conference on Massive Storage Systems and Technology (MSST 2017), Santa Clara, CA, USA, 12–15 May 2017.

33. Kaddani, S.; Vanderpooten, D.; Vanpeperstraete, J.-M.; Aissi, H. Weighted sum model with partial preference information: Application to multi-objective optimization. *Eur. J. Oper. Res.* **2017**, *260*, 665–679. [CrossRef]

34. Song, B.; Kang, S. A Method of Assigning Weights Using a Ranking and Nonhierarchy Comparison. *Adv. Decis. Sci.* **2016**, *2016*, 8963214. [CrossRef]

35. Al-Sayed, H.M.A.; Ahmed, A.R. Utilization of watermelon rinds and sharlyn melon peels as a natural source of dietary fiber and antioxidants in cake. *Ann. Agric. Sci.* **2013**, *58*, 83–95. [CrossRef]

36. Jiang, L.; Shang, J.; He, L.; Qian, W. Comparisons of Microwave-Assisted and Conventional Heating Extraction of Pectin from Seed Watermelon Peel. *Adv. Mat. Res.* **2012**, *550–553*, 1801–1806. [CrossRef]

37. Prakash Maran, J.; Sivakumar, V.; Thirugnanasambandham, K.; Sridhar, R. Microwave assisted extraction of pectin from waste Citrullus lanatus fruit rinds. *Carbohydr. Polym.* **2014**, *101*, 786–791. [CrossRef]

38. Persia, M.E.; Parsons, C.M.; Schang, M.; Azcona, J. Nutritional evaluation of dried tomato seeds. *Poult. Sci.* **2003**, *82*, 141–146. [CrossRef]

39. Silva-Beltrán, N.P.; Ruiz-Cruz, S.; Chaidez, C.; Ornelas-Paz, J.d.J.; López-Mata, M.A.; Márquez-Ríos, E.; Estrada, M.I. Chemical constitution and effect of extracts of tomato plants byproducts on the enteric viral surrogates. *Int. J. Environ. Health Res.* **2015**, *25*, 299–311. [CrossRef] [PubMed]

40. Taveira, M.; Ferreres, F.; Oliveira, L.; Valentão, P.; Andrade, P. Fast determination of bioactive compounds from Lycopersicon esculentum Mill. leaves. *Food Chem.* **2012**, *135*, 748–755. [CrossRef] [PubMed]

41. Strati, I.F.; Oreopoulou, V. Recovery of carotenoids from tomato processing by-products—A review. *Food Res. Int.* **2014**, *65*, 311–321. [CrossRef]

42. Khan, A.S.; Singh, Z.; Abbasi, N.; Swinny, E. Pre- or post-harvest applications of putrescine and low temperature storage affect fruit ripening and quality of 'Angelino' plum. *J. Sci. Food Agric.* **2008**, *88*, 1686–1695. [CrossRef]

43. Wang, T.; Li, X.; Zhou, B.; Li, H.; Zeng, J.; Gao, W. Anti-diabetic activity in type 2 diabetic mice and α-glucosidase inhibitory, antioxidant and anti-inflammatory potential of chemically profiled pear peel and pulp extracts (*Pyrus* spp.). *J. Funct. Foods* **2015**, *13*, 276–288. [CrossRef]

44. Dincel Kasapoglu, E.; Kahraman, S.; Tornuk, F. Apricot juice processing byproducts as sources of value-added compounds for food industry. *Eur. Food Sci. Eng.* **2020**, *1*, 18–23.

45. Paraskevopoulou, C.; Vlachos, D.; Bechtsis, D.; Tsolakis, N. An assessment of circular economy interventions in the peach canning industry. *Int. J. Prod. Econ.* **2022**, *249*, 108533. [CrossRef]

46. Wu, H.; Shi, J.; Xue, S.; Kakuda, Y.; Wang, D.; Jiang, Y.; Ye, X.; Li, Y.; Subramanian, J. Essential oil extracted from peach (*Prunus persica*) kernel and its physicochemical and antioxidant properties. *LWT Food Sci. Technol.* **2011**, *44*, 2032–2039. [CrossRef]

47. Nowicka, P.; Wojdyło, A. Content of bioactive compounds in the peach kernels and their antioxidant, anti-hyperglycemic, anti-aging properties. *Eur. Food Res. Technol.* **2019**, *245*, 1123–1136. [CrossRef]

48. Doğantürk, M.; Seçilmiş Canbay, H. Oil Ratio and Fatty Acid Composition of Cherry Seed Oil. *Turk. J. Health Sci. Life* **2019**, *2*, 21–24.

49. Poguberović, S.S.; Krčmar, D.M.; Maletić, S.P.; Kónya, Z.; Pilipović, D.D.T.; Kerkez, D.V.; Rončević, S.D. Removal of As(III) and Cr(VI) from aqueous solutions using "green" zero-valent iron nanoparticles produced by oak, mulberry and cherry leaf extracts. *Ecol. Eng.* **2016**, *90*, 42–49. [CrossRef]

50. Yukui, R.; Wenya, W.; Rashid, F.; Qing, L. Fatty Acids Composition of Apple and Pear Seed Oils. *Int. J. Food Prop.* **2009**, *12*, 774–779. [CrossRef]

51. Barreira, J.C.M.; Arraibi, A.A.; Ferreira, I.C.F.R. Bioactive and functional compounds in apple pomace from juice and cider manufacturing: Potential use in dermal formulations. *Trends Food Sci. Technol.* **2019**, *90*, 76–87. [CrossRef]

52. Bonarska-Kujawa, D.; Cyboran, S.; Oszmiański, J.; Kleszczyńska, H. Extracts from apple leaves and fruits as effective antioxidants. *J. Med. Plant Res.* **2011**, *5*, 2339–2347.

53. Guneser, B.A.; Yilmaz, E. Bioactives, Aromatics and Sensory Properties of Cold-Pressed and Hexane-Extracted Lemon (*Citrus limon* L.) Seed Oils. *J. Am. Oil Chem. Soc.* **2017**, *94*, 723–731. [CrossRef]

54. Burnett, C.L.; Bergfeld, W.F.; Belsito, D.V.; Hill, R.A.; Klaassen, C.D.; Liebler, D.C.; Marks, J.G.; Shank, R.C.; Slaga, T.J.; Snyder, P.W.; et al. Safety Assessment of Citrus Fruit-Derived Ingredients as Used in Cosmetics. *Int. J. Toxicol.* **2021**, *40*, 5S–38S. [CrossRef]

55. Giwa, S.; Muhammad, M.; Giwa, A. Utilizing orange peels for essential oil production. *J. Eng. Appl. Sci.* **2018**, *13*, 17–27.

56. Willner, I.; Baron, R.; Willner, B. Growing Metal Nanoparticles by Enzymes. *Adv. Mater.* **2006**, *18*, 1109–1120. [CrossRef]

57. Shankar, S.S.; Rai, A.; Ankamwar, B.; Singh, A.; Ahmad, A.; Sastry, M. Biological synthesis of triangular gold nanoprisms. *Nat. Mater.* **2004**, *3*, 482–488. [CrossRef]

58. Basavegowda, N.; Rok Lee, Y. Synthesis of silver nanoparticles using Satsuma mandarin (*Citrus unshiu*) peel extract: A novel approach towards waste utilization. *Mater. Lett.* **2013**, *109*, 31–33. [CrossRef]

59. Roselló-Soto, E.; Barba, F.J.; Parniakov, O.; Galanakis, C.M.; Lebovka, N.; Grimi, N.; Vorobiev, E. High Voltage Electrical Discharges, Pulsed Electric Field, and Ultrasound Assisted Extraction of Protein and Phenolic Compounds from Olive Kernel. *Food Bioproc. Tech.* **2015**, *8*, 885–894. [CrossRef]

60. Difonzo, G.; Squeo, G.; Pasqualone, A.; Summo, C.; Paradiso, V.M.; Caponio, F. The challenge of exploiting polyphenols from olive leaves: Addition to foods to improve their shelf-life and nutritional value. *J. Sci. Food Agric.* **2021**, *101*, 3099–3116. [CrossRef] [PubMed]

61. Perumalla, A.; Hettiarachchy, N. Green tea and grape seed extracts—Potential applications in food safety and quality. *Food Res. Int.* **2011**, *44*, 827–839. [CrossRef]

62. Ferhi, S.; Santaniello, S.; Zerizer, S.; Cruciani, S.; Fadda, A.; Sanna, D.; Dore, A.; Maioli, M.; D'hallewin, G. Total Phenols from Grape Leaves Counteract Cell Proliferation and Modulate Apoptosis-Related Gene Expression in MCF-7 and HepG2 Human Cancer Cell Lines. *Molecules* **2019**, *24*, 612. [CrossRef]

63. Pinela, J.; Carvalho, A.M.; Ferreira, I.C.F.R. Wild edible plants: Nutritional and toxicological characteristics, retrieval strategies and importance for today's society. *Food Chem. Toxicol.* **2017**, *110*, 165–188. [CrossRef] [PubMed]

64. Chiellini, E.; Cinelli, P.; Chiellini, F.; Imam, S.H. Environmentally Degradable Bio-Based Polymeric Blends and Composites. *Macromol. Biosci.* **2004**, *4*, 218–231. [CrossRef]

65. Liang, S.; McDonald, A.G. Chemical and Thermal Characterization of Potato Peel Waste and Its Fermentation Residue as Potential Resources for Biofuel and Bioproducts Production. *J. Agric. Food Chem.* **2014**, *62*, 8421–8429. [CrossRef]

66. Mitrus, M.; Moscicki, L. Physical properties of thermoplastic starches. *Int. Agrophysics* **2009**, *23*, 305–308.

67. Wang, K.; Yin, J.; Shen, D.; Li, N. Anaerobic digestion of food waste for volatile fatty acids (VFAs) production with different types of inoculum: Effect of pH. *Bioresour. Technol.* **2014**, *161*, 395–401. [CrossRef]

68. Bhatia, S.K.; Yang, Y.H. Microbial production of volatile fatty acids: Current status and future perspectives. *Rev. Environ. Sci.* **2017**, *16*, 327–345. [CrossRef]

69. Agler, M.T.; Wrenn, B.A.; Zinder, S.H.; Angenent, L.T. Waste to bioproduct conversion with undefined mixed cultures: The carboxylate platform. *Trends Biotechnol.* **2011**, *29*, 70–78. [CrossRef] [PubMed]

70. Baumann, I.; Westermann, P. Microbial Production of Short Chain Fatty Acids from Lignocellulosic Biomass: Current Processes and Market. *Biomed Res. Int.* **2016**, *2016*, 8469357. [CrossRef] [PubMed]

71. Zacharof, M.-P.; Lovitt, R.W. Complex Effluent Streams as a Potential Source of Volatile Fatty Acids. *Waste Biomass Valorization* **2013**, *4*, 557–581. [CrossRef]

72. Bioplastics Market Data. European Bioplastics. Available online: https://www.european-bioplastics.org/market/ (accessed on 20 May 2023).

73. Insect Protein Market by Insect Type, Form & Region for 2021–2029 | Global Sales Analysis and Opportunity—2029 | FMI future-marketinsights.com), Insect Protein Market Size, Share & Covid analysis. Available online: https://www.fortunebusinessinsights.com/industry-reports/insect-based-protein-market-100780 (accessed on 21 May 2023).

74. International Platform of Insects for Food and Feed—IPIFF, Insects as Feed EU Legislation—Aquaculture, Poultry & Pig Species. Available online: https://ipiff.org/insects-eu-legislation/ (accessed on 21 May 2023).

75. Białkowska, A. Strategies for efficient and economical 2,3-butanediol production: New trends in this field. *World J. Microbiol. Biotechnol.* **2016**, *32*, 200. [CrossRef]

76. Markou, G.; Georgakakis, D. Cultivation of filamentous cyanobacteria (blue-green algae) in agro-industrial wastes and wastewaters: A review. *Appl. Energy* **2011**, *88*, 3389–3401. [CrossRef]

77. Lundquist, T.; Woertz, I.; Quinn, N.; Benemann, J. A Realistic Technology and Engineering Assessment of Algae Biofuel Production. *Energy* **2010**, *1*.

78. Sirakov, I.; Velichkova, K.; Stoyanova, S.; Staykov, Y. The importance of microalgae for aquaculture industry. Review. *Int. J. Fish. Aquat. Sci.* **2015**, *2*, 81–84.

79. Brown, M.R. Nutritional value of microalgae for aquaculture. Advances in Aquaculture nutrition. In Proceedings of the VI International Symposium on Nutrition Aquaculture, Cancun, Quintana Roo, Mexico, 3–6 September 2002.

80. European Biochar Certificate—Guidelines for a Sustainable Production of Biochar EBC (2012–2022). European Biochar Foundation (EBC), Arbaz, Switzerland. Version 10.1 from 10 January 2022. Available online: http://european-biochar.org (accessed on 20 May 2023).

81. Europe Biochar Market Trends, Size, Share, Analysis, Forecast 2022–2030. Available online: https://inkwoodresearch.com/reports/europe-biochar-market/ (accessed on 20 May 2023).

waste

Article

Twin-Screw Extrusion Mechanical Pretreatment for Enhancing Biomethane Production from Agro-Industrial, Agricultural and Catch Crop Biomasses

Arthur Chevalier [1,2], Philippe Evon [1], Florian Monlau [3], Virginie Vandenbossche [1] and Cecilia Sambusiti [2,*]

1 Laboratoire de Chimie Agro-Industrielle (LCA), Université de Toulouse, INRAE, Toulouse INP, 31030 Toulouse, France
2 TotalEnergies, CSTJF, Avenue Larribau, 64018 Pau, France
3 TotalEnergies, PERL—Pôle D'Etudes et de Recherche de Lacq, Pôle Economique 2, BP 47–RD 817, 64170 Lacq, France
* Correspondence: cecilia.sambusiti@totalenergies.com

Abstract: This study aimed to evaluate the effects of mechanical treatment through twin-screw extrusion for the enhancement of biomethane production. Four lignocellulosic biomasses (i.e., sweetcorn by-products, whole triticale, corn stover and wheat straw) were evaluated, and two different shear stress screw profiles were tested. Chemical composition, particle size reduction, tapped density and cellulose crystallinity were assessed to show the effect of extrusion pretreatment on substrate physico-chemical properties and their biochemical methane production (BMP) capacities. Both mechanical pretreatments allowed an increase in the proportion of particles with a diameter size less than 1 mm (from 3.7% to 72.7%). The most restrictive profile also allowed a significant solubilization of water soluble coumpounds, from 5.5% to 13%. This high-shear extrusion also revealed a reduction in cellulose crystallinity for corn stover (i.e., 8.6% reduction). Sweetcorn by-products revealed the highest BMP values (338–345 NmL/gVS), followed by corn stover (264–286 NmL/gVS), wheat straw (247–270 NmL/gVS) and whole triticale (233–247 NmL/gVS). However, no statistical improvement in maximal BMP production was provided by twin-screw extrusion. Nevertheless, BMP kinetic analysis proved that both extrusion pretreatments were able to increase the specific rate constant (from 13% to 56% for soft extrusion and from 66% to 107% for the high-shear one).

Keywords: anaerobic digestion; kinetics; lignocellulosic biomass; mechanical pretreatment; methane potential; twin-screw extrusion

Citation: Chevalier, A.; Evon, P.; Monlau, F.; Vandenbossche, V.; Sambusiti, C. Twin-Screw Extrusion Mechanical Pretreatment for Enhancing Biomethane Production from Agro-Industrial, Agricultural and Catch Crop Biomasses. *Waste* **2023**, *1*, 497–514. https://doi.org/10.3390/waste1020030

Academic Editors: Catherine N. Mulligan, Dimitris P. Makris and Vassilis Athanasiadis

Received: 17 March 2023
Revised: 4 May 2023
Accepted: 12 May 2023
Published: 22 May 2023

1. Introduction

The increase in the world's population correlated with the increase in energy demand has raised concerns about the problem of the supply of fossil resources such as oil, gas or coal, and their harmful effects, contributing strongly to global warming. Renewable and alternative energy sources may be able to solve these problems, but they require major investments and innovative technologies to do so [1].

One of the most promising and profitable biotechnologies for replacing fossil energy with renewable energy, such as bioethanol or biodiesel production, is anaerobic digestion (AD) as methane produced by the bioconversion of numerous sources of organic materials can be used to generate heat, electricity and fuel, while the digestate generated in the process may be disposed of as an organic amendment for agricultural soils [2]. AD is a biological process characterized by four successive metabolic pathways, i.e., hydrolysis, acidogenesis, acetogenesis and methanogenesis, and involves the gradual conversion of high molecular weight compounds such as carbohydrates, proteins and lipids into biomethane [3].

Lignocellulosic biomass represents one of the most attractive sources of organic matter for biogas production since it is widely available through by-products or wastes generated by activities such as agriculture or agro-industry [4]. Lignocellulose consists of three main biopolymers, which are associated with each other to form a complex network. Cellulose, the most abundant natural polymer, is a polysaccharidic homopolymer made of glucose units linked by β-1,4 glycosidic bonds. Hemicelluloses are also polysaccharides, but these heteropolymers are made from different monomers such a pentoses (e.g., xylose and arabinose), hexoses (e.g., mannose, glucose and galactose) and uronic acids. Finally, lignin is, unlike the previous biopolymers, an amorphous polyphenolic heteropolymer composed of three different monolignols (i.e., coumaryl, coniferyl and sinapyl alcohols). [5].

Among lignocellulosic biomasses, agro-industrial or agricultural by-products are key carbon resources available in vast quantities since the four most farmed plants in the world, namely sugar cane, corn, rice and wheat, already generate more than 2.55 billion tons of plant fiber annually [6]. Catch crops such as grass, clover, hemp or triticale also represent an appealing answer to the competitive use of agricultural land, and they can be efficiently valorized through biogas production [7].

However, as the complex structure and physico-chemical properties of cell walls inside lignocellulosic substrates constitute a natural barrier against pests such as insects or microorganisms, they also limit the hydrolysis stage of AD, and therefore the substrate's natural digestibility, which is the main limiting factor for their utilization [8].

Therefore, to break this complex structure of lignocellulose and access the monomeric sugars forming cellulose and hemicelluloses, which are consumed by microorganisms during AD, a pretreatment step is essential [9]. Biological, mechanical, physical and/or chemical pretreatments are the main pretreatments that can be applied [10–12]. For the mechanical pretreatments, such as grinding, milling or shredding, the main issue is their energy requirement, especially when coupled with thermal pretreatment [13]. However, these are able to efficiently disrupt the cell wall structure, thus improving the accessible contact surface area between anaerobic microorganisms and the substrate. Physical pretreatments, mainly ultrasound or microwave, or physico-chemical pretreatments such as steam explosion, are also able to disrupt the cell wall structure, and they are more cost-effective than mechanical pretreatments. However, the former lack their industrial maturity, whereas the latter may generate toxic compounds [10,14,15]. Chemical pretreatments, mainly alkaline and acid pretreatments, while often very efficient in increasing biogas production through the removal of either hemicelluloses or lignin, have the main disadvantages of being pretty expensive and not very environmentally friendly, and they can also generate AD inhibitors in the process such as furans and phenols compounds [16]. On the other hand, biological pretreatments, which mainly use enzymes or fungi, are environmentally friendly, and do not generate inhibitors, but they are expensive and constitute pretty slow processes [17].

Twin-screw extrusion is a continuous mechanical process allowing biomass disruption with strong mixing and shearing forces generated at least by the intermeshing of the two rotating screws [18] and, in most cases, by the use of specific shearing screw elements along the screw profile [19]. Twin-screw extrusion is considered a promising process to simultaneously apply mechanical and thermal pretreatments for lignocellulosic substrates, explaining why it was investigated since the 1990s [20]. This technology presents key parameters that can be set according to the required purposes, i.e., configuration parameters such as screw elements and module types, and operational parameters such as screw rotation speed, temperature profile, solid-to-liquid ratio and feeding rates [21]. Twin-screw extrusion processes can be implemented from ambient to hot temperatures, and they can adapt to many types of biomasses. Twin-screw extrusion also has the possibility of combining mechanical pretreatment with other categories of pretreatment, either thanks to the addition of chemicals in the case of reactive extrusion [22] or enzymes in the case of bioextrusion [23,24]. Finally, twin-screw extrusion is an easy scalable technology with excellent repeatability results when transferred from a laboratory scale to pilot and industrial ones [25,26].

Over the past decade, several studies have especially focused their efforts on assessing the effect of twin-screw extrusion pretreatment for AD. Several biomasses including rice, wheat or corn straw, grass, sprout stem, vine shoot or miscanthus were proven to generate better methane yields (from a 16% to 72% increase) after mechanical twin-screw extrusion pretreatment [27–30]. Triticale, when harvested fresh at the milky stage as a catch crop, is another biomass of interest for AD as it presents a lower lignin content than other cereal straws harvested at the fully ripe stage [31] and, to the best of our knowledge, has not yet been pretreated through twin-screw extrusion. Furthermore, for all the references listed above, a particle size reduction up to a 2–40 mm range has been conducted beforehand, which may minimize the real efficiency of mechanical pretreatment through twin-screw extrusion. Lastly, a 60–100 °C temperature range has also been set inside the extruder barrel, which could add thermal pretreatment in addition to the mechanical one.

The aim of this study was to assess the effect of twin-screw extrusion mechanical pretreatment on biomethane production from different agro-industrial, agricultural and catch crop biomasses. Two different screw profiles were tested on four biomasses (i.e., sweetcorn by-product, triticale, corn stover, wheat straw), and twin-screw extrusion treatment was conducted at ambient temperature while avoiding particle size reduction before extrusion. Then, the influence of twin-screw extrusion pretreatment on their physico-chemical properties and their performances in AD were evaluated.

2. Materials and Methods

2.1. Feedstocks and Inoculum

Four feedstocks were selected among agro-industrial wastes, agricultural by-products or energy catch crops. These were chosen from among lignocellulosic biomasses available in the south-west part of France, and in sufficient quantities for potential large-scale exploitation through biogas plants. Sweetcorn by-product (i.e., husk and cob) recovered from Soleal–Bonduelle factory (Bordères-et-Lamensans, France) is referred to as "SB". Whole fresh triticale harvested in a plot located in the city of Pavie (France) is referred to as "WT". Corn stover harvested in the city of Cescau (France) is referred to as "CS". Lastly, wheat straw, also harvested in the city of Cescau, is referred to as "WS". All four biomasses were comb milled using a hammer mill (Goulu N, Electra, Poudenas, France). Then, the only WS was also processed through a crushing mill (Bro140, Electra, Poudenas, France) equipped with a 15 mm grid. Comb or crush milled samples before extrusion are referred to as "CM". WT was then dried at 50 °C during 24 h for conservation purposes, and then rehydrated at initial humidity (i.e., 69% moisture content) by mixing it with water in a concrete mixer before extrusion.

Anaerobic digester inoculum was recovered from an industrial biogas plant from TotalEnergies located in the city of Bénesse-Maremne (France). Inoculum was maintained in a 5 L glass bioreactor at 38 °C under stirring and anaerobic conditions before performing Biochemical Methane Potential (BMP) tests. Main inoculum parameters were 3.9 ± 0.1 TS, 2.7 ± 0.1 VS, 7.8 ± 0.0 pH, 1.5 ± 0.0 gCH$_3$COOH/L, 4.1 ± 1.4 gN-NH4$^+$/L and 0.17 ± 0.0 FOS/TAC ratio.

2.2. Extrusion Pretreatment

A BC 45 twin-screw extruder (Clextral, Firminy, France) was used for extrusion pretreatment (Figure 1). The machine is made of 7 consecutive modules, each 200 mm in length, and its screw elements have a diameter of 55 mm and lengths of either 50 or 100 mm. Three different screw profiles were investigated (Figure 2). The first screw profile (A) is referred to as soft extrusion (SE), and it was tested on all four biomasses. It consists of trapezoidal double-thread screw elements (T2F) on module 1, followed by conveying double-thread screw elements (C2F) on modules 2 to 6, and then 5 kneading (i.e., bilobal) elements (BL22) with a −45° angle at the early beginning of module 7, followed by C2F elements up to the extruder outlet. The second screw profile (B) is referred to as high-shear extrusion (HE), and it was only used on SB and WT. It also starts with T2F elements on module 1, followed by C2F elements on modules 2 to 4, then conveying single-thread screw elements (C1F) in modules 5 and 6, and lastly reverse single-thread screw elements with grooves (CF1C) and C1F elements on module 7. For this screw profile, a filter section was positioned at the level of module 6 to allow the collection of a liquid separately from the solid and thus avoid waterlogging the machine over time. It consisted of six semi-cylindrical grids with eight per square centimeter conical holes with 1 mm inside diameter and 2 mm outside diameter. The third and last screw profile (C), referred to as high-shear extrusion with rehydration (HER), is actually only a variation of the second one. It was needed as CS and WS required rehydration at the moment of high-shear extrusion to prevent the machine from blocking. It only differs from profile B by the addition of 5 BL22 elements with a +45° angle on module 4 to favor intimate mixing between the solid and water. The latter was injected thanks to a piston pump at the beginning of module 3. HE and HER screw profiles are expected to generate the same mechanical pretreatment as the intense mechanical shear zone in module 7, made of CF1C elements, and are perfectly identical.

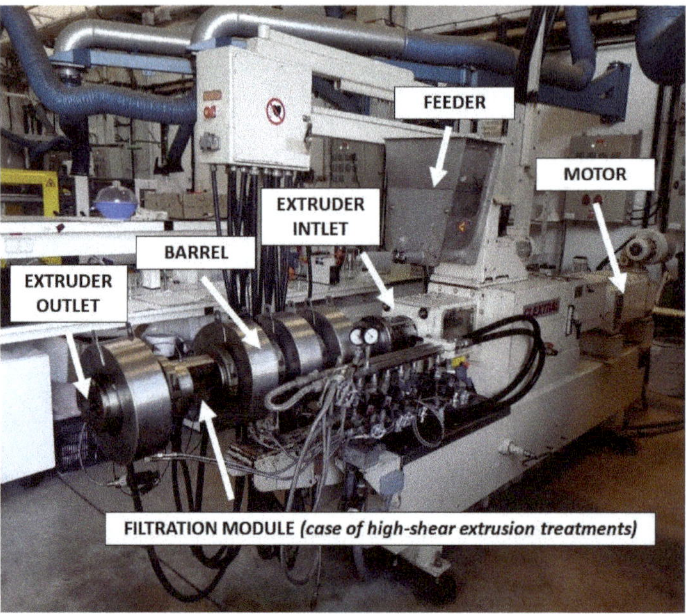

Figure 1. Photograph of the Clextral BC 45 twin-screw extruder that has been used in this study.

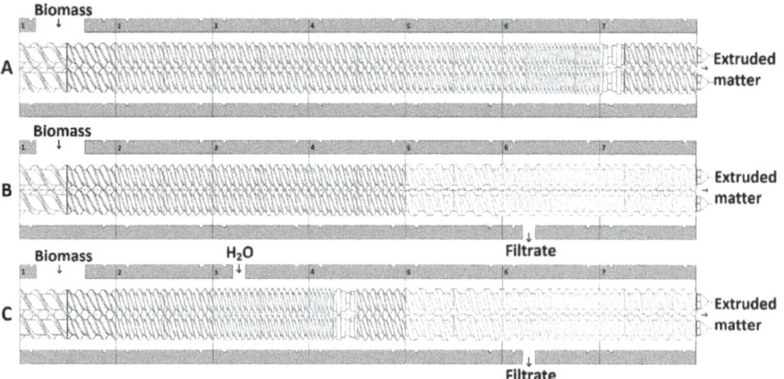

Figure 2. Schematic representation of the three screw profiles tested in the Clextral BC 45 twin-screw machine. (**A**) Soft extrusion, (**B**) high-shear extrusion and (**C**) high-shear extrusion with rehydration.

During extrusion tests, neither heating nor temperature regulation were applied but material temperature was measured at the level of the stress zone (i.e., module 7). Screw rotation speed was set at 60 rpm. Biomass supply was fulfilled either manually in the case of SB, WT, CS and WS for SE, and in the case of SB and WT for HE, or with a volumetric twin-screw feeder in the specific case of CS and WS for HER. CS and WS were slightly rehydrated by mixing them with water in a concrete mixer up to a 70% total solid content (TS) and 75% TS, respectively, before SE as they were too dry to be extruded as such. Extrudate samples and filtrate ones for the HE and HER configurations were collected after reaching the extruder stability. For each test, three samplings were conducted during 4 min, and this enabled the calculation of mean outlet flow rates. The latter were then used through a material balance to calculate the raw material inlet flow rate, based on the dry matter contents of raw and extruded solid samples plus that of the filtrate for HE and HER.

2.3. Biochemical Methane Potential (BMP)

Extruded samples were dried at 40 °C for 24 h before BMP. Batch BMP values were assessed in 569 mL glass flask with a working volume 269 mL at 37 °C with a stirring speed of 75 RPM for a duration of 32 days corresponding to time needed to reach a daily biomethane production <1%. An 18 mL macro element solution (Na_2HPO_4 22.4 g/L; NH_4Cl 10.6 g/L; KH_2PO_4 5.4 g/L; $MgCl_2$ 2 g/L; $CaCl_2$ 1.5 g/L; $FeCl_2$ 0.4 g/L), 0.3 mL oligoelement solution ($CoCl_2$ 0.1 g/L; $MnCl_2$ 0.05 g/L: $NiCl_2$ 0.01 g/L; $ZnCl_2$ 0.005 g/L; H_3BO_3 0.005 g/L; Na_2SeO_3 0.005 g/L; $CuCl_2$ 0.003 g/L; $NaMoO_4$ 0.001 g/L) and 15 mL bicarbonate buffer solution ($NaHCO_3$ 50 g/L) were added the medium to ensure good bacterial development. Carboxymethylcellulose was used as positive control and sole inoculum as negative control (BMP produced was then subtracted from results). Extrudate and filtrate samples from HE/HER were reassembled as one sample keeping the same proportions as at the extrusion outlet. A 0.5 gVS substrate/gVS inoculum ratio was set, and 7–8.5 pH range check were realized to avoid initial medium acidification. Flasks were flushed with nitrogen at the beginning of the trials to ensure anaerobic condition, then were sealed with impermeable red butyl rubber septum-type stoppers. BMP tests were performed in duplicate.

Biogas analyses were performed by using an Agilent 990 Micro GC with two columns: one at 80 °C and 200 kPa with argon as carrier phase for H_2, O_2, N_2 and CO_2, one at 60 °C and 150 kPa with helium as carrier phase CH_4 and one for CO_2 and H_2S. Injector temperature was 80 °C. Biogas production was calculated from pressure increase measured with a manometer.

Elemental analyses ($C_aH_bN_cS_dO_e$) of dried samples were performed to calculate their maximal theorical BMP (BMP_{th}) using Buswell equation [32,33]:

$$BMP_{th}\left[\frac{NmL}{gVS}\right] = \frac{4a + b - 2c - 3d - 3e}{12a + 2 + 16c + 14d + 32e} \times 1000 \tag{1}$$

Khongchamnan et al.'s [34] elemental analysis of lignin was also used to calculate lignin maximal theorical BMP (621 $NmLCH_4/gVS$), which was then used to determine samples' adjusted theorical maximal BMP ($BMP_{th\ adjusted}$) using the following formula:

$$BMP_{th\ adjusted}\left[\frac{NmL}{gVS}\right] = BMP_{th\ substrate} - BMP_{th\ lignin} \times \%lignin_{substrate} \tag{2}$$

The biodegradability index (BI) was finally calculated using the following equation:

$$BI\ (\%) = \frac{BMP_{experiemental}}{BMP_{th\ adjusted}} \tag{3}$$

2.4. Analytical Methods

2.4.1. Sample Preparation for Analysis

Parts of raw and extruded samples were dried at 105 °C until constant weight for total solid content (TS) determination, and they were then mineralized at 550 °C during 8 h for ash and volatile solid content (VS) determination according to the National Renewable Energy Laboratory (NREL) procedures [35]. The rest of raw and extruded samples were dried at 40 °C during 24 h for further characterizations. Parts of dried samples were used as such for granulometry and tapped density measurements, and the remaining dried samples were milled with a 1 mm grid on a microfine grinder drive (MF 10 basic, IKA Werke, Staufen im Breisgau, Germany). Two mechanical sieves (aperture sizes of 0.8 and 0.18 mm, respectively) and a bottom plate were used on a vibratory sieve shaker (AS 200, Retsch, Hann, Germany) during 10 min at a 3 mm amplitude to recover sample fractions depending on their particle sizes. Sample fractions between 0.8 and 0.18 mm were used for lignocellulosic composition analysis, whereas fractions under 0.18 mm were used for cellulose crystallinity assessment.

2.4.2. Fiber Composition

Fiber composition of lignocellulosic samples was determined in triplicate according to an adapted protocol [36] from the National Renewable Energy Laboratory (NREL) procedures [37]. Water and then 96% (v/v) ethanol extractions were performed on an extraction system (Fibertec FT 122, Foss, Hillerød, Denmark) using 1 g of dry sample and 100 mL of boiled solvent at 100 °C and atmospheric pressure for 1 h. Cellulose, hemicelluloses and lignin contents were assessed with a two-step hydrolysis using a 72% (w/w) sulphuric acid at 30 °C for 1 h and then a 4% (w/w) solution after dilution during 1 h at 121 °C, followed by filtration. Acid-insoluble lignin in residues was determined by weight loss after calcination during 8 h at 450 °C. Part of the extracts was then used for acid-soluble lignin on UV spectrophotometer (UV-1800, Shimadzu, Kyoto, Japan) at 320 nm for CS samples and at 240 nm for the other ones, and the rest of the extracts were neutralized with calcium carbonate until reaching pH 5–7 and then filtered using a 0.2 μm cellulose acetate filter. Analysis of sugar monomers (i.e., arabinose, glucose, galactose, xylose and mannose) was performed on a Dionex (Sunnyvale, CA, USA) ICS-3000 type ion chromatography HPLIC system with a pumping device, an auto-injector, an electrochemical detector with a gold electrode and an Ag/AgCl reference electrode. A pre-column (4 × 50 mm, Dionex) connected to a Carbopac PA1 column (4 × 250 nm, Dionex) was used for the stationary phase with a 1 mM sodium hydroxide solution as an eluent. A total of 25 μL of samples were injected automatically with separation of sugars carried out at a flow rate of 1 mL/min at 25 °C. A range of standards was made from 1 to 100 mg/L to undertake external calibration for the quantification of sugar monomers. Chromeleon analysis was conducted with the 6.8 version of Dionex processing software.

2.4.3. Granulometry and Tapped Density

Seven mechanical sieves (aperture sizes of 4.0, 2.0, 1.0, 0.8, 0.5, 0.25 and 0.125 mm) and a bottom plate were used to measure, in triplicate, the particle size distribution using the Retsch AS 200 vibratory sieve shaker during 10 min at a 3 mm amplitude.

A bulk density tapping instrument (Densi-Tap, Ma.Tec, Novara, Italy), modified to support a 1000 mL graduated cylinder, with a cam shaft speed of 250 rpm and a stroke travel of 3.2 mm, was used to determine tapped density. A total of 1000 taps were repeated until the tapped volume did not change between two consecutive cycles. The final tapped volume was read on the graduated cylinder. All determinations were conducted in triplicate.

2.4.4. Cellulose Crystallinity

Cellulose crystallinity was determined using an X-ray diffraction (XRD) instrument (D8 Advance, Brucker, San Jose, CA, USA) with a 0.154 nm wavelength, Cu/Kα radiation at 40 kV and 40 mA tube current. Samples were implemented with a speed of $1°/min$, in a range of 2θ varying from $6°$ to $30°$, and a step size of $0.0303°$ at room temperature.

The crystallinity index (Cr) was determined using the following equation [38]:

$$Cr = \frac{I_{002} - I_{amorphous}}{I_{002}} \times 100 \tag{4}$$

where:

I_{002} is the intensity of the crystalline portion of the biomass (cellulose) at $2\theta = 22°$;
$I_{amorphous}$ is the peak of the amorphous portion at $2\theta = 16°$.
Analyses were conducted in triplicate.

2.5. Data Analyses

Kinetic study of the biomethane production was assessed by applying a model based on the modified Gompertz equation [39]:

$$B = BMP_\infty \exp\left\{ -\exp\left[\frac{R_m \times e}{BMP_\infty} \times (\lambda - t) + 1\right]\right\} \tag{5}$$

where:
B: cumulative biomethane production (NmL/gVS);
BMP_∞: maximal biomethane production (NmL/gVS);
R_m: specific biomethane production rate (NmL/gVS.day);
λ: lag phase time (day).

Kinetic parameters were determined by minimizing the sum of the least squares between the observed and predicted values.

2.6. Statistical Analyses

For statistical analyses made on chemical composition, tapped density, cellulose crystallinity and experimental BMP, Student's tests were conducted with statistical significance level of $p < 0.05$ on the Microsoft Office Excel software (Microsoft, Albuquerque, NM, USA). Data are expressed as means $\pm$ standard deviations (s.d.).

3. Results and Discussion

3.1. Description of the Twin-Screw Extrusion Pretreatments

Usually, when working on laboratory scale extruders, biomasses are first reduced to smaller particle sizes (within a 1 mm to 2 cm range) before extrusion to enable stable device feeding [25,40,41]. In this study, in order to be as close as possible to pilot or industrial reality, biomasses were extruded fresh and they were just comb milled to avoid any influence of additional mechanical pretreatment before extrusion [4]. Measured parameters during samplings and material balances are given in Table 1.

Dry inlet flow rates were intended to be set to 5 kg/h. However, from a practical point of view, it was not possible to achieve precisely this 5 kg/h dry inlet flow rate as substrates were manually fed into the machine for the SE and HE configurations. In the case of SE, dry inlet flow rates were 5.3 kg/h, 5.6 kg/h and 5.1 kg/h, respectively, for SB, WT and WS. For CS, it was only 3.7 kg/h due to there being coarser and less flexible solid particles in the starting material. When using the HE configuration, trials conducted from SB and WT revealed rather different inlet flow rates, i.e., 7.2 kg/h and 3.7 kg/h, respectively. Regarding HER configuration, which was applied to CS and WS only, rehydration should have been set to a 75/25 water-to-dry matter ratio. However, this could not be performed as such because when this rehydration rate was applied, it generated non-homogeneous rehydration in the extruder and a too low moisture content, resulting in the blocking of the machine. CS and WS were instead rehydrated to higher water-to-dry matter ratios, which were respectively 85/15 and 82/18.

Table 1. Measured parameters and material balances of the different twin-screw extrusion pretreatments.

Raw Material		SB		WT		CS		WS	
	Screw Profile	SE	HE	SE	HE	SE	HER	SE	HER
Measured parameters	Material temperature range in module 7 (°C)	18–22	31–36	26–27	30–34	32–34	28–42	27–32	58–63
	Motor current range (A)	14–15	16–25	27–35	41–47	11–19	27–37	13–21	27–30
Substrate	Dry matter content (%)	27.8 ± 0.2	28.2 ± 1.0	21.1 ± 0.2	21.1 ± 0.2	69.3 ± 1.0	89.6 ± 0.1	77.0 ± 1.1	88.5 ± 0.1
	Inlet flow rate (kg/h)	19.2 ± 3.1	26.0 ± 3.3	26.5 ± 3.8	17.5 ± 2.8	5.3 ± 0.5	1.0 ± 0.1	6.6 ± 0.3	2.0 ± 0.1
	Dry inlet flow rate (kg/h)	5.3 ± 0.9	7.2 ± 0.9	5.6 ± 0.8	3.7 ± 0.6	3.6 ± 0.3	0.9 ± 0.1	5.1 ± 0.2	1.8 ± 0.1
Liquid	Water inlet flow rate (kg/h)	-	-	-	-	-	5.0	-	8.5
	Water-to-dry matter ratio	-	-	-	-	-	85/15	-	82/18
Extrudate	Dry matter content (%)	30.6 ± 0.3	43.6 ± 1.2	23.6 ± 0.3	63.4 ± 0.3	75.3 ± 0.6	82.3 ± 0.1	81.6 ± 0.4	76.3 ± 0.3
	Outlet flow rate (kg/h)	17.4 ± 2.8	13.4 ± 1.8	23.6 ± 3.2	4.9 ± 0.2	4.9 ± 0.4	1.0 ± 0.1	6.3 ± 0.3	2.2 ± 0.1
	Dry outlet flow rate (kg/h)	5.3 ± 0.9	5.8 ± 0.8	5.6 ± 0.8	3.1 ± 0.1	3.6 ± 0.3	0.8 ± 0.1	5.1 ± 0.2	1.7 ± 0.1
Filtrate	Dry matter content (%)	-	12.7 ± 0.8	-	7.3 ± 0.0	-	1.9 ± 0.0	-	1.9 ± 0.0
	Outlet flow rate (kg/h)	-	11.0 ± 2.5	-	8.1 ± 0.6	-	4.7 ± 0.2	-	7.6 ± 0.3
	Dry outlet flow rate (kg/h)	-	1.4 ± 0.3	-	0.6 ± 0.0	-	0.1 ± 0.0	-	0.1 ± 0.0

As no temperature regulation was applied during the extrusion pretreatment, the temperature in the stress zone (i.e., module 7) was probably too low to allow good rehydration according to the literature [42,43]. However, this solution has the advantage of being less energy-consuming and therefore less expensive. Extrudate-to-filtrate ratios expressed in terms of dry mass for the HE/HER configurations were 80/20 for SB, 84/16 for WT, 90/10 for CS and 92/8 for WS.

The decrease in the proportion of dry matter in the filtrate in the case of the HER configuration in need of rehydration at the moment of twin-screw extrusion seems to correlate with the hypothesis of superficial rehydration. SB, WT, CS and WS initial dry matter contents increased by 2.8%, 2.6%, 6.0% and 4.6%, respectively, using the SE configuration, and they increased much more, i.e., by 15.4%, 42.3%, 67.0% and 59.0%, respectively, using the HE/HER ones. It has to be noted here that for the HER configuration, water injected in the extruder at the moment of the pretreatment was taken into account for calculating the above-mentioned increases in dry matter content for the CS and WS solid materials.

The high compression action of the CF1C reverse screw elements used in the HE and HER screw profiles reflects the increase in mechanical shear applied to biomasses in comparison with the less restrictive bilobal elements used during soft extrusion. This compression action appeared to be more intense for CS and WS as illustrated by the higher values of the extrudate's dry matter contents.

Moreover, when comparing the SE and HE/HER configurations with each other, temperature ranges in the stress zone increased from 18–22 °C to 31–36 °C for SB, from 26–27 °C to 30–34 °C for WT, from 32–34 °C to 28–42 °C for CS and from 27–32 °C up to 58–63 °C for WS, which illustrates once again the mechanical stress increase using the high-shear extrusion conditions, especially with WS and, to a lesser extent, with CS. In the same way, the motor current range increased between SE and HE/HER from 14–15 A to 16–25 A for SB, from 27–35 A to 41–47 A for WT, from 11–19 A to 27–37 A for CS and from 12–21 A to 27–30 A for WS. These amperage increases were further proof of the increase in mechanical shear applied to the biomass using CF1C reverse screws instead of BL22 kneading elements.

These results are consistent with those found in the literature where the reverse screw elements used in HE/HER are presented as stronger flow-restricting elements than the kneading elements mounted with reverse pitch used in SE [19,44].

3.2. Effect of the Twin-Screw Extrusion Mechanical Pretreatment on the Chemical Composition of Biomasses

The chemical compositions of biomasses before and after extrusion are shown in Table 2. SB is characterized by 30%TS of cellulose, 17%TS of hemicelluloses and 15.5%TS of Klason lignin, which is perfectly in accordance with the literature data as Lallement et al. [45] showed 27–30% content for cellulose, 16–21% for hemicelluloses and 13–16% for Klason lignin for maize residues. WT is characterized by 27%TS of cellulose, 24.5%TS of hemicelluloses and 18%TS of Klason lignin, which is similar to the already published data for lignin and hemicelluloses but a bit lower for cellulose: e.g., 36%TS, 25%TS and 16%TS for cellulose, hemicelluloses and lignin, respectively, in Pronyk et al. [46], and 35%TS, 23%TS and 17.5%TS in Tamaki and Mazza [47].

As triticale was harvested fresh, its lignocellulosic composition may differ from the literature data in which it was harvested at a late stage [48]. In addition, triticale is not a species that is as documented as others such as wheat or maize. Thus, its composition could also differ from those of other varieties as triticale includes over 320 species in the European catalogue [49]. CS contained 32%TS of cellulose, 19%TS of hemicelluloses and 17%TS of Klason lignin, which is similar to the already published data, i.e., 28–44% TS, 13–25%TS and 14–26%TS, respectively, for cellulose, hemicelluloses and lignin [50]. WS contained 34%TS of cellulose, 19%TS of hemicelluloses and 22.5%TS of Klason lignin, which is also quite similar to the already published data, i.e., 23–36%, 11–31% and 10–23%, respectively, for cellulose, hemicelluloses and lignin [51], although presenting high cellulose and Klason lignin contents.

Overall, there are slight variations regarding SB before and after extrusion even if a small but statistically significant reduction in the content of extractables was observed, especially after HE, due to their partial solubilization and then their removal by filtration in the case of HE. Regarding WT, for all families of molecules quantified, the differences in composition between CM, SE and HE are systematically significant from a statistical point of view. Only a small solubilization of extractables (5%), leading to an increase in cellulose (5%) and in lignin (1.5%), simultaneously with a small decrease in hemicelluloses (3%), is observed after SE. However, a much more significant solubilization of extractables (15%) is observed after HE, also leading to the increases in both cellulose (8%) and lignin (6%) and a very small decrease in hemicelluloses (1%).

As shown earlier with the motor's amperage, HE was more intense for WT than for SB, which led to a much higher proportion of extractables removed by filtration and thus to a much more significant reduction in the content of extractables inside the WT-based extrudate. For CS, there is a moderate difference in the biomass chemical composition after SE. However, there is again an important solubilization of extractables (13.5%) after HER, leading as well to statistically significant increases in cellulose (9%), hemicelluloses (3.5%) and lignin (3%). Finally, WS shows similar trend as CS with no drastic modification in chemical composition after SE but a solubilization of extractables (5.5%), leading, at the same time, to the increases in cellulose (2%), hemicelluloses (1.5%) and lignin (2%) after HER. Moreover, for each biomass, the lower extractable content measured after HE/HER was always statistically different compared to SE.

Even if Menardo et al. [26] used a mixture of rice straw silage, maize silage and triticale silage in their study, no literature was found about whole triticale or triticale straw extrusion. However, as a rye and wheat hybrid [52], it might be approximated as such. Vandenbossche et al. [53] reported slight composition variations as well when extruding wheat straw with reverse single-thread screw elements despite an internal heating temperature of 80 °C. Moreover, Zheng et al. [54] showed no difference in corn cob composition after using conveying, kneading or reverse elements through twin-screw extrusion. Finally, in the case of rehydrated corn stover, which was mechanically treated with a twin-screw extruder, Wang et al. [55] showed a slight increase for cellulose (1%), and slight decreases for hemicelluloses (1%) and, especially, lignin (0.2%) contents. Lastly, in their very recent study, Elalami et al. [56] showed no statistical difference in the chemical composition of corn stover before and after extrusion.

Table 2. Chemical compositions of solid samples before and after twin-screw extrusion pretreatment.

	SB			WT			CS			WS		
	CM	SE	HE	CM	SE	HE	CM	SE	HER	CM	SE	HER
TS[2] (%FM[1])	28.2 ± 0.3^{a}	28.8 ± 0.3^{a}	43.6 ± 1.1^{b}	21.1 ± 0.2^{a}	23.6 ± 0.3^{b}	63.4 ± 0.3^{c}	89.6 ± 0.1^{a}	75.3 ± 0.6^{b}	82.3 ± 0.1^{c}	88.5 ± 0.1^{a}	81.6 ± 0.4^{b}	76.3 ± 0.3^{c}
VS[3] (%FM[1])	27.2 ± 0.3^{a}	27.8 ± 0.3^{a}	42.5 ± 1.1^{b}	18.9 ± 0.2^{a}	21.5 ± 0.2^{b}	59.6 ± 0.3^{c}	86.5 ± 0.1^{a}	72.7 ± 0.6^{b}	80.8 ± 0.1^{c}	85.0 ± 0.1^{a}	78.4 ± 0.4^{b}	75.1 ± 0.3^{c}
Ash (%FM[1])	1.0 ± 0.0^{a}	1.0 ± 0.0^{a}	1.2 ± 0.0^{b}	2.2 ± 0.0^{a}	2.1 ± 0.0^{b}	3.8 ± 0.0^{c}	3.1 ± 0.0^{a}	2.6 ± 0.0^{b}	1.5 ± 0.2^{c}	3.5 ± 0.1^{a}	3.2 ± 0.1^{b}	1.2 ± 0.0^{c}
Extractables (%TS[2])	28.4 ± 0.7^{a}	25.8 ± 0.3^{b}	22.3 ± 0.2^{c}	25.2 ± 0.0^{a}	20.5 ± 0.1^{b}	10.0 ± 0.1^{c}	18.6 ± 0.1^{a}	18.1 ± 0.3^{a}	5.3 ± 0.0^{b}	10.7 ± 0.0^{a}	9.9 ± 0.1^{b}	5.2 ± 0.3^{c}
Cellulose (%TS[2])	29.9 ± 0.7^{a}	29.6 ± 0.5^{a}	32.0 ± 0.4^{b}	27.1 ± 0.1^{a}	31.6 ± 0.2^{b}	34.9 ± 1.8^{c}	31.7 ± 1.6^{a}	33.3 ± 0.0^{a}	40.7 ± 1.0^{b}	34.3 ± 0.2^{a}	33.6 ± 0.1^{b}	36.2 ± 0.1^{c}
Hemicelluloses (%TS[2])	17.0 ± 0.7^{a}	19.7 ± 0.6^{b}	18.0 ± 0.4^{c}	24.6 ± 0.1^{a}	21.3 ± 0.1^{b}	23.4 ± 1.1^{c}	18.8 ± 0.5^{a}	18.6 ± 0.0^{a}	22.3 ± 0.6^{b}	19.3 ± 0.1^{a}	19.2 ± 0.0^{a}	20.9 ± 0.1^{b}
Klason lignin (%TS[2])	15.5 ± 0.2^{a}	16.2 ± 0.1^{b}	16.7 ± 0.2^{c}	17.8 ± 0.5^{a}	19.1 ± 0.2^{b}	24.2 ± 0.0^{c}	17.3 ± 0.3^{a}	17.5 ± 0.1^{a}	20.5 ± 0.3^{b}	22.6 ± 0.1^{a}	22.9 ± 0.1^{b}	24.6 ± 0.2^{c}
Residual chemicals (%TS[2])	9.2 ± 0.2^{a}	8.8 ± 0.1^{b}	10.9 ± 0.2^{c}	5.4 ± 0.4^{a}	7.5 ± 0.2^{b}	7.4 ± 0.0^{b}	13.7 ± 0.3^{a}	12.5 ± 0.1^{b}	11.3 ± 0.3^{c}	13.1 ± 0.1^{a}	14.3 ± 0.1^{b}	13.3 ± 0.2^{a}

[1] FM: fresh matter, [2] TS: total solids, [3] VS: volatile solids. For each biomass, means $\pm$ s.d. followed by different letters are significantly different ($p < 0.05$).

In conclusion, although HE/HER configuration allowed partial solubilization of extractables, which significantly impacted the extrudate's composition in lignocellulosic compounds from a statistical point of view, mechanical extrusion only slightly altered the biomass lignocellulosic composition.

3.3. Effect of the Twin-Screw Extrusion Mechanical Pretreatment on Granulometry

The samples' cumulated granulometries before and after extrusion treatments are shown in Figure 3 and expressed in % of total particle weight (%wt). Overall, SE and HE had quite similar effects regarding SB: only 2.3%wt of particles were smaller than 0.25 mm before extrusion instead of 5.7%wt after SE and 4.3%wt after HE. In the same way, for bigger particles, 6.4% of them were smaller than 0.8 mm before extrusion instead of 24.3% after SE and 30.4%wt after HE, and 16.8%wt of particles were smaller than 2 mm before extrusion instead of 59.3%wt after SE and 65.9%wt after HE.

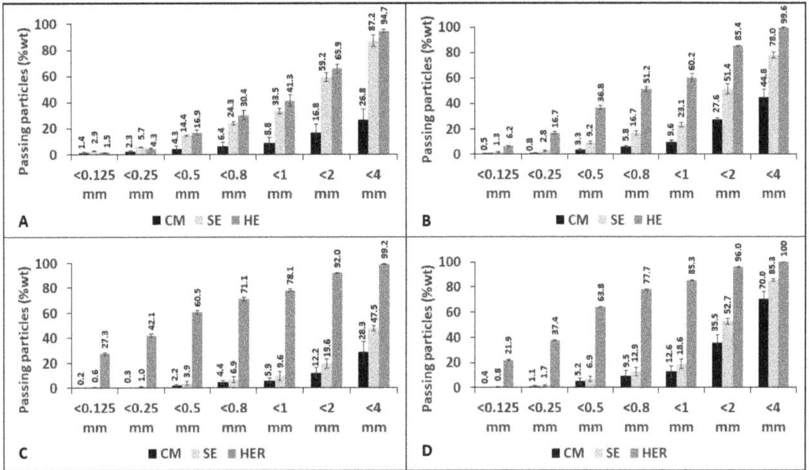

Figure 3. Cumulated granulometries of solid samples before and after twin-screw extrusion pretreatment: (**A**) SB, (**B**) WT, (**C**) CS and (**D**) WS.

For WT, the results obtained are really different: only 0.8%wt of particles were smaller than 0.25 mm before extrusion instead of 2.8%wt after SE and 16.7%wt after HE. In the same way, 5.8%wt of particles were smaller than 0.8 mm before extrusion instead of 16.7%wt after SE and 51.2%wt after HE, and 27.6% of particles were smaller than 2 mm before extrusion instead of 51.4%wt after SE and 85.4%wt after HE. In the case of WT, the SE effect in particle size reduction was less important than for SB. However, the HE one was much more important, with the high-shear extrusion process contributing to a significant additional reduction in particle size in comparison with the soft one.

Concerning CS, only 0.3%wt of particles were smaller than 0.25 mm before extrusion, and this mass content increased to 1%wt after SE and to 42.1%wt after HER. Identically, 4.4%wt of particles were smaller than 0.8 mm before extrusion instead of 6.9%wt after SE and 71.1%wt after HER. For particles smaller than 2 mm, their mass content was 12.2%wt before extrusion, 19.6%wt after SE and 92%wt after HER. The important reduction in the particle size after HER in comparison with SE is therefore even more pronounced for CS than for WT. Oppositely, the SE effect in particle size reduction was less important than for SB and WT.

Quite the same conclusions can be made for WS. Only 1.1%wt of particles were smaller than 0.25 mm before extrusion instead of 1.7%wt after SE and 37.4%wt after HER. In the same way, 9.5%wt of particles were smaller than 0.8 mm before extrusion instead of 12.9%wt after SE and 77.7%wt after HER, and 35.5%wt of particles were smaller than 2 mm before extrusion instead of 52.7%wt after SE and 96%wt after HE. SE and HER effects in particle size reduction are thus similar for WS and CS samples. However, in contrast to the other substrates, one should be aware that WS had to be ground once more through a 15 mm grid before extrusion, which must have already reduced its particle size.

In their study, Duque et al. [57] showed a small (i.e., 20%) reduction of particles with a size that was more important than the 3.14 mm on extruded barley straw processed with reverse screws as samples were crush milled with a 5 mm mesh before extrusion, while Zheng et al. [54] showed a clear reduction in particle size after extrusion in the case of corn cob but no difference depending on the type of screws used along the profile (i.e., conveying, kneading or reverse elements). However, in that study, corn cob particles were already ground to particle sizes between 0.6 and 0.76 cm before extrusion, meaning that the effect of reducing the size of large particles more or less at the moment of the extrusion pretreatment depending on the screw elements used may probably not have been illustrated properly. In contrast, Garuti et al. [58] showed a net reduction of 7.1% of particles with a size of more than 5 mm, simultaneously with a 19.6% increase in particles less than 0.3 mm in diameter after extrusion conducted on an agricultural waste mix.

In the present study, the absence of fine milling before extrusion, although being more restrictive for the application of the mechanical pretreatment in the extruder as shown earlier, showed more realistic particle size reduction at the moment of the only pretreatment. This reduction in particle size was evidenced for the four biomasses treated, and, even if both SE and HE pretreatments resulted in the same size reduction for SB, the high-shear extrusion pretreatment was much more restrictive for WT, and especially for CS and WS. This important reduction in size with the HER configuration (i.e., CS and WS feedstocks) is probably the consequence of superficial rehydration inside the extruder. With cell walls within the particles being less moist and therefore more rigid, a reduction in size was probably favored.

3.4. Effect of the Twin-Screw Extrusion Mechanical Pretreatment on Tapped Density

The evolutions of tapped densities before and after extrusion are shown in Table 3, and the results are in general agreement with those of particle size distribution presented previously (Figure 3). After SE, the tapped densities of all substrates increased significantly, with increases of 2.7, 4.8, 2.4 and 2.6 times for SB, WT, CS and WS, respectively, indicating particle refining even with soft extrusion. After HE/HER, tapped densities increased even more with increases of 5.3, 5.4, 7.0 and 8.5 times, respectively, for SB, WT, CS and WS. This increase in tapped densities was thus consistent with the particle size reduction discussed earlier, as smaller particle sizes led to a better stacking of the particles between them after compaction, and therefore a reduction in the inter-particle voids and, as a result, higher densities. This was also confirmed by Chen et al. [27] who showed that the bulk density of rice straw increased by 2.2-fold after extrusion. However, when looking at SB results, and taking into account that particle sizes were quite similar after SE and HE pretreatments (Figure 3), it would have been expected for values of tapped densities to be closer to each other. Likewise, WT tapped density after HE would have been expected to be much higher than that after SE. In this case, one possible explanation would be the formation of aggregates during sample drying after extrusion that would have been separated after particle size distribution measurements but not in the case of the tapped density ones.

Table 3. Tapped densities and cellulose crystallinities of solid samples before and after twin-screw extrusion pretreatment.

	SB			WT			CS			WS		
	CM	SE	HE	CM	SE	HE	CM	SE	HER	CM	SE	HER
Tapped density (kg/m^3)	21.0 ± 0.9 [a]	57.5 ± 0.8 [b]	110.6 ± 1.1 [c]	11.6 ± 1.4 [a]	55.3 ± 1.4 [b]	62.9 ± 1.3 [c]	28.7 ± 0.8 [a]	69.2 ± 1.1 [b]	200.3 ± 1.7 [c]	14.7 ± 0.2 [a]	38.7 ± 0.0 [b]	124.0 ± 0.5 [c]
Cellulose crystallinity (%)	29.2 ± 1.2 [a]	27.3 ± 0.5 [a]	28.0 ± 0.1 [a]	52.4 ± 0.4 [a]	52.3 ± 1.6 [a]	53.9 ± 0.4 [a]	67.5 ± 2.1 [a]	64.2 ± 0.8 [ab]	58.9 ± 1.4 [b]	55.5 ± 0.6 [a]	53.7 ± 1.1 [a]	56.4 ± 1.3 [a]

For each biomass, means $\pm$ s.d. followed by different letters are significantly different ($p < 0.05$).

3.5. Effect of the Twin-Screw Extrusion Mechanical Pretreatment on Cellulose Crystallinity

As shown in Table 3, no cellulose crystallinity change was observed after extrusion for SB, WT and WS. In contrast, the 8.6% decrease in cellulose crystallinity of CS after HER pretreatment was statistically significant. Regarding the results for CS, the most logical explanation would be that HER pretreatment was more impactful on it than on the other biomasses as it was previously illustrated by its high decrease in particle size (Figure 3) and its much higher tapped density value (Table 3). In the literature, mechanical pretreatments are generally known to be efficient in the decrystallization of cellulose [59]. However, it is not always the case with extrusion pretreatments. While Zhang et al. [60] achieved an impressive 48.4% cellulose crystallinity reduction on rice straw with only extrusion, Zhang et al. [22] instead did not observe a decrease in cellulose crystallinity for corn stover after reactive alkali extrusion. Here, the most likely reason for the cellulose crystallinity reduction of CS after HER is that its more rigid morphological structure compared to other biomasses was more affected by the high-shear twin-screw extrusion mechanical pretreatment.

3.6. Effect of the Twin-Screw Extrusion Mechanical Pretreatment on BMP Results

Figure 4 shows experimental BMP and the BMP kinetics using the modified Gompertz equation model, while their parameters are gathered in Table 4. The experimental BMP tests showed methane production ranges of 338–345 NmL/gVS for SB, 233–247 NmL/gVS for WT, 264–286 NmL/gVS for CS and 247–270 NmL/gVS for WS, which is in accordance with standard values for lignocellulosic biomasses [61,62]. SB shows the highest methanogenic potential with a 40–48% increase compared to WT, a 20–28% increase compared to CS and a 27–37% increase compared to WS, which definitely makes it the best candidate for biomethane production followed by CS.

However, no BMP increase or BMP decrease provided by SE or HE/HER pretreatments was observed from a statistical point of view, and this for all the biomasses tested. Therefore, extrusion had no impact on the experimental maximal BMP. Wahid et al. [63] and Victorin et al. [64] also reported no statistical increase or statistical decrease in final BMP after the extrusion of wheat straw. Hjorth et al. [18] also reported no statistical increase in BMP after 90 days on extruded barley straw. This confirms that the mechanical treatment provided by extrusion does not generate or allow a better degradation of the compounds usually non-valorized during AD, which also correlates with the lack of cellulose crystallinity variation shown earlier.

Adjusted maximal theorical BMPs are proposed in this study as a way to calculate a more realistic estimation of the real biodegradability of the substrates as the monolignols constituting the native lignin are not consumed during AD [65]. SB, WT, CS and WS achieved around 89%, 66%, 76% and 74%, respectively, of adjusted theoretical biodegradability after CM, SE and HE/HER pretreatments. Therefore, a BMP improvement from 11% to 34% is still accessible for these biomasses according to the adjusted theorical biodegradability. The higher biodegradability shown by SB compared to the other biomass is in agreement with the 132 AD feedstock database established by Lallement et al. [45], who reported maximal theoretical biodegradabilities of 72% for agro-industrial residues and 56% for lignocellulosic matter, and this correlates with its higher experimental CH_4 production.

BMP kinetics showed that most of the biomethane was produced within the first 10–20 days of BMP tests for all biomasses, which is usual according to the literature data [63,64]. A modified Gompertz model was used to model BMP kinetics as it is one of the most accurate models for anaerobic digestion processes [66]. The predicted BMP showed methane production ranges of 330–339 NmL/gVS for SB, 229–244 NmL/gVS for WT, 259–284 NmL/gVS for CS and 242–265 NmL/gVS for WS. These results are slightly lower to the experimental ones but nonetheless are very close as proven by an R^2 correlation coefficient of more than 0.99 for all kinetics, and, this is expected when using a modified Gompertz model [67–69].

Lag phases were very short with an initial 0.2–1.3 day range, and they varied slightly with a 0.2–1.3 day range after SE and with a 0.01–1.4 day range after HE while retaining their short durations, which shows excellent adequacy between used inocula and substrates, and illustrates the easily digested biomass. This result is also correlated with the higher biodegradability of SB compared to WT, which, respectively, had the shortest and longest initial lag phases. The impact of pretreatment seems more impactful in the case of a longer initial lag phase as demonstrated by Tsapekos et al. [70] who managed to reduce the grass lag phase from 3.4 to 2.7 days thanks to a mechanical pretreatment.

The specific rate constant increased for all biomasses after SE and even more after HE or HER. For SB, it increased by 26% after SE and by 81% after HE. For WT, it increased by 56% after SE and by 67% after HE. For CS, it increased by 13% after SE and by 107% after HER. Lastly, for WS, it increased by 21% after SE and by 72% after HER. This demonstrates that the morphological modifications of the biomasses induced by the twin-screw extrusion mechanical pretreatments, especially the high-shear one, allowed better accessibility and degradability of methanizable compounds during digestion by the anaerobic microorganisms, which correlates with the particle size reduction and increase in tapped densities discussed earlier. Chen et al. [27] also reported a decrease in anaerobic digestion time after rice straw extrusion as well as Pérez-Rodrígez et al. [71] after the extrusion of corn cob. This improvement in the specific rate constant is a great indicator of how a substrate would react in the case of continuous biogas production [26], and the impact of SE and especially HE/HER pretreatments on the four biomasses evidenced in this study indicates promising perspectives in reducing their hydraulic residence times [72]. As a direct consequence, a smaller sizing of new biogas plant bioreactors could be considered, thus allowing a diminution in their capital expenditures (CAPEX) and their operating expenditures (OPEX) [73].

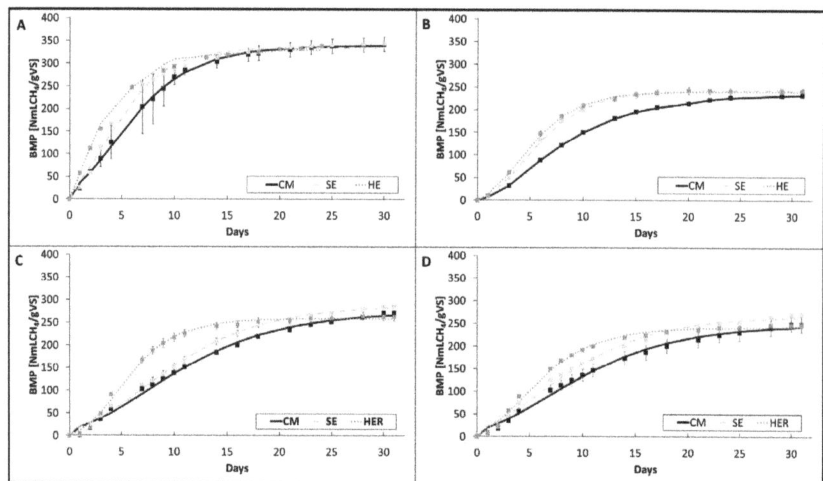

Figure 4. Experimental BMP (square symbols) and BMP kinetics using the modified Gompertz equation model (lines) before and after twin-screw extrusion pretreatment: (**A**) SB, (**B**) WT, (**C**) CS and (**D**) WS. BMPs from HE/HER samples were assessed after mixing extrudate and filtrate samples in the same proportions as at the extrusion outlet.

Table 4. Experimental BMP, maximal and adjusted maximal theorical BMP and kinetics parameters of modified Gompertz equation model before and after twin-screw extrusion pretreatment.

Biomass	SB			WT			CS			WS		
Pretreament	CM	SE	HE	CM	SE	HE	CM	SE	HER	CM	SE	HER
Experimental BMP (NmL/gVS)	345 ± 16 [a]	345 ± 7 [a]	338 ± 2 [a]	233 ± 4 [a]	247 ± 2 [a]	242 ± 6 [a]	272 ± 5 [ab]	286 ± 3 [a]	264 ± 9 [b]	249 ± 17 [ab]	270 ± 2 [a]	247 ± 0 [b]
BMP_{th} adjusted (NmL/gVS)	395	388	375	356	346	362	366	369	350	352	343	346

Table 4. *Cont.*

Biomass	SB			WT			CS			WS		
Pretreament	CM	SE	HE	CM	SE	HE	CM	SE	HER	CM	SE	HER
BI (%)	87	89	90	66	66	67	74	78	75	71	79	71
B (NmL/gVS)	339	336	330	229	244	241	268	284	259	244	265	242
R_m (NmL/gVS.day)	31	39	56	18	28	30	15	17	31	14	17	24
λ (day)	0.19	0.20	0.01	1.28	1.33	1.03	0.83	1.08	1.43	0.46	0.29	0.80
R^2	>0.99	>0.99	>0.99	>0.99	>0.99	>0.99	>0.99	>0.99	>0.99	>0.99	>0.99	>0.99

For each biomass, means $\pm$ s.d. followed by different letters are significantly different ($p < 0.05$).

4. Conclusions

The effect of the mechanical pretreatment induced through twin-screw extrusion was able to greatly impact the physico-chemical properties of all four biomasses tested, at different levels depending on the selected screw profile as illustrated by the increase in the proportion of smaller size particles correlated with the impressive increase in tapped densities. The more intense mechanical shear effect of the high-shear extrusion pretreatments also allowed a significant solubilization of water-soluble compounds just as the reduction in the corn stover cellulose crystallinity by 8.6% showed, without affecting those of other biomasses. With soft extrusion, no modification of chemical composition and cellulose crystallinity was observed. Sweetcorn by-products revealed the highest BMP values (i.e., 338–345 NmL/gVS), followed by corn stover (264–286 NmL/gVS), wheat straw (247–270 NmL/gVS), and lastly, whole triticale (233–247 NmL/gVS), thereby illustrating their great potential as inputs for biogas production. Even if none of the physico-chemical modifications induced by the two applied extrusion pretreatments were proven to statistically improve the maximal BMP production, kinetic analysis revealed that both extrusion pretreatments, and especially the high-shear one, were able to increase the specific rate constant by up to 56% for soft extrusion and even 106% for high-shear extrusion. Further studies on either bioextrusion (i.e., addition of enzymes) and/or reactive extrusion (i.e., addition of chemicals) as additional pretreatments, known to improve biomethane production, are recommended to attain a synergistic effect with the only mechanical pretreatment.

Author Contributions: Conceptualization, A.C., P.E., F.M., V.V. and C.S.; methodology, A.C., P.E., F.M., V.V. and C.S.; validation, P.E., F.M., V.V. and C.S.; formal analysis, A.C.; investigation, A.C.; resources, P.E. and F.M.; writing—original draft preparation, A.C.; writing—review and editing, P.E., F.M., V.V. and C.S.; supervision, P.E., F.M., V.V. and C.S.; project administration, P.E. and C.S.; funding acquisition, P.E. and C.S. All authors have read and agreed to the published version of the manuscript.

Funding: This research was funded by ANRT (National Association for Research and Technology in France) under grant number 2021/0737.

Data Availability Statement: All the data are described in the Figures and Tables.

Acknowledgments: The authors would like to express their sincere gratitude to Ovalie Innovation (Auch, France) for supplying the batch of whole triticale used for the purpose of this study. They would also like to deeply thank the PERL and CIRIMAT laboratory teams for their respective help with BMP and cellulose crystallinity assessments.

Conflicts of Interest: The authors declare no conflict of interest.

References

1. Abas, N.; Kalair, A.; Khan, N. Review of fossil fuels and future energy technologies. *Futures* **2015**, *69*, 31–49. [CrossRef]
2. Li, Y.; Chen, Y.; Wu, J. Enhancement of methane production in anaerobic digestion process: A review. *Appl. Energy* **2019**, *240*, 120–137. [CrossRef]
3. Ampese, L.C.; Sganzerla, W.G.; Di Domenico Ziero, H.; Mudhoo, A.; Martins, G.; Forster-Carneiro, T. Research progress, trends, and updates on anaerobic digestion technology: A bibliometric analysis. *J. Clean. Prod.* **2022**, *331*, 130004. [CrossRef]
4. Konan, D.; Koffi, E.; Ndao, A.; Peterson, E.C.; Rodrigue, D.; Adjallé, K. An Overview of Extrusion as a Pretreatment Method of Lignocellulosic Biomass. *Energies* **2022**, *15*, 3002. [CrossRef]
5. Hendriks, A.T.W.M.; Zeeman, G. Pretreatments to enhance the digestibility of lignocellulosic biomass. *Bioresour. Technol.* **2009**, *100*, 10–18. [CrossRef]

6. Tye, Y.Y.; Lee, K.T.; Wan Abdullah, W.N.; Leh, C.P. The world availability of non-wood lignocellulosic biomass for the production of cellulosic ethanol and potential pretreatments for the enhancement of enzymatic saccharification. *Renew. Sustain. Energy Rev.* **2016**, *60*, 155–172. [CrossRef]

7. Gissén, C.; Prade, T.; Kreuger, E.; Nges, I.A.; Rosenqvist, H.; Svensson, S.-E.; Lantz, M.; Mattsson, J.E.; Börjesson, P.; Björnsson, L. Comparing energy crops for biogas production—Yields, energy input and costs in cultivation using digestate and mineral fertilisation. *Biomass Bioenergy* **2014**, *64*, 199–210. [CrossRef]

8. Xu, N.; Liu, S.; Xin, F.; Zhou, J.; Jia, H.; Xu, J.; Jiang, M.; Dong, W. Biomethane Production from Lignocellulose: Biomass Recalcitrance and Its Impacts on Anaerobic Digestion. *Front. Bioeng. Biotechnol.* **2019**, *7*, 191. [CrossRef]

9. Atelge, M.R.; Atabani, A.E.; Banu, J.R.; Krisa, D.; Kaya, M.; Eskicioglu, C.; Kumar, G.; Lee, C.; Yildiz, Y.Ş.; Unalan, S.; et al. A critical review of pretreatment technologies to enhance anaerobic digestion and energy recovery. *Fuel* **2020**, *270*, 117494. [CrossRef]

10. Alvira, P.; Tomás-Pejó, E.; Ballesteros, M.; Negro, M.J. Pretreatment technologies for an efficient bioethanol production process based on enzymatic hydrolysis: A review. *Bioresour. Technol.* **2010**, *101*, 4851–4861. [CrossRef]

11. Carrere, H.; Antonopoulou, G.; Affes, R.; Passos, F.; Battimelli, A.; Lyberatos, G.; Ferrer, I. Review of feedstock pretreatment strategies for improved anaerobic digestion: From lab-scale research to full-scale application. *Bioresour. Technol.* **2016**, *199*, 386–397. [CrossRef]

12. Mankar, A.R.; Pandey, A.; Modak, A.; Pant, K.K. Pretreatment of lignocellulosic biomass: A review on recent advances. *Bioresour. Technol.* **2021**, *334*, 125235. [CrossRef]

13. Kasinath, A.; Fudala-Ksiazek, S.; Szopinska, M.; Bylinski, H.; Artichowicz, W.; Remiszewska-Skwarek, A.; Luczkiewicz, A. Biomass in biogas production: Pretreatment and codigestion. *Renew. Sustain. Energy Rev.* **2021**, *150*, 111509. [CrossRef]

14. Gałązka, A.; Szadkowski, J. Enzymatic hydrolysis of fast-growing poplar wood after pretreatment by steam explosion. *Cellul. Chem. Technol.* **2021**, *55*, 637–647. [CrossRef]

15. Thamizhakaran Stanley, J.; Thanarasu, A.; Senthil Kumar, P.; Periyasamy, K.; Raghunandhakumar, S.; Periyaraman, P.; Devaraj, K.; Dhanasekaran, A.; Subramanian, S. Potential pre-treatment of lignocellulosic biomass for the enhancement of biomethane production through anaerobic digestion—A review. *Fuel* **2022**, *318*, 123593. [CrossRef]

16. Monlau, F.; Sambusiti, C.; Barakat, A.; Quéméneur, M.; Trably, E.; Steyer, J.-P.; Carrère, H. Do furanic and phenolic compounds of lignocellulosic and algae biomass hydrolyzate inhibit anaerobic mixed cultures? A comprehensive review. *Biotechnol. Adv.* **2014**, *32*, 934–951. [CrossRef]

17. Naik, G.P.; Poonia, A.K.; Chaudhari, P.K. Pretreatment of lignocellulosic agricultural waste for delignification, rapid hydrolysis, and enhanced biogas production: A review. *J. Indian Chem. Soc.* **2021**, *98*, 100147. [CrossRef]

18. Hjorth, M.; Gränitz, K.; Adamsen, A.P.S.; Møller, H.B. Extrusion as a pretreatment to increase biogas production. *Bioresour. Technol.* **2011**, *102*, 4989–4994. [CrossRef]

19. Evon, P.; Vandenbossche, V.; Candy, L.; Pontalier, P.-Y.; Rouilly, A. Twin-Screw Extrusion: A Key Technology for the Biorefinery. In *Biomass Extrusion and Reaction Technologies: Principles to Practices and Future Potential*; ACS Symposium Series; American Chemical Society: Washington, DC, USA, 2018; Volume 1304, pp. 17–33. ISBN 978/0/8412/3371/3.

20. Turick, C.; Peck, M.; Chynoweth, D.; Jerger, D.; White, E.; Zsuffa, L.; Kenney, W. Methane fermentation of woody biomass. *Bioresour. Technol.* **1991**, *37*, 141–147. [CrossRef]

21. Zheng, J.; Rehmann, L. Extrusion pretreatment of lignocellulosic biomass: A review. *Int. J. Mol. Sci.* **2014**, *15*, 18967–18984. [CrossRef]

22. Zhang, S.; Keshwani, D.R.; Xu, Y.; Hanna, M.A. Alkali combined extrusion pretreatment of corn stover to enhance enzyme saccharification. *Ind. Crop. Prod.* **2012**, *37*, 352–357. [CrossRef]

23. Vandenbossche, V.; Brault, J.; Hernandez-Melendez, O.; Evon, P.; Barzana, E.; Vilarem, G.; Rigal, L. Suitability assessment of a continuous process combining thermo-mechano-chemical and bio-catalytic action in a single pilot-scale twin-screw extruder for six different biomass sources. *Bioresour. Technol.* **2016**, *211*, 146–153. [CrossRef]

24. Gatt, E.; Rigal, L.; Vandenbossche, V. Biomass pretreatment with reactive extrusion using enzymes: A review. *Ind. Crop. Prod.* **2018**, *122*, 329–339. [CrossRef]

25. Cha, Y.-L.; Yang, J.; Park, Y.; An, G.H.; Ahn, J.-W.; Moon, Y.-H.; Yoon, Y.-M.; Yu, G.-D.; Choi, I.-H. Continuous alkaline pretreatment of *Miscanthus sacchariflorus* using a bench-scale single screw reactor. *Bioresour. Technol.* **2015**, *181*, 338–344. [CrossRef]

26. Menardo, S.; Cacciatore, V.; Balsari, P. Batch and continuous biogas production arising from feed varying in rice straw volumes following pre-treatment with extrusion. *Bioresour. Technol.* **2015**, *180*, 154–161. [CrossRef] [PubMed]

27. Chen, X.; Zhang, Y.; Gu, Y.; Liu, Z.; Shen, Z.; Chu, H.; Zhou, X. Enhancing methane production from rice straw by extrusion pretreatment. *Appl. Energy* **2014**, *122*, 34–41. [CrossRef]

28. Khor, W.C.; Rabaey, K.; Vervaeren, H. Low temperature calcium hydroxide treatment enhances anaerobic methane production from (extruded) biomass. *Bioresour. Technol.* **2015**, *176*, 181–188. [CrossRef] [PubMed]

29. Frydendal-Nielsen, S.; Hjorth, M.; Baby, S.; Felby, C.; Jørgensen, U.; Gislum, R. The effect of harvest time, dry matter content and mechanical pretreatments on anaerobic digestion and enzymatic hydrolysis of miscanthus. *Bioresour. Technol.* **2016**, *218*, 1008–1015. [CrossRef] [PubMed]

30. Pérez-Rodríguez, N.; García-Bernet, D.; Domínguez, J.M. Faster methane production after sequential extrusion and enzymatic hydrolysis of vine trimming shoots. *Environ. Chem. Lett.* **2018**, *16*, 295–299. [CrossRef]

31. Niu, W.; Huang, G.; Liu, X.; Chen, L.; Han, L. Chemical Composition and Calorific Value Prediction of Wheat Straw at Different Maturity Stages Using Near-Infrared Reflectance Spectroscopy. *Energy Fuels* **2014**, *28*, 7474–7482. [CrossRef]

32. Buswell, A.M.; Mueller, H.F. Mechanism of Methane Fermentation. *Ind. Eng. Chem.* **1952**, *44*, 550–552. [CrossRef]
33. Achinas, S.; Euverink, G.J.W. Theoretical analysis of biogas potential prediction from agricultural waste. *Resour. Effic. Technol.* **2016**, *2*, 143–147. [CrossRef]
34. Khongchamnan, P.; Wanmolee, W.; Laosiripojana, N.; Champreda, V.; Suriyachai, N.; Kreetachat, T.; Sakulthaew, C.; Chokejaroenrat, C.; Imman, S. Solvothermal-Based Lignin Fractionation from Corn Stover: Process Optimization and Product Characteristics. *Front. Chem.* **2021**, *9*, 697237. [CrossRef]
35. Sluiter, A. Determination of Total Solids in Biomass and Total Dissolved Solids in Liquid Process Samples: Laboratory Analytical Procedure (LAP). *Tech. Rep.* **2008**, *9*. Available online: https://www.nrel.gov/docs/gen/fy08/42621.pdf (accessed on 2 May 2022).
36. Beaufils, N.; Boucher, J.; Peydecastaing, J.; Rigal, L.; Vilarem, G.; Villette, M.-J.; Candy, L.; Pontalier, P.-Y. The effect of time and temperature on the extraction of xylose and total phenolic compounds with pressurized hot water from hardwood species used for pulp and paper production in the South of France. *Bioresour. Technol. Rep.* **2021**, *16*, 100832. [CrossRef]
37. Sluiter, A.; Hames, B.; Ruiz, R.; Scarlata, C.; Sluiter, J.; Templeton, D.; Crocker, D. Determination of structural carbohydrates and lignin in biomass, in: Laboratory Analytical Procedure (LAP). *Natl. Renew. Energy Lab.* **2008**, *1617*. Available online: https://www.nrel.gov/docs/gen/fy13/42618.pdf (accessed on 2 May 2022).
38. Segal, L.; Creely, J.J.; Martin, A.E.; Conrad, C.M. An Empirical Method for Estimating the Degree of Crystallinity of Native Cellulose Using the X-ray Diffractometer. *Text. Res. J.* **1959**, *29*, 786–794. [CrossRef]
39. Zwietering, M.H.; Jongenburger, I.; Rombouts, F.M.; van't Riet, K. Modeling of the Bacterial Growth Curve. *Appl. Environ. Microbiol.* **1990**, *56*, 1875–1881. [CrossRef]
40. Evon, P.; Vandenbossche, V.; Pontalier, P.-Y.; Rigal, L. New thermal insulation fiberboards from cake generated during biorefinery of sunflower whole plant in a twin-screw extruder. *Ind. Crop. Prod.* **2014**, *52*, 354–362. [CrossRef]
41. Heredia-Olea, E.; Pérez-Carrillo, E.; Montoya-Chiw, M.; Serna-Saldívar, S.O. Effects of Extrusion Pretreatment Parameters on Sweet Sorghum Bagasse Enzymatic Hydrolysis and Its Subsequent Conversion into Bioethanol. *BioMed Res. Int.* **2015**, *2015*, 325905. [CrossRef]
42. Mazaheri Tehrani, M.; Ehtiati, A.; Sharifi Azghandi, S. Application of genetic algorithm to optimize extrusion condition for soy-based meat analogue texturization. *J. Food Sci. Technol.* **2017**, *54*, 1119–1125. [CrossRef] [PubMed]
43. Brishti, F.H.; Chay, S.Y.; Muhammad, K.; Ismail-Fitry, M.R.; Zarei, M.; Saari, N. Texturized mung bean protein as a sustainable food source: Effects of extrusion on its physical, textural and protein quality. *Innov. Food Sci. Emerg. Technol.* **2021**, *67*, 102591. [CrossRef]
44. Choudhury, G.S.; Gautam, A. Comparative study of mixing elements during twin-screw extrusion of rice flour. *Food Res. Int.* **1998**, *31*, 7–17. [CrossRef]
45. Lallement, A.; Peyrelasse, C.; Lagnet, C.; Barakat, A.; Schraauwers, B.; Maunas, S.; Monlau, F. A Detailed Database of the Chemical Properties and Methane Potential of Biomasses Covering a Large Range of Common Agricultural Biogas Plant Feedstocks. *Waste* **2023**, *1*, 195–227. [CrossRef]
46. Pronyk, C.; Mazza, G.; Tamaki, Y. Production of Carbohydrates, Lignins, and Minor Components from Triticale Straw by Hydrothermal Treatment. *J. Agric. Food Chem.* **2011**, *59*, 3788–3796. [CrossRef]
47. Tamaki, Y.; Mazza, G. Measurement of structural carbohydrates, lignins, and micro-components of straw and shives: Effects of extractives, particle size and crop species. *Ind. Crop. Prod.* **2010**, *31*, 534–541. [CrossRef]
48. Armstrong, D.G.; Cook, H.; Thomas, B. The lignin and cellulose contents of certain grassland species at different stages of growth. *J. Agric. Sci.* **1950**, *40*, 93–99. [CrossRef]
49. Plant Variety Catalogues, Databases & Information Systems. Available online: https://food.ec.europa.eu/plants/plant-reproductive-material/plant-variety-catalogues-databases-information-systems_en (accessed on 5 March 2023).
50. Pordesimo, L.O.; Hames, B.R.; Sokhansanj, S.; Edens, W.C. Variation in corn stover composition and energy content with crop maturity. *Biomass Bioenergy* **2005**, *28*, 366–374. [CrossRef]
51. Collins, S.R.; Wellner, N.; Martinez Bordonado, I.; Harper, A.L.; Miller, C.N.; Bancroft, I.; Waldron, K.W. Variation in the chemical composition of wheat straw: The role of tissue ratio and composition. *Biotechnol. Biofuels* **2014**, *7*, 121. [CrossRef]
52. Zillinsky, F.J. The Development of Triticale. In *Advances in Agronomy*; Brady, N.C., Ed.; Academic Press: Cambridge, MA, USA, 1974; Volume 26, pp. 315–348.
53. Vandenbossche, V.; Doumeng, C.; Rigal, L. Thermomechanical and Thermo-mechano-chemical Pretreatment of Wheat Straw using a Twin-screw Extruder. *BioResources* **2014**, *9*, 1519–1538. [CrossRef]
54. Zheng, J.; Choo, K.; Rehmann, L. The effects of screw elements on enzymatic digestibility of corncobs after pretreatment in a twin-screw extruder. *Biomass Bioenergy* **2015**, *74*, 224–232. [CrossRef]
55. Wang, Z.; He, X.; Yan, L.; Wang, J.; Hu, X.; Sun, Q.; Zhang, H. Enhancing enzymatic hydrolysis of corn stover by twin-screw extrusion pretreatment. *Ind. Crop. Prod.* **2020**, *143*, 111960. [CrossRef]
56. Elalami, D.; Aouine, M.; Monlau, F.; Guillon, F.; Dumon, C.; Hernandez Raquet, G.; Barakat, A. Enhanced enzymatic hydrolysis of corn stover using twin-screw extrusion under mild conditions. *Biofuels Bioprod. Biorefining* **2022**, *16*, 1642–1654. [CrossRef]
57. Duque, A.; Manzanares, P.; Ballesteros, I.; Negro, M.J.; Oliva, J.M.; González, A.; Ballesteros, M. Sugar production from barley straw biomass pretreated by combined alkali and enzymatic extrusion. *Bioresour. Technol.* **2014**, *158*, 262–268. [CrossRef]
58. Garuti, M.; Sinisgalli, E.; Soldano, M.; Fermoso, F.G.; Rodriguez, A.J.; Carnevale, M.; Gallucci, F. Mechanical pretreatments of different agri-based feedstock in full-scale biogas plants under real operational conditions. *Biomass Bioenergy* **2022**, *158*, 106352. [CrossRef]

59. Zheng, Y.; Zhao, J.; Xu, F.; Li, Y. Pretreatment of lignocellulosic biomass for enhanced biogas production. *Prog. Energy Combust. Sci.* **2014**, *42*, 35–53. [CrossRef]
60. Zhang, Y.; Chen, X.; Gu, Y.; Zhou, X. A physicochemical method for increasing methane production from rice straw: Extrusion combined with alkali pretreatment. *Appl. Energy* **2015**, *160*, 39–48. [CrossRef]
61. Monlau, F.; Sambusiti, C.; Barakat, A.; Guo, X.M.; Latrille, E.; Trably, E.; Steyer, J.-P.; Carrere, H. Predictive Models of Biohydrogen and Biomethane Production Based on the Compositional and Structural Features of Lignocellulosic Materials. *Environ. Sci. Technol.* **2012**, *46*, 12217–12225. [CrossRef]
62. Thomsen, S.T.; Spliid, H.; Østergård, H. Statistical prediction of biomethane potentials based on the composition of lignocellulosic biomass. *Bioresour. Technol.* **2014**, *154*, 80–86. [CrossRef]
63. Wahid, R.; Hjorth, M.; Kristensen, S.; Møller, H.B. Extrusion as Pretreatment for Boosting Methane Production: Effect of Screw Configurations. *Energy Fuels* **2015**, *29*, 4030–4037. [CrossRef]
64. Victorin, M.; Davidsson, Å.; Wallberg, O. Characterization of Mechanically Pretreated Wheat Straw for Biogas Production. *BioEnergy Res.* **2020**, *13*, 833–844. [CrossRef]
65. Khan, M.U.; Ahring, B.K. Lignin degradation under anaerobic digestion: Influence of lignin modifications—A review. *Biomass Bioenergy* **2019**, *128*, 105325. [CrossRef]
66. Velázquez-Martí, B.; Meneses Quelal, O.; Gaibor, J.; Niño, Z. Review of Mathematical Models for the Anaerobic Digestion Process. In *Anaerobic Digestion*; IntechOpen: London, UK, 2018; ISBN 978/1/83881/849/4.
67. Mancini, G.; Papirio, S.; Lens, P.N.L.; Esposito, G. Increased biogas production from wheat straw by chemical pretreatments. *Renew. Energy* **2018**, *119*, 608–614. [CrossRef]
68. Lee, J.; Park, K.Y. Impact of hydrothermal pretreatment on anaerobic digestion efficiency for lignocellulosic biomass: Influence of pretreatment temperature on the formation of biomass-degrading byproducts. *Chemosphere* **2020**, *256*, 127116. [CrossRef]
69. Sieborg, M.U.; Jønson, B.D.; Larsen, S.U.; Vazifehkhoran, A.H.; Triolo, J.M. Co-Ensiling of Wheat Straw as an Alternative Pre-Treatment to Chemical, Hydrothermal and Mechanical Methods for Methane Production. *Energies* **2020**, *13*, 4047. [CrossRef]
70. Tsapekos, P.; Kougias, P.G.; Angelidaki, I. Mechanical pretreatment for increased biogas production from lignocellulosic biomass; predicting the methane yield from structural plant components. *Waste Manag.* **2018**, *78*, 903–910. [CrossRef]
71. Pérez-Rodríguez, N.; García-Bernet, D.; Domínguez, J.M. Extrusion and enzymatic hydrolysis as pretreatments on corn cob for biogas production. *Renew. Energy* **2017**, *107*, 597–603. [CrossRef]
72. Meegoda, J.N.; Li, B.; Patel, K.; Wang, L.B. A Review of the Processes, Parameters, and Optimization of Anaerobic Digestion. *Int. J. Environ. Res. Public. Health* **2018**, *15*, 2224. [CrossRef]
73. Janke, L.; Weinrich, S.; Leite, A.F.; Terzariol, F.K.; Nikolausz, M.; Nelles, M.; Stinner, W. Improving anaerobic digestion of sugarcane straw for methane production: Combined benefits of mechanical and sodium hydroxide pretreatment for process designing. *Energy Convers. Manag.* **2017**, *141*, 378–389. [CrossRef]

Review

Value Chain Analysis of Rice Industry by Products in a Circular Economy Context: A Review

W. A. M. A. N. Illankoon [1,*], Chiara Milanese [2], Maria Cristina Collivignarelli [3] and Sabrina Sorlini [1]

[1] Department of Civil, Environmental, Architectural Engineering, and Mathematics (DICATAM), University of Brescia, Via Branze 43, 25123 Brescia, Italy

[2] Department of Chemistry, University of Pavia & Center for Colloid and Surface Science, Viale Taramelli 16, 27100 Pavia, Italy

[3] Department of Civil Engineering and Architecture, University of Pavia, Via Ferrata 3, 27100 Pavia, Italy

* Correspondence: a.wijepalaabeysi@unibs.it

Abstract: The quantity of organic waste generated by agricultural sectors is continually increasing due to population growth and rising food demand. Rice is the primary consumable food in Asia. However, many stakeholders follow a linear economic model such as the "take–make–waste" concept. This linear model leads to a substantial environmental burden and the destruction of valuable resources without gaining their actual value. Because these by-products can be converted into energy generating and storage materials, and into bio-based products by cascading transformation processes within the circular economy concept, waste should be considered a central material. This review examines the composition of rice straw, bran, and husks, and the procedures involved in manufacturing value-added goods, from these wastes. Moreover, starting with an extensive literature analysis on the rice value chains, this work systematizes and displays a variety of strategies for using these by-products. The future development of agricultural waste management is desirable to capitalize on the multi-functional product by circulating all the by-products in the economy. According to the analysis of relevant research, rice straw has considerable potential as a renewable energy source. However, there is a significant research gap in using rice bran as an energy storage material. Additionally, modified rice husk has increased its promise as an adsorbent in the bio-based water treatment industry. Furthermore, the case study of Sri Lanka revealed that developing countries have a huge potential to value these by-products in various sectors of the economy. Finally, this paper provides suggestions for researchers and policymakers to improve the current agriculture waste management system with the best option and integrated approach for economic sustainability and eco- and environmental solution, considering some case studies to develop sustainable waste management processes.

Keywords: rice straw; rice bran; rice husk; agricultural waste; valorization; circular economy; biomass; bioeconomy

Citation: Illankoon, W.A.M.A.N.; Milanese, C.; Collivignarelli, M.C.; Sorlini, S. Value Chain Analysis of Rice Industry by Products in a Circular Economy Context: A Review. *Waste* **2023**, *1*, 333–369. https://doi.org/10.3390/waste1020022

Academic Editors: Catherine N. Mulligan, Dimitris P. Makris and Vassilis Athanasiadis

Received: 15 December 2022
Revised: 10 January 2023
Accepted: 8 February 2023
Published: 4 April 2023

1. Introduction

Rice is an annual plant crop mainly cultivated in areas with high rainfall and rice is a primary edible food and crop in Asia [1,2]. Many South Asian countries have an agrarian economy and have been producing significant amounts of agricultural waste related to the rice industry. As a food crop, rice ranks first in consumption, second in total output, and third in total cultivation [1]. Most (around 90%) of the world's rice supply is grown in Asia; the world's two major rice producers, China and India, provide more than half of the world's rice supply [1,3]. The world increase in cultivated area from 120 million to 163 million ha (0.5%) each year and the increase in paddy yield from 1.8 to 4.6 t/ha (1.6%) between 1960 and 2018 caused a more than threefold increase in global rice output, from 221 million to 745 million metric tons (2.1% per year) [4] (Figure 1). The Green Revolution dramatically increased rice production, which helped stave off famines, feed millions of people, alleviate poverty and hunger, and improve the lives of countless people throughout Asia [1,5]. Since there is a limit on how much land may be devoted to rice cultivation, any future increases in output will have to come from higher yields. Most of the world's population relies on rice for sustenance. With a global per capita consumption of 64 kg per year, milled rice accounts for 19% of global daily calorie intake [1,3,4,6].

Agricultural wastes are biomass residues that can be grouped into two categories, i.e., crop residues and agro-industrial residues [7]. However, traditional approaches handling agricultural waste have been linked to environmental damage and financial losses. Many farmers and other agro-industry stakeholders engage in open field burning or open dumping to clear their land for future cultivation rather than extracting their total value. In developing countries, burning crop straws and other agricultural wastes in the open air or in the kitchen is a significant contributor to dangerously high levels of air pollution. According to estimates provided by the World Bank and the Institute for Health Metrics and Evaluation (2016), the welfare losses caused by exposure to air pollution cost the global economy around $5.11 trillion in 2013 [8]. For sake of example and comparison, the size of welfare losses in Belarus, Bulgaria, India, Romania, Kazakhstan, and Bangladesh as a share of regional gross domestic product (GDP) are as follows: 9.25%, 8.85%, 7.69%, 7.21%, and 6.81%. The cost of air pollution caused by open-air straw burning in mainland China in 2004 was estimated to be over 19.65 billion CNY or around 0.14% of the country's GDP [9]. Thus, it is possible to minimize these adverse effects by considering the valorization process for these potential feedstocks and using them as a valuable material or source for the national economy [10].

Considering their waste quantities and physical and chemical properties, rice industry by-products have a high potential for generating energy and extracting nutrients, minerals, and biochemicals through different valorization processes [10]. However, their potential still needs to be explored due to issues relating to their supply chain, appropriate technologies for pretreatment, and cost-effective methods. Therefore, it is necessary to combine them with feasible business models to facilitate the valorization methods for the rice industry by-products in many developed and developing countries [7,10].

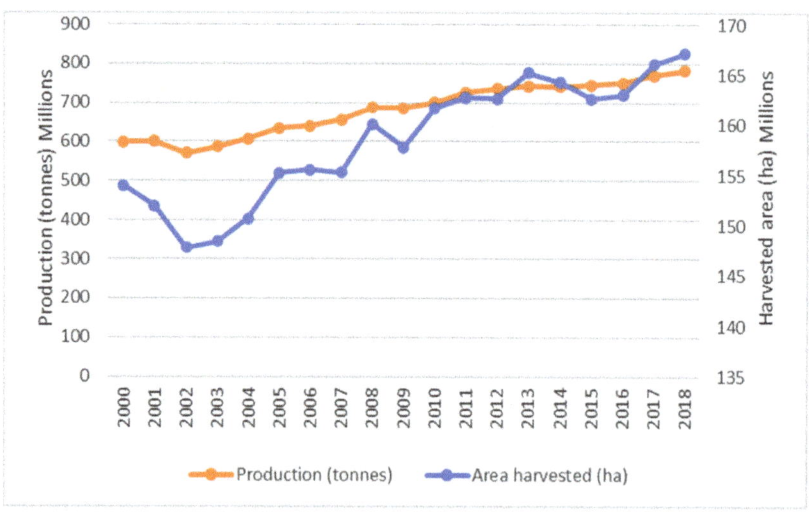

Figure 1. Rice production and area harvested, adopted from [11].

However, if wastes are valorized as raw materials to complete parts of equipment through disassembly and remanufacturing, they can create value for entire supply chains, allowing additional markets for used components beyond the raw materials market and developing new specialized professions. These openings are crucial because they establish an indirect market where resources, energy, and components from various waste streams in the rice value chain produce income for the communities [12].

Therefore, the primary purpose of this study is to consider current waste management practices for the main types of waste in the rice industry, namely rice straws, rice bran, and rice husk, including rice husk ash that remains from processes involving energy generation. Additionally, this review presents the strategies and potentialities of these by-products for developing full utilization in the economy. Therefore, the study is focused on the potentiality of all types of products, such as high-value low-volume products, high-volume low-value products, and value-added products (Figure 2). However, the implementation of these valorization options will be determined by practical feasibility. In addition, these options will be analyzed for compatibility with other factors such as economic feasibility and social acceptance.

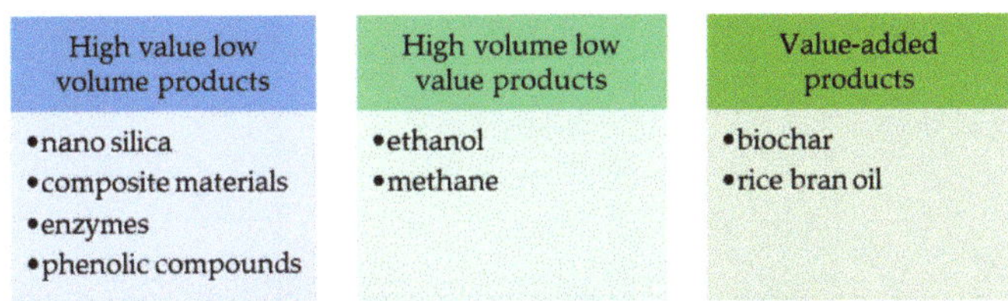

Figure 2. Focused valorization options.

2. Rice Industry Value Chain

Agricultural food processing consists of a variety of value chains and generates different types of agricultural waste through the value chain from farm to fork [13,14] (Figure 3). Large quantities of valuable wastes are produced during the harvesting and processing stages, and these wastes should be studied and analyzed to extract their valuable parts sustainably. For example, paddy straws are produced in large quantities during paddy harvesting. These straws are commonly found in the field and are often used as fodder for animals and as bedding for livestock [15]. The remaining part is burned by the farmers when ready for subsequent cultivation. Additionally, rice husk frequently becomes a material for burning and is discarded in landfills [16]. The destruction of this valuable biomass without proper use will cause irreversible damage to the environment and all living beings on the planet. Bran is another by-product of the rice processing value chain (Figure 3). After harvesting and milling, the by-products of the rice industry can be subjected to industrial symbiosis to exploit their full value [2,13].

Rice farming is integrated with geography, soil type, water availability, harvesting and processing techniques, and market behavior [17,18]. Farmers grow different types of rice in different regions because each grows best in specific soil types and climates. Additionally, access to water and proper irrigation systems boost agricultural production [19,20]. Many small and medium rice farmers still depend on labor [14]. Traditional techniques are slow and dependent on workability and experience. Modern technology and equipment are helpful in rice-producing areas but expensive [13,14,21,22]. Consequently, different production conditions affect the rice industry's waste [21,22]. Therefore, waste calculation should include unused rice and other waste due to harvesting and processing errors [2,13]. Consumer behavior and attitudes will determine how much rice is wasted at the end of the food value chain [23–25]. Edible rice wastage is due to bad taste, rotting, and leftovers [23–25].

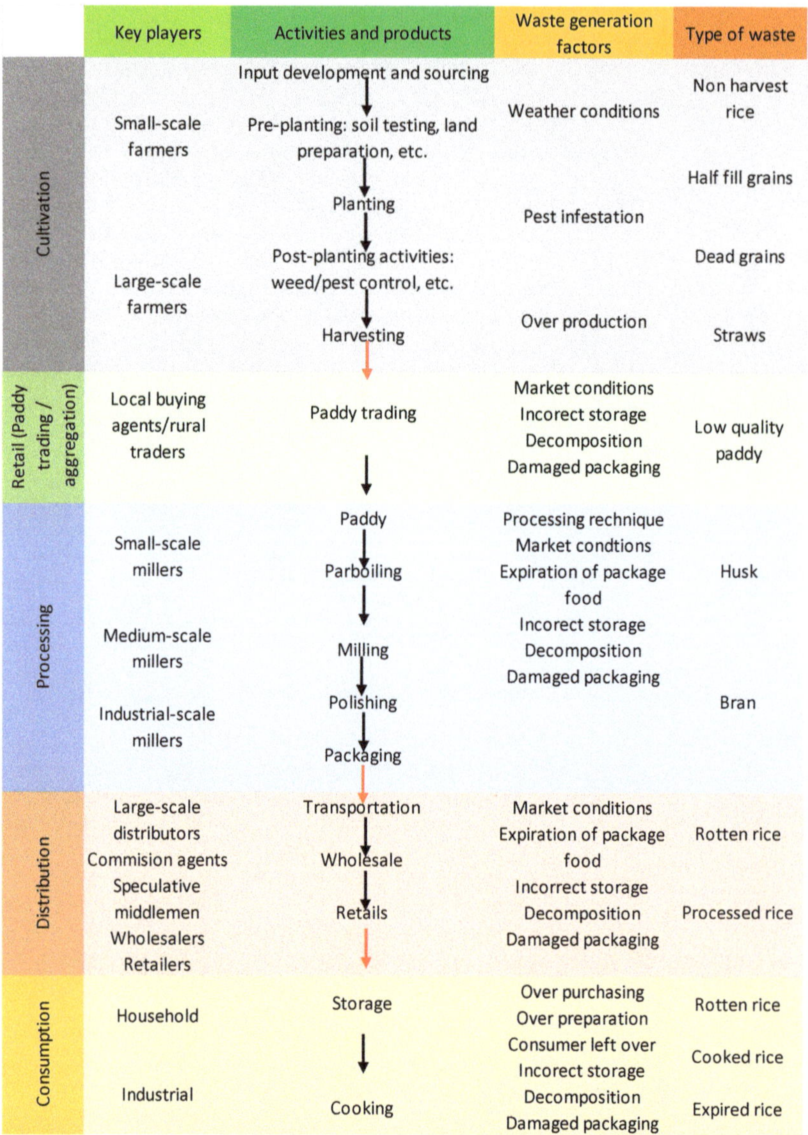

Figure 3. Conceptual model of the rice waste generated all along the food chain (red arrows represent the interaction points of each stage throughout the value chain) adopted from [14,26–29].

Waste Production throughout the Value Chain

After completion of the harvesting process, the paddy is transferred to the rice mills to be processed into white or brown rice. The paddy is subjected to a series of operational procedures during the rice milling process to remove straw particles, half-filled seeds, husks, bran, and germ. Several milling processes exist, such as one-stage milling and multi-stage milling [2,30]. Compared to multi-stage mills, the one-stage milling technique produces fewer by-products. The large-scale industrial milling process has several steps, such as cleaning (removing chaff, dead seeds, seeds that are only half full, and stones), parboiling, de-husking, peeling, polishing, and grading [2,30,31]. In addition, specific varieties of rice will be washed in hot water for a certain amount of time to remove the husk, enhance its size, and obtain a better shape of the grains [31].

There are several ways to remove rice husks from rice seeds. The germ particles and outer bran are removed after the husking in a series of huller reels and pearling cones, where the waxy cuticle is sheared off by friction between the high-speed abrasive cone and its casing [30–32]. As a result, rice bran is generated as a by-product [33]. The milling gap between the cone and the cover can be changed. Therefore, the grinding ratio can be changed by raising or lowering the cone [30,31]. Typically, in most rice mills, the rice passes through several cones, each with a higher milling rate than the previous one [30]. Since the milling space between the cone and the casing is adjustable [31], the milling rate can be varied by raising or lowering the cone [29]. The bran from the different stages is usually quantified as one product [33]. Next, rice from the pearlier is passed through polishers to get a finer appearance to the rice grains [2,26]. In this process, some parts of the starchy kernel are removed. This by-product is called fine bran if it is included within the inner bran layer. Finally, the mixture of whole and broken rice from the polishers are subjected to the sieving process and graded according to the standard at which the rice is sold [27].

According to previous research, the ratio of useable products to by-products is shown in Figure 4. Pollards are often a blend of polishings and bran. However, all these by-products are generated during the rice milling process, and their amount is roughly 60% rice husk, 35% rice bran, and 5% polishing from the whole rice mill waste stream [16,34–39].

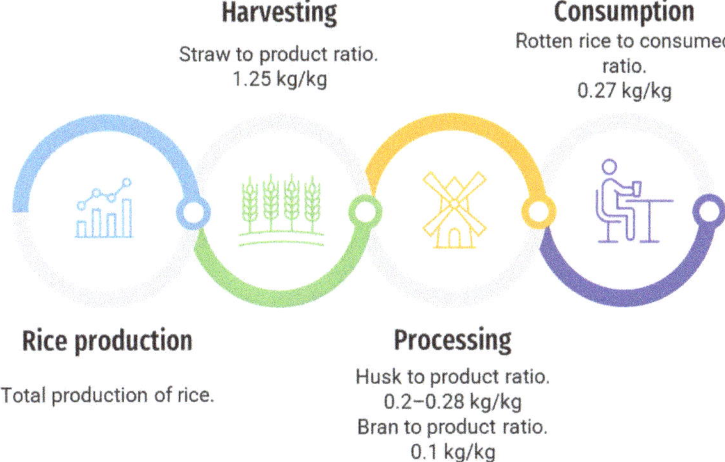

Figure 4. Quantity of rice processing by-products throughout the value chain adopted from [16,34–39].

Total rice consumption worldwide from 2008/2009 to 2021/2022 (in 1000 metric tons) is shown in Figure 5. The FAO Agricultural Outlook predicts that paddy production will rise to 52603 metric tons by 2027 compared to 2018 [40]. Due to factors including the increase in agriculturally usable land, technological advancements, and faster population growth in recent decades, global agricultural output has expanded dramatically [40].

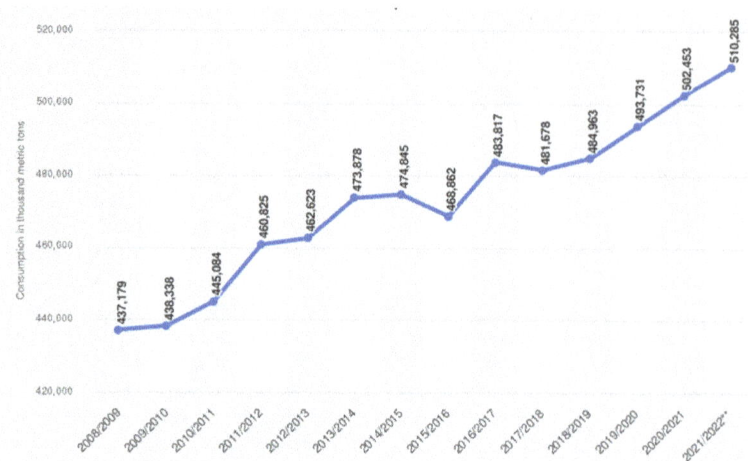

Figure 5. Total rice consumption worldwide from 2008/2009 to 2021/2022 (in 1000 metric tons) adopted from [41]. ** estimate as of January 2022.

As agricultural waste generates economic benefits, agricultural waste recycling is not meant to degrade value like other industrial waste recycling does [42]. Due to the nature of systemically implemented operations, recycling must be compared to materials that remain the same or lose performance when recycled. Due to their inherent propensity for rapid spoiling, agri-food supply chain management may need to be more sustainable and efficient [43]. Having a systemic vision and viewpoint that prioritizes the concepts of complexity and networks is essential for solving this challenge [13,42–44]. According to this method of thinking, a system is a collection of interconnected individuals whose behaviors are determined by their connections. When all of these elements are considered, they form a holistic system with more worth than just the sum of its individual parts. From this point of view, designing the agri-food scenario using a systemic approach is a viable method to begin a paradigm shift that entails switching from linear to circular structures.

3. Analysis of Rice Supply Chain Waste

3.1. Rice Straw

Rice straw is the vegetative part of the rice plant (*Oryza sativa* L.). Rice straw consists of the plant's stem, leaves, and pods and is generated after being cut off during harvest. Rice straw comprises cellulose, lignin, waxes, silicates, and minerals. In general, animals are often fed with rice straw, and rice straw can be utilized for creating compost, paper, cow bedding, and crafts; it also offers energy to specific industries and covers agricultural areas [45,46]. The rice straw of the current year is usually burned before the subsequent plowing to prepare the field.

Composition of Rice Straw

Variety, cultivated area, seasons, nitrogen fertilizer, plant maturity, plant health, and several other environmental and human variables significantly impact the chemical composition of any biomass [15,47]. Changes in chemical and physical parameters affect the yield and quality of the final product. Heterogeneity is thus seen as detrimental to the manufacturing process. Additionally, this impacts how by-products are used at the end of their life cycle. Therefore, compositional analysis and structural characterization should be considered to enhance the effectiveness of the valorization options. Tables 1–3 provide the average values of various important parameters describing raw and processed rice straw (based on the energy, nutritional, and fertilizer properties, respectively) as obtained by the chemical analyses. Rice straw has a greater silica concentration but less lignin than the straws of other cereals [48]. In order to maximize silica amount in the stem ratio, it is advised that the rice straw be shortened as much as possible [2]. Cell walls may contain silica, or silica may be soluble in water. They are eliminated with urine, where they sometimes crystallize. Since rice straw has a high oxalate content (1–2% of dry matter) and is known to lower Ca concentrations, adding supplemental Ca is often recommended [49]. Variety, time duration between harvest and storage, amount of nitrogen fertilizer used, plant maturity (lignin content increases with maturity), plant health, and environmental conditions affect the quality of rice straw [15]. Although rice straw is a rich energy source, it contains only 2–7% protein and is indigestible due to its high silica content. Therefore, it is considered a coarse and low-quality food source [50]. Minerals such as sulfur may be a limiting factor when considering it as fodder [51]. Other conditions usually involve:

- Excessive amounts of neutral detergent fiber (NDF) lead to decreased feed consumption and fat-corrected milk output [52].
- There is not enough P, Cu, Zn, Ca, and NaCl to meet the needs of animals [53].
- In comparison to corn silage, it contains less energy, has an unpleasant taste, and uses nitrogen less effectively [54].

Table 1. Energy characteristics of rice straw [2,47,55–59].

Properties	Average Values (MJ/kg)				
	Rice Straw	Rice Straw Urea Treated	Rice Straw Ammonia Treated	Rice Straw NaOH Treated	Other
Higher heating value (HHV)	-	-	-	-	15.5
Lower heating value (LHV)	-	-	-	-	14.43
	Average values (% wt)				
Fixed carbon	-	-	-	-	14.6
Volatile matter	-	-	-	-	60.28
Ash	-	-	-	-	16.83
Moisture	7.2	6.00	29.1	30.6	8.47
Carbon	-	-	-	-	38.58
Oxygen	-	-	-	-	35.79
Hydrogen	-	-	-	-	4.97
Nitrogen	-	-	-	-	1.12
Sulphur	-	-	-	-	0.145

Table 2. Fodder characteristics of rice straw [2,47,55–59].

Properties	Average Values (% wt)				
	Rice Straw	Rice Straw Urea Treated	Rice Straw Ammonia Treated	Rice Straw NaOH Treated	Other
Dry matter	92.8	94.0	70.9	69.4	-
Crude protein	4.2	7.9	11.0	2.9	-
Crude fiber	35.1	34.2	-	36.2	39.00
Neutral detergent fiber (NDF)	69.1	68.6	67.5	63.4	67.10
Acid detergent fiber (ADF)	42.4	42.3	-	37.7	-
Lignin	4.8	5.2	-	-	12.50
Ether extract	1.4	1.3	-	0.9	-
Ash	18.1	19.3	15.2	19.0	18.6
Gross energy	15.5	16.0	-	15.2	-

Table 3. Fertilizer characteristics of rice straw [2,47,55–59].

Properties	Average Values (g/kg)				
	Rice Straw	Rice Straw Urea Treated	Rice Straw Ammonia Treated	Rice Straw NaOH Treated	Other
Nitrogen	6.72	12.64	17.6	4.64	11.2
Phosphorus	0.9	1.5	-	0.4	1.2
Potassium	18	17.5	-	-	20.2
Calcium	2.9	3.2	-	2.1	3.3
Magnesium	1.9	1.7	-	-	2.0
Sulphur	-	-	-	-	0.7
Silica	-	-	-	-	68.8
Manganese	454	387	-	-	-
Zinc	34	34	-	-	-
Copper	6	3	-	-	-
Iron	355	-	-	-	956.3

3.2. Rice Bran

A significant waste product in the value chain of rice processing is rice bran. It is mainly used as animal feed and is regarded as a healthy source of fiber for pets because of its high nutritional content. Additionally, farmers may get it at a significant discount because of its availability. Due to the high fat and fiber content of rice bran, up to 40% of it is added to the diets of cattle, dogs, pigs, and chickens [60–62]. Additionally, rice bran is a valuable feed for many animals since it contains 14–18% oil. Therefore, dehulled rice bran may be utilized in more value-added processes than ordinary rice bran [2].

Composition of Rice Bran

The composition of rice bran has a significant role in defining its possible valorization options. Rice bran's physical and chemical properties are influenced by several aspects concerning the grain and the milling procedure [63]. Rice variety, environmental circumstances, grain size and form, distribution, chemical components, strength of the outermost layer, and breaking resistance are the primary elements affecting rice grain [64]. Additionally, the type of grinding machine is the main factor related to the processing conditions, and the grinding process of different layers of rice grain at different depths shows different chemical compositions [63,64].

Rice bran contains various nutrients, including carbohydrates, proteins, minerals, and lipids. It has a high carbohydrate (cellulose and hemicellulose) content and is simple to employ to create microbial products with added value [65]. As a result, before the valuation procedure, it is required to assess the composition. Because rice bran is employed in value-added goods as a microbial product or as a food additive, it is generated during several phases of the rice milling process, which are eventually combined and discharged as rice bran. As a consequence, the chemical composition varies significantly [2]. In addition, the chemical composition of raw rice bran and de-oiled rice bran varies in fiber concentration [55]. Tables 4–7 reflect the chemical analyses regarding its energy-related parameters, fertilizer-related features, feed-related parameters, and bioactive-component qualities.

Rice bran stands out compared to other cereal grains due to the tocotrienol, tocopherol, γ-oryzanol, and β-sitosterol contents [66]. This is significant since there is mounting evidence that these substances may help to lower levels of total plasma cholesterol, triglycerides, and low-density lipoprotein while raising levels of high-density lipoprotein [66]. In addition, ferulic acid and soluble fiber (including β-glucan, pectin, and gums) are found in the indigestible cell walls of rice bran. While the United States Department of Agriculture (USDA) nutritional database values for crude rice bran are often utilized in animal diet formulation [67], caution must be taken since they may not account for changes across rice cultivars [68].

The variety of rice bran utilized determines the chemical content and quality of the end product. According to Hong and his co-workers [69], the fatty-acid content of rice bran oil varies based on the type of rice bran utilized. According to the same paper, rice bran oil, which includes a high concentration of free fatty acids, has several drawbacks when used as fuel in diesel engines in the winter season.

Table 4. Energy characteristics of rice bran [2,55,56,63–65,70–74].

Parameters	Average Values (MJ/kg)							
	Fiber < 4%	Fiber 4–11%	Fiber 11–20%	Fiber > 20%	Defatted Fiber < 11%	Defatted Fiber 11–20%	Defatted Fiber > 20%	Other
HHV	-	-	-	-	-	-	-	15.29
LHV	-	-	-	-	-	-	-	14.17
	Average Values (% wt)							
Fixed carbon	-	-	-	-	-	-	-	19.53
Volatile Mmatter	-	-	-	-	-	-	-	61.83 (organic)
Moisture	10.0	9.9	9.8	8.3	10.3	11.0	8.4	-
Carbon	-	-	-	-	-	-	-	38.92
Oxygen	-	-	-	-	-	-	-	36.77
Hydrogen	-	-	-	-	-	-	-	5.12
Nitrogen	-	-	-	-	-	-	-	0.55
Sulphur	-	-	-	-	-	-	-	0.0

Table 5. Fodder characteristics of rice bran [2,55,56,63–65,70–74].

Parameters	Average Values (% wt)							
	Fiber < 4%	Fiber 4–11%	Fiber 11–20%	Fiber > 20%	Defatted Fiber < 11%	Defatted Fiber 11–20%	Defatted Fiber > 20%	Other
Dry matter	90.0	90.1	90.2	91.7	89.7	89.0	91.6	-
Crude protein	14.2	14.8	12.7	8.8	16.0	17.1	6.7	-
Crude fiber	4.1	8.6	16.3	28.3	9.8	14.8	30.8	-
NDF	12.4	25.2	34.4	48.7	26.6	32.6	51.7	-
ADF	3.2	11.2	19.6	32.7	12.5	18.0	35.4	25.2
Lignin	1.2	4.1	6.8	11.0	4.5	6.3	11.8	10.0
Ether extract	13.2	17.2	14.4	10.3	4.1	1.0	4.8	-
Ash	6.9	9.4	12.4	13.6	12.3	14.2	19.1	19.0
Starch	42.0	28.8	22.4	14.7	32.2	26.4	14.2	-
Total sugars	3.8	2.8	2.8	1.0	2.7	3.0	1.6	-
Gross energy	20.5	21.2	20.2	19.3	17.9	17.1	17.0	-
Lipids	-	-	-	-	-	-	-	-

Table 6. Fertilizer characteristics of rice bran [2,55,56,63–65,70–74].

Parameters	Average Values							
	Fiber < 4%	Fiber 4–11%	Fiber 11–20%	Fiber > 20%	Defatted Fiber < 11%	Defatted Fiber 11–20%	Defatted Fiber > 20%	Other
Nitrogen	22.7	23.68	2032	14.08	25.6	27.36	10.72	-
Phosphorus	13.9	17.0	13.8	7.4	12.1	19.2	4.9	-
Potassium	10.8	14.9	12.3	6.3	8.5	7.4	7.3	-
Calcium	0.6	0.7	0.7	4.7	0.8	2.5	1.0	-
Magnesium	6.1	7.8	6.5	2.1	4.6	4.4	2.4	-
Sulphur	-	-	-	-	-	-	-	-
	Average Values (mg/kg)							
Manganese	-	211.0	138.0	-	221.0	164.0	157.0	-
Zinc	-	63.0	55.0	-	80.0	80.0	34.0	-
Copper	-	8.0	9.0	-	14.0	13.0	7.0	-
Iron	-	106.0	-	-	297.0	556.0	443.0	-

Table 7. Biochemical characteristics of rice bran [2,55,56,63–65,70–75].

Parameters	Average Values (% wt)							
	Fiber < 4%	Fiber 4–11%	Fiber 11–20%	Fiber > 20%	Defatted Fiber < 11%	Defatted Fiber 11–20%	Defatted Fiber > 20%	Other
Protein	-	-	-	-	-	-	-	6.4
Aminoamides	% Protein							
Alanine	5.9	6.4	5.8	-	6.0	5.7	6.2	-
Arginine	7.7	6.6	7.2	-	7.0	6.2	7.4	-
Aspartic acid	7.9	9.0	9.3	-	8.7	8.8	8.1	-
Cystine	1.1	1.2	1.7	-	1.7	1.7	1.2	-
Glutamic acid	13.5	13.0	12.7	-	15.5	12.6	12.7	-
Glycine	4.9	5.3	5.2	-	5.1	5.0	5.4	-
Histidine	2.6	2.6	2.4	-	2.5	2.3	2.4	-
Isoleucine	5.8	5.9	5.3	-	4.8	4.2	6.7	-
Leucine	6.7	6.7	7.0	-	7.2	7.0	7.5	-
Lysine	4.5	4.7	4.4	-	4.4	3.9	4.6	-
Methionine	2.3	2.2	1.9	-	2.4	1.9	2.1	-
Phenylalanine	4.6	4.4	4.4	-	4.9	4.7	4.8	-
Proline	4.7	5.3	4.6	-	5.1	5.6	6.1	-
Serine	4.3	4.6	4.0	-	4.8	4.5	4.3	-

3.3. Rice Husk

Rice husk is the outer covering of the rice grain and is produced as a by-product of the rice milling process. It is also called hull and chaff [39,76]. In agricultural nations, this is the most prevalent agricultural by-product. In particular, rice husk is utilized as the primary source of energy in rice mills, poultry farming, and silica-rich cement [56,77,78]. Additionally, small quantities are used as construction materials and fertilizers [79]. However, most rice husks eventually wind up in landfills or are burned in the open air, significantly polluting the environment. The calorific value of rice husk is considerably high, roughly 16,720 kJ/kg [80]. As previously stated, many millers directly burn or gasify rice husk as their primary energy source [16]. Rice husk ash is another type of waste produced during this burning procedure. This additional waste, which makes up around 25% of the original volume of rice husks, has a significant adverse effect on the environment [38].

Composition of Rice Husk

Due to photosynthesis and biochemical interactions, silica and a barrier layer are formed on the rice plant's stem and husk surfaces [81]. These layers have developed to shield the rice plant and its grains from environmental changes such as temperature variations, excessive water evaporation, and microbial assault [81]. Approximately 20–30% of the rice husk is made up of mineral components, including silica and metallic residues containing magnesium (Mg), iron (Fe), and sodium (Na). Calcium (Ca), manganese (Mn), and potassium (K) are further examples of trace elements [82]. Rice husk mainly comprises organic compounds, including cellulose, lignin, and hemicellulose, making up around 70–80% of the total weight [37,83]. Rice husk is maturing into a raw material prospective in the manufacturing sector. However, when rice husk accumulates to the point that it poses a severe threat to the local ecosystem, it is classified as agro-waste. As a result, these adverse effects on the environment must be softened via a process of valorization or value addition. Therefore, it is crucial to conduct a physicochemical investigation and determine the composition of the material. Tables 8–11 show the characteristics of rice husks in terms of energy, fodder, fertilizer, and biochemical properties. The chemical components of rice husk ash are shown in Table 12.

Table 8. Energy characteristics of rice husks [2,16,35,55,56,84–86].

Parameter	Average Value (MJ/kg)
HHV	13.18
LHV	12.01
Parameter	**Average Value (% wt)**
Fixed carbon	24.62
Volatile matter	46.13
Ash	19.77
Moisture	9.01
Carbon	38.52
Oxygen	35.37
Hydrogen	4.79
Nitrogen	0.39
Sulphur	0.14

Table 9. Fodder characteristics of rice husks [2,16,35,55,56,84–86].

Parameter	Average Value (% wt)
Dry matter	91.9
Crude protein	3.7
Crude fiber	42.6
NDF	75.7
ADF	52.2
Lignin	25.05
Ether extract	1.5
Ash	20.52
Starch	5.3
Gross energy	16.3

Table 10. Fertilizer characteristics of rice husks [2,16,35,55,56,84–86].

Parameter	Average Value (g/kg)
Nitrogen	4.47
Phosphorus	0.74
Potassium	4.19
Calcium	1.96
Magnesium	0.43
Sulfur	1.84
Silica	95.4
	Average Value (mg/kg)
Manganese	442
Zinc	43
Copper	2
Iron	139.4

Table 11. Biochemical characteristics of rice husk in different rice varieties [2,55,56].

Bio-Active Compounds	Average Value			
	Gladio Variety	**Carolina Variety**	**Creso Variety**	**Scirocco Variety**
Total polar phenol (TPP) (mgGAE/kg)	$27{,}898 \pm 803$	$25{,}487 \pm 1038$	$24{,}155 \pm 797$	$19{,}662 \pm 334$
p-coumaric acid (mg/kg)	6367 ± 146	5692 ± 308	5565 ± 109	4879 ± 122
ferulic acid (mg/kg)	2037 ± 110	1752 ± 76	1771 ± 103	1510 ± 86

Table 12. Chemical composition of rice husk ash [87–89].

Parameter	Average Value (% wt)
SiO_2	93.4
Al_2O_3	0.05
Fe_2O_3	0.06
CaO	0.31
MgO	0.35
K_2O	1.4
Na_2O	0.1
P_2O_5	0.8

4. Valorization Potential of Rice Industry By-Products

4.1. Valorization of Rice Straw

Rice straw could also be valorized for four different purposes: energy production, animal feed, fertilizer, and other uses. By pyrolyzing rice straw, bio-oil, biochar, and syngas may be generated. Numerous chemical substances are found in rice straw bio-oil, including alcohols, acids, furans, aromatics, ketones, phenols, and pyranoglucose [47]. Alcohol and pyranoglucose are created as a consequence of the pyrolysis of cellulose, while hemicelluloses are used to create ketones [47]. The metabolic process through which carbohydrates are changed into alcohols or acids is known as fermentation, as shown in Equations (1) and (2). Second-generation biofuels are made from cellulose feedstock (Equation (1)). Physical, chemical, or biological pretreatment and fermentation are all viable routes to their production. While its lack of competition from other feedstock substrates is an advantage, its need for highly efficient lignohemicellulose enzymatic breakdown is a drawback. Although the commercialization of second-generation ethanol facilities shows promise, the longevity of these plants will primarily rely on the market availability of the feedstocks at affordable costs [90]. Bacteria convert carbohydrates into lactic acid (Equation (2)). Numerous chemical or physical pretreatments are required, followed by enzymatic hydrolysis to convert fermentable sugars from lignocellulosic materials into ethanol or lactic acid. In addition to its many uses in the food and beverage industries, lactic acid and its derivatives also have a wide range of applications in the pharmaceutical, cosmetic, and manufacturing industries [91,92]. Numerous studies have demonstrated that rice straw can be utilized to make second-generation biofuels [47,93–96]. Typically, bacteria and yeast turn carbohydrates into lactic acid and sugars into alcohol. *Trichoderma reesei*, which was derived from decaying rice straw waste, produces cellulases that break down cellulose in the rice straw to glucose, which is then fermented with yeasts such as *S. cerevisiae* to make ethanol [93,97,98].

$$C_6H_{12}O_6 \rightarrow 2C_2H_5OH + 2CO_2 \tag{1}$$

$$C_6H_{12}O_6 \rightarrow 2CH_3CHOHCOOH \text{ (lactic acid)} \tag{2}$$

The anaerobic digestion process may convert rice straw into biogas [75,99,100]. Anaerobic digestion is a sustainable process that converts organic waste into usable energy. Generating green energy from rice straw is an effective way to lessen the effects of global warming [75]. Around the globe, rice straw is utilized directly as an energy source for heating rooms by direct burning, firing clay pots, and cooking [101]. Additionally, small grids in certain nations such as Nigeria have a higher potential for using rice husks and straw as a source of rural power [102,103]. Umar and co-workers [102,103] claimed that rice straws have the potential to generate 1.3 million MWhy-1 energy in a country such as Nigeria. A 36 MW power plant in Sutton, Ely, Cambridgeshire, was constructed in 2000, producing more than 270 GWh annually while using 200,000 tons of rice straw [104]. Another work shows that Sri Lanka has a total energy capacity of 2129.24 ktoe/year of primary energy from rice straw and rice husk and a capacity of 977 Mwe, allowing it to produce 5.65 TWh of electricity per year [16].

According to literature, rice straw may be used efficiently for composite preparation [105–111]. Furthermore, rice straw microfibrils at 5% increase the characteristics of rice straw polypropylene composites [107]. Another study highlights many uses of rice-straw cement bricks for load-bearing walls [106]. Rice straw can also be used to lower the price of cement bricks with sufficient thermal insulation, appropriate mechanical qualities, and fire resistance [112–115]. Furthermore, rice straw-based composites with adhesives generated from starch can be used as ceiling panels and bulletin boards [109]. Finally, following proper pretreatment, rice straw could also be utilized to produce fiberboard [116].

Rice straw may be used to produce many kinds of enzymes in an industrial setting [47,57,58]. *Trichoderma harsanum* SNRS3 can generate cellulase and xylanase using alkali-pretreated rice straw [58]. According to this research, rice straw is a more effective inducer of the formation of cellulase and xylanase and does not need the inclusion of other chemicals. Lactic acid can be produced by using pretreated rice straw [117,118]. A Naviglio extractor and trifluoroacetic acid could transform rice straw into a unique bioplastic that can be used as shrink films, sheets, or for shape memory effects. Its mechanical characteristics are equivalent to polystyrene in the dry state, while in the wet state, the cast bioplastic performs equal to plasticized poly(vinyl chloride) [119].

Rice straw has a low value as a feeding material, despite its use as bedding for cattle [15]. In contrast to ruminants, which depend on symbiotic bacteria to break down cellulose in the gastrointestinal system, all vertebrates lack the enzymes necessary to dissolve β-acetyl bonds [52]. Additionally, dried rice straw contains low nutrient value owing to its low amount of protein and high amounts of lignin and silica. However, this may be addressed by pretreating it with ammonia or urea [48]. To increase the nutrient availability of rice straw, it can be converted into silage. Therefore, some researchers have focused on improving rice straw harvesting technologies for silage production [120]. Other studies have explored several practical examples of silage processing, including using different additives to enhance fermentation quality and adding yeasts such as *Candida tropicalis* [121–123]. Feed intake, digestibility, rumen fermentation, and microbial N synthesis efficiency are improved after urea treatment of rice straw [124].

Rice straw has been proposed as a low-cost adsorbent for purifying contaminated water [125]. However, straw surface composition and metal speciation significantly impact the adsorption capacity, which changes with metal ions and water pH [126,127]. On the adsorbent surface of rice straw, methyl/methylene, hydroxyl, quaternary ammonium, ether, and carbonyl groups predominate; adding additional quaternary ammonium or incorporating carboxyl groups enhances its adsorption capability [128]. The existence of these groups is supported by the ATR–FTIR spectrum shown in Figure 6. However, competing cations and chelators in the solution are likely to result in decreased sorption capacities [129]. Furthermore, most heavy metal ions exhibit maximal adsorption capacities around pH 5. In contrast, very acidic circumstances promote Cr adsorption [130], which could be the result of the reduction of Cr(VI) to Cr(III). Moreover, cellulose phosphate derived from rice straw that has been treated with NaOH and then reacted with phosphoric acid in the presence of urea has a more remarkable ability to absorb heavy metals. This ability is increased when microwave heating is used to produce it [131]. The addition of epoxy and amino compounds to rice straw by reacting with epichlorohydrin and trimethylamine results in a high sulfate adsorption efficiency, demonstrating the material's anion exchangeability [132]. Like rice husk, straw can be used as an adsorbent for different water contaminants, such as alkali and phenolic chemicals, that can usually be recovered using anionic species [133]. Various adsorbents from rice straws have also been developed to remove dyes from wastewater. An example of cationic dye application is rice straw treated with citric acid, which increases the specific surface area and pore size. These treated straws have been used to absorb crystal violet or methylene blue from an aqueous phase [134]. It has been observed that the addition of activated rice straw causes a significant reduction in microalgae in water, which has been attributed to the synergistic effects of humic chemicals and H_2O_2 created by the straw breakdown [135]. According to a different investigation, water and methanol extracts from rice straw controlled the cyanobacterium *Anabaena* sp. but promoted *Chlorella* sp. To prevent the development of *Anabaena* sp., rice straw extraction is an economical and ecologically beneficial option, but it may not work as well on other cyanobacteria and microalgae [136].

Rice straw is utilized as organic fertilizer for various crops in many places throughout the globe. It can also be used as a soil conditioner to replace the organic matter in the soil [112]. In addition, rice straw is also a growth medium for mushrooms [137]. Adding biochar derived from rice straw to the soil makes it possible to enhance the characteristics of the soil by lowering its pH, cation exchange capacity (CEC), nutrient availability, and nitrate leaching [138–140]. Figure 7 displays an overview of all rice straw value-adding possibilities [2,47,53,57,58,75,100,106,110,130,137–139,141–144].

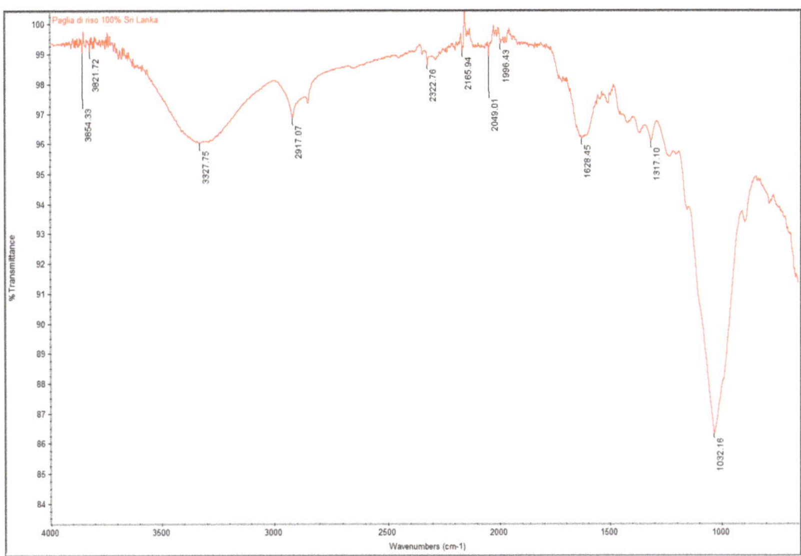

Figure 6. ATR–FTIR spectrum of Sri Lankan rice straw (own source) as-received sample acquired at room temperature using a Nicolet FTIR iS10 spectrometer (Nicolet, Madison, WI, USA) equipped with a Smart iTR with diamond plate. Straw was dried at 45 °C, milled, and sieved with a 1000 μm mesh sieve before analysis. Thirty-two scans in the 4000–600 cm^{-1} range at 4 cm^{-1} resolutions were co-added.

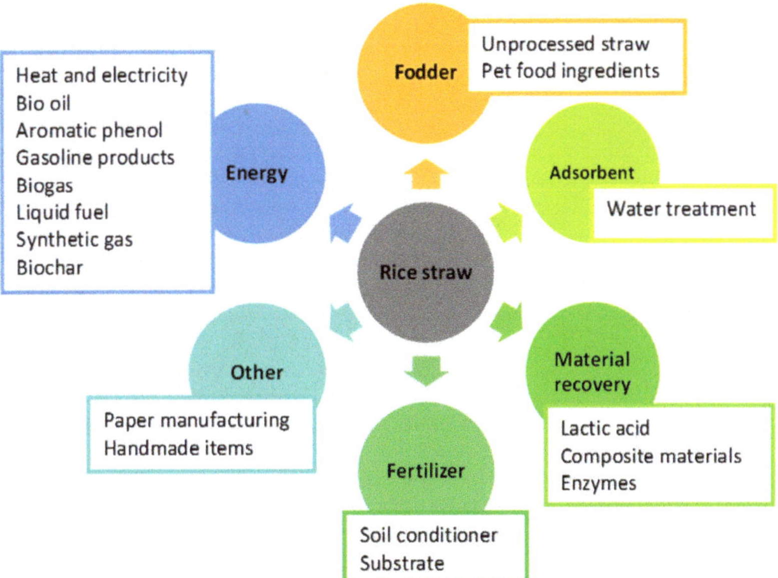

Figure 7. Summary of possible valorization options for rice straw.

4.2. Valorization of Rice Bran

Considering the concept of circular economy and green product technology, the biorefinery plan would be the best option for managing and utilizing rice bran [145]. In addition to providing more nutrients than other cereal grains, rice bran has more lipids, protein, and calories [146–148] (Table 5). Rice bran is vulnerable to oxidative rancidity; thus, heat stabilization is necessary to avoid spoilage and rancidity [149]. Rice bran oil is widely recommended around the world due to the presence of several beneficial natural and healthy bioactive ingredients. Companies have been encouraged to manufacture stabilized rice bran and rice bran products to improve the health of organisms because of the unique mix of lipids, minerals, and nutrients found in rice bran, including calcium, phosphorus, and magnesium [65]. In addition, several researchers have found that the manufacturing of de-oiled rice bran and rice bran oil is in great demand worldwide [62,150,151].

The production of biodiesel from rice bran is actively marketed all over the globe. However, rice bran oil must be removed from the rice bran to produce biodiesel via a transesterification process [152]. Several methods have purportedly been utilized to produce biodiesel from rice bran oil, including acid-catalyzed and base-catalyzed transesterification and lipase-catalyzed transesterification. However, each technique has different environmental effects as well as technological and economic benefits and drawbacks [153–156].

Some researchers have examined bioethanol synthesis from rice by-products such as rice bran, defatted rice bran, and rice washing drainage [157–160]. After pretreating stripped rice bran with diluted acid and detoxifying it, the *Pichia stipitis* NCIM 3499 strain generated an ethanol concentration of 12.47 g/L [161]. Additionally, another study found that biological pretreatment with the fungus *Aspergillus niger* increased ethanol output [162]. Numerous scientists have attempted to manufacture lactic acid from dehulled rice bran using various microbes [163–166]. Another study discovered that many *Bacillus coagulans* isolates could grow in denatured rice bran enzymatic hydrolysates without adequate nutrients, with the majority producing concentrations of lactic acid more significant than 65 g/L and yields greater than 0.85 g/g [163]. They stressed in the same paper that manufacturing lactic acid from dehulled rice bran might be economically viable.

Due to its numerous similarities to gasoline, biobutanol is the most ecologically benign substitute for traditional fossil fuels. Additionally, when HCl and enzyme treatments are used together, they can remove 41.18 g/L of sugar from dehulled rice bran and 36.2 g/L of sugar from rice bran [167]. Another study reported that both defatted rice bran hydrolysates and rice bran hydrolysates could be fermented in bioreactors with nutrients to make butanol at a rate of 12.24 g/L and 11.4 g/L, respectively [163].

Rice bran can be used as an adsorbent for polluting substances because it has a granular shape, is chemically stable, does not dissolve in water, and is easy to get. Its surface has several active sites that can remove pollutants [39]. How well these sites work depends on the chemical nature of the solution and whether or not there are other ions in the solution besides the ones to be trapped. Additionally, various functional groups on the surface of rice bran, such as hydroxyl and carbonyl groups, are responsible for its high adsorption effectiveness [39]. The existence of these groups is supported by the ATR–FTIR spectrum shown in Figure 8, which exhibits rice-straw-like peaks. Some researchers have tried to figure out the best way to remove arsenic from water using a fixed-bed column system made of rice bran. The objective of this study was to look at how different design parameters, such as flow rate, bed height, and initial concentration affected the adsorption process. The uptake capacities of As(III) and As(V) were found to be 66.95 µg/g and 78.95 µg/g, respectively [168].

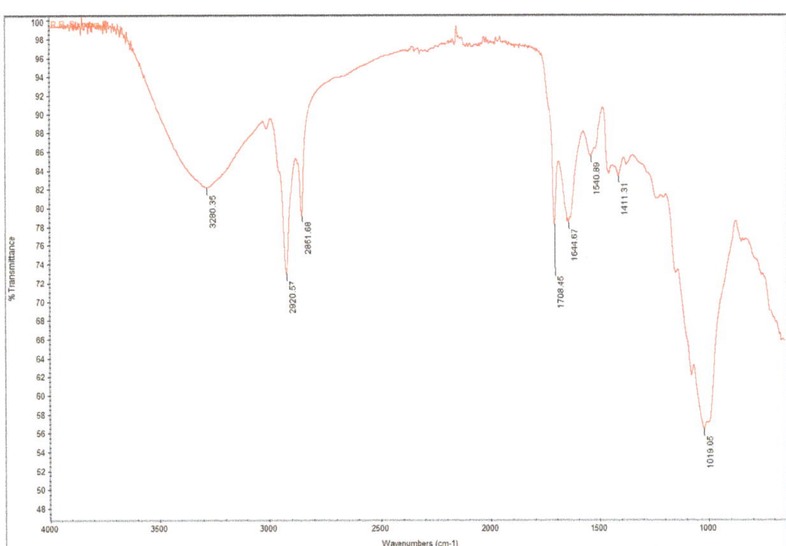

Figure 8. ATR–FTIR spectrum of Sri Lankan rice bran (own source) as-received sample acquired at room temperature using a Nicolet FTIR iS10 spectrometer (Nicolet, Madison, WI, USA) equipped with a Smart iTR with diamond plate. Bran was dried at 45 °C, milled, and sieved with a 1000 μm mesh sieve before analysis. Thirty-two scans in the 4000–600 cm^{-1} range at 4 cm^{-1} resolutions were co-added.

Another excellent substitute for conventional fossil fuels is hydrogen, which, when oxidized, merely produces water vapor (H_2O). Additionally, hydrogen has a higher energy content for mass units than traditional fuels, ranging from 112 to 142 kJ/g [168,169]. Photo and dark fermentation and their combination are all capable of producing bio-hydrogen [170]. Some studies have investigated hydrogen generation from rice bran and defatted rice bran using isolated bacteria from the same substrates. They identified *E.ludwigli* IF2SW-B4 as the most promising strain. When rice bran was utilized as a substrate, 545 mL/L of bio-hydrogen was produced [171]. The whole biotechnology process will be more economical once enzymes are produced utilizing low-cost ingredients. For the environmentally friendly and more effective release of fermentable sugars from different affordable and sustainable biomasses such as rice bran, enzymatic hydrolysis is used [171]. Researchers have conducted several investigations to synthesize enzymes from defatted rice bran and rice bran [172–174]. Figure 9 displays an overview of all rice bran value-adding possibilities.

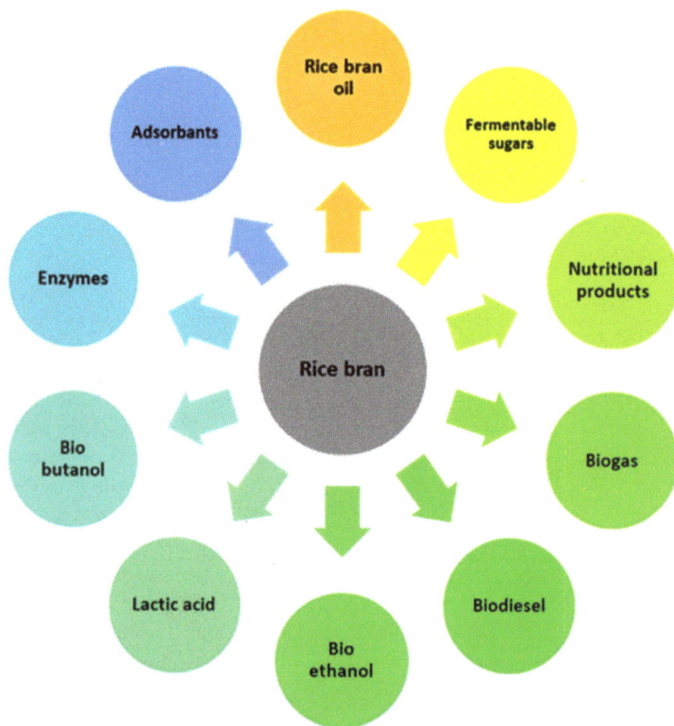

Figure 9. Possible valorization options for rice bran [39,145,146,150–153,157–174].

4.3. Valorization of Rice Husk

South Asian countries such as India, Pakistan, Bangladesh, and Sri Lanka were among the best in the world in utilizing rice husk from 1970 to 1985 [175]. In addition, governments and other organizations engaged in rice farming and the post-harvest process have provided essential direction and strong support for rice husk management. Rice husk differs from other agricultural wastes in several important physicochemical aspects, including high silica concentration, low density, high porosity, and a significant outer surface area [176]. Because of these qualities, rice husk is more valuable than other waste materials. As a result, it covers a range of industrial applications.

In water treatment, using activated carbon for the adsorption process to remove heavy metals from industrial effluents is appealing. Numerous functional groups, including hydroxyl, methyl/methylene, ether, and carbonyl are present on the rice husk's adsorbent surface, contributing to the material's enhanced adsorbent efficiency (Figure 10) [39]. The existence of these groups is supported by the ATR–FTIR spectrum shown in Figure 10, which mainly exhibits bands at 3307 cm^{-1}, 2921 cm^{-1}, 2000–2500 cm^{-1}, 1654 cm^{-1}, 1034 cm^{-1}, and 788 cm^{-1} representing hydroxyl groups, C–H groups, C≡C or C≡N bonds, C=O groups, C–O and C–H bonds, and Si–O bonds, respectively.

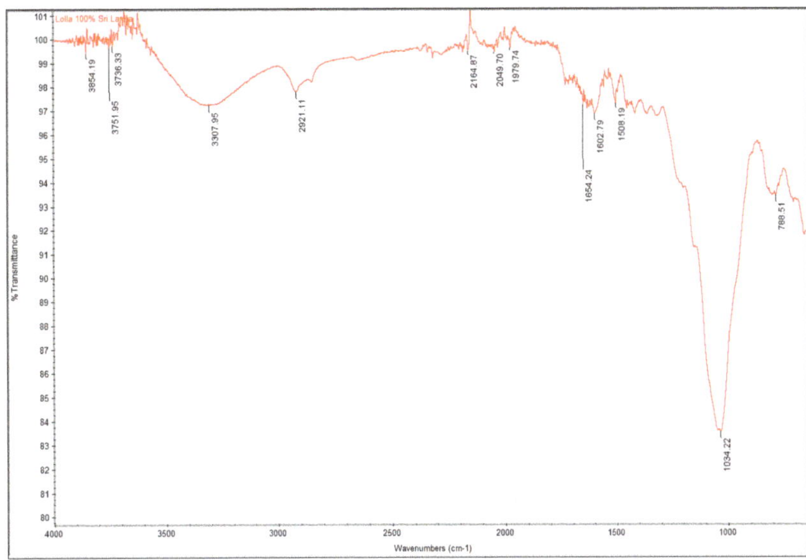

Figure 10. ATR–FTIR spectra of Sri Lankan rice husk (own source) as-received sample acquired at room temperature using a Nicolet FTIR iS10 spectrometer (Nicolet, Madison, WI, USA) equipped with a Smart iTR with diamond plate. Husk was dried at 45 °C, milled, and sieved with a 1000 μm mesh sieve before analysis. Thirty-two scans in the 4000–600 cm^{-1} range at 4 cm^{-1} resolutions were co-added.

The presence of several types of polar groups on the surface of rice husks results in a significant cation exchange capacity, indicating a potential efficacy in physisorption mode [177]. Rice husk treated with H_3PO_4 showed enhanced copper absorption capacity [178]. Some studies found that chemically treated rice husk can absorb cationic dyes such as methylene blue [179,180] and malachite green [181]. To study the absorption of fluoride from aqueous solutions, some researchers produced rice husk by chemically impregnating it with nitric acid, followed by physical activation [182]. According to their findings, the highest absorption of fluoride was 75% at a pH of 2, and the ability to absorb fluoride decreased as the pH rose from 2 to 10. Ahmaruzzaman and Gupta [183] confirmed this conclusion. When modified rice husk is cross-linked with poly(methyl methacrylate-co-maleic anhydride), nanoparticles are formed that can be used to absorb heavy metal ions (such as Pb(II)) and dyes (such as crystal violet) [184]. Researchers have discovered that novel green ceramic hollow fiber membranes made from rice husk ash can act as an adsorber and separator to remove heavy metals from water effectively [185]. Treating rice husk with H_2SO_4 and NaOH prior to heating enhances the product's capacity to absorb phenol [186].

Biomass derived from agricultural waste has been identified as a rich source of feedstock for biochar production; however, at present, farmers, and other stakeholders such as millers, practice open field burning or open dumping to dispose of these by-products. Compared to low-cost traditional treatment procedures (boiling, chlorination, sand filtration, and solar disinfection), biochar adsorbent offers various advantages. It is also suitable for low-income countries because of its availability, cheap cost, and accessible technology. Low-cost conventional approaches mainly destroy pathogens, while biochar can remove a wide range of pollutants from drinking water. Existing processes, such as chlorination, emit carcinogenic by-products, and boiling concentrates chemical contaminants. Pyrolysis temperature, vapor residence time, and other chemical and physical alteration variables influence the properties of biochar (Figure 11). Compared to conventionally activated biochar, rice husk biochar activated by a single phase of KOH-catalyzed pyrolysis under CO_2 has a larger surface area and greater capacity for phenol adsorption [187]. The gold-thiourea complex can be effectively adsorbed using biochar derived from rice husks that have been heated to 300 °C and have particular silanol groups and oxygen functional groups [188]. Rice husk-activated carbons are effective in removing phenol [189], chlorophenols [190], basic dyes [191,192], and acidic dyes [193,194] from water, as well as heavy metal ions such as Cr(VI) at low pH [195,196], Cu(II), and Pb(II) [197].

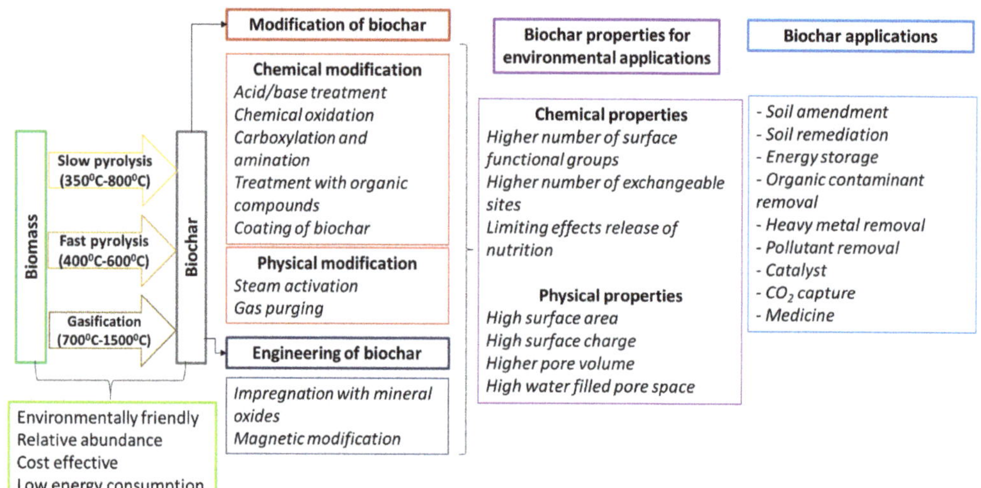

Figure 11. Strategic schematic diagram for biochar production, modification/engineering, characteristics, and water treatment applications adopted from [198–201].

Rice husk pellets provide an alternative to diesel oil and coal for energy generation in small-scale power plants. Through pyrolysis and gasification processes, they may also be used to produce biodiesel [202]. Rice husk, subjected to a thermochemical conversion process, may provide an inexhaustible supply of gaseous and liquid fuel. Thermochemical and biochemical processes are shown in Figure 12 as the two ways rice husk can be converted to energy. Thermochemical processes such as combustion, gasification, and pyrolysis are often regarded as the primary means of producing secondary energy substances. Fermentation and transesterification are also critical biochemical steps in ethanol and biodiesel production [175,203,204]. The briquettes made from rice husks with starch or gum arabic as binders burn stronger and more efficiently than timbers [205]. Another study describes a reactor that uses rice husk combined with sawdust or charcoal to generate high-grade fuel [206]. In order to obtain charcoal, which has a comparatively high calorific content, rice husk is subjected to carbonation using starch as a binder and either ferrous sulfate or sodium hypophosphite, which promote ignition [135]. Economically viable primary pyrolysis oil, suitable as boiler fuel oil and for the manufacture of catalytically treated, upgraded liquid products, can be obtained by fluidized-bed rapid pyrolysis with the catalytic treatment of rice husk [207].

Materials derived from rice husks have been used in the world's most advanced technical equipment and industries. For example, the Indian space agency has figured out how to extract high-quality silica from rice husk ash. This high-purity silica might also increase its use in the information technology sector [175]. In addition, the same publication has stated that other scientists have discovered how to extract and purify silica from rice husk ash to produce semiconductors. In addition, several researchers have pointed out the prospects of using silicon-based compounds extracted from rice husk and ash in various industries [208,209].

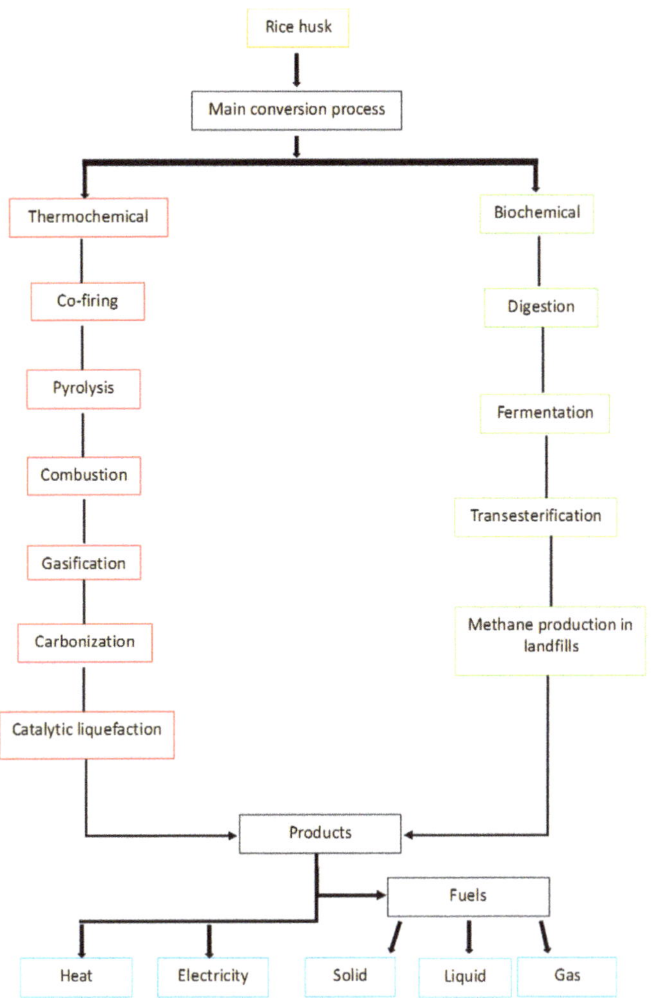

Figure 12. Energy conversion process of rice husk, adopted from [176,203,204].

Li et al. [210] stated that KOH-activated rice husk char could make porous carbons for CO_2 capture at low pressures. Low activation temperature and a small KOH/char ratio favor high CO_2 absorption and CO_2-over-N_2 selectivity. This is presumed to be due to the micropores' narrow size distribution. A similar investigation has been conducted using KOH-activated rice husk biochar for hydrogen storage. It revealed that 77 K/6 bars have a hydrogen storage capacity of 2.3%wt [211].

Due to its microscopic particle size, high solution pH, and low supportive electrolyte content, rice husk ash is an effective adsorbent for heavy metals, including lead and mercury [212,213]. In addition, the fluoride-absorption capacity of rice husk ash treated with aluminum hydroxide is enormous [214]. Both the effective removal of phenol from aqueous solutions and the adsorption of various dyes, including indigo carmine, Congo red, and methylene blue, has been accomplished using rice husk ash [180,215–218]. Due to its high silica concentration and the existence of mesopores and macropores, rice husk ash is a promising adsorbent for removing contaminants from biodiesel [219]. Zou and Yang [220] examined different approaches for generating silica and silica aerogel from rice husk ash.

Epoxy paints can use rice husk ash as a filler, and the inclusion of rice husk ash can improve a variety of qualities, including wear resistance, elongation, and scratch resistance [221]. In addition, a paper's printing quality might be enhanced by using rice husk ash. Because rice husk ash contains more silica, it can improve the paper's surface quality, and the coating layer it generates reduces the quantity of ink penetrating the paper [222]. Additionally, some researchers have studied pigments made from rice husk and ash [175].

Rice husk ash can improve the properties of cementitious materials such as concrete in resistance to corrosion. With the addition of rice husk ash, the cement particles are encased in a calcium silicate hydrate gel, making the cement denser and less porous. These characteristics are also helpful in protecting concrete against cracking, corrosion, and chemical breakdown caused by leaching agents [89,223–226]. The use of powdered rice husk ash derived under controlled burning conditions as a reinforcing filler for different rubbers has been researched. The authors discovered that substantially reinforced rice husk ash had no adverse effect on the vulcanization or aging behavior of certain rubber types, such as natural rubber, styrene–butadiene rubber, and ethylene–propylene–diene elastomers [227]. Moreover, the use of rice husk ash as a raw material in cement production has the potential to reduce production costs. Figure 13 displays an overview of all rice husk and rice husk ash value-adding possibilities.

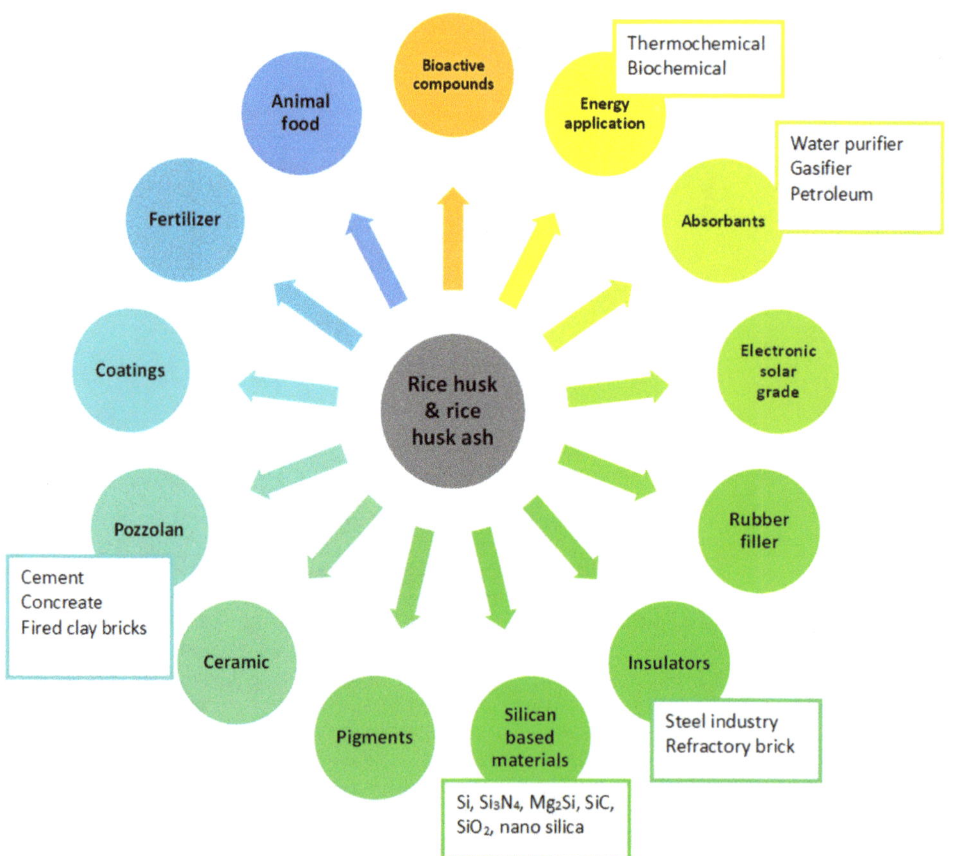

Figure 13. Possible valorization options for rice husk and its ash [176–202,205–210,212–214,216–222, 224–226,228,229].

5. Circular Economy

The circular economy (CE) has gained attention as a way to overcome the current production and consumption model of "take, make and throw away" or the linear model based on continuous progress and increasing resource output [230]. CE aims to optimize resource usage and achieve an equilibrium between economy, environment, and society by supporting closed manufacturing processes [231,232]. Numerous studies have focused on political, environmental [233], economic, and corporate issues [13]. The "reduce, reuse and recycle" (3R) concept has nine steps from recovery to recovery (Figure 14). Industrial ecology, environmental economics, and environmental policy have influenced CE [13,232,233]. Some authors have claimed that broad systems theory is where CE first emerged [13,232]. Modern concepts including "sustainable design", "performance economy", "cradle-to-grave", "biomimicry", and "blue economy" are associated with developing CE [234,235]. CE was first introduced to Europe in 1976 with Germany's Waste Disposal Act [13,232]. Later, the European Union promoted CE through the Waste Directive 2008/98/EC and the Circular Economy Package [236]. "Reduce, reuse and recycle" is part of the European Waste Directive 2008/98/EC and has been part of the US Solid Waste Agenda since 1989 [236,237]. "CE includes corporate-level sustainable production practices, increasing producer and consumer awareness and responsibility, using renewable technology and materials (where possible), and adopting appropriate, consistent and clear policies and systems" [238].

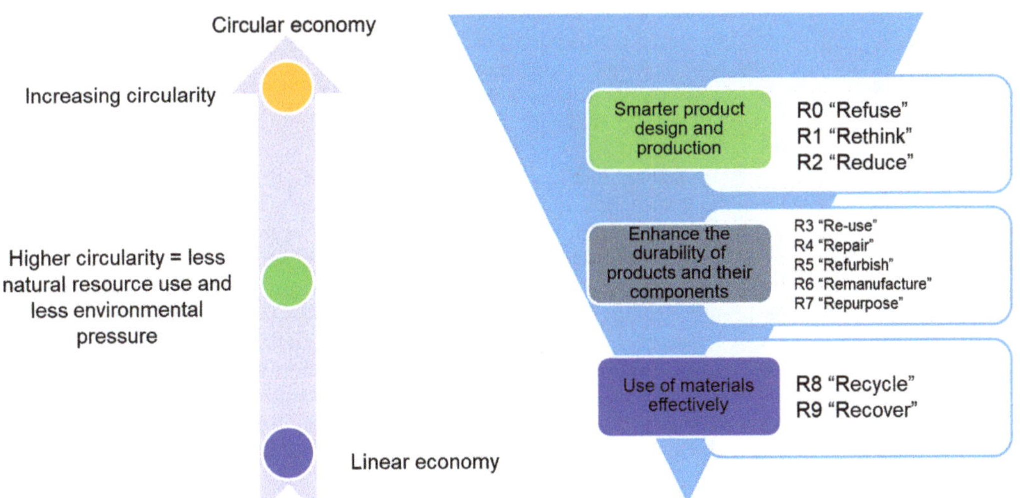

Figure 14. Priority order for circularity techniques within the production chain, adopted from [230–232].

A systemic perspective requires new solutions focusing on environmental processes and stakeholders in the relevant sectors. Agri-food waste management aims to increase resource efficiency and protect the environment. Innovative waste management solutions are needed to reduce waste or transform it into new raw materials. These management practices are part of the CE system, an industrial framework meant to be restorative or regenerative (producing no waste or pollution). Several studies have demonstrated the potential to produce bioenergy, biodegradable polymers, alcohols, and antioxidants from the food supply chain to manage agricultural wastes effectively. Thus, agricultural waste is a source of macronutrients, including proteins, carbohydrates, and fats, as well as micronutrients and bioactive chemicals used to generate new products. Biorefineries use agricultural waste to produce value-added energy and industrial goods. The scientific community considers this concept a sustainable alternative.

If a circular economy-based waste management system is successfully implemented, waste can become a source of wealth for a community or country. Sweden is a prime instance of this. They have made significant investments in infrastructure and imported a large percentage of Norway's waste to convert it into energy (electricity and heat). Thanks to this decision, Sweden can turn Norway's waste into money for its people. Consequently, it charges Norway for waste treatment, generates sufficient energy (electricity and heat) from Norway's waste to meet demand, and recycles or sells the metals it extracts from the bottom ash. In addition, the remaining bottom ash is taken to use in public infrastructure and precast concrete products, drastically reducing the need for mining operations. Because of its innovative approach to waste management within the context of the circular economy, Sweden is a standout among countries [12].

6. Case Study: Sri Lanka

6.1. Paddy and Rice Value Chain in Sri Lanka

Since 800 BC, rice has been grown in Sri Lanka [239]. Rice agriculture has grown throughout the nation due to ideal climatic conditions and geographic locations for paddy production (Figure 15). The low country dry zone has the highest rice production as this zone has had a well-planned irrigation system since ancient times. According to the Sri Lanka Rice Research Center, rice consumption per capita in 2019 was close to 107 kg. The Yala season (March to August) and the Maha season (September to December) are the two primary rice harvesting seasons in Sri Lanka. The paddy and rice value chain in Sri Lanka comprises public and private partners connecting rice producers such as small, medium, and industrial-scale farmers, millers (cooperative millers, rice marketing boards, and private millers), food processors, and consumers. In specific Sri Lankan mills, just one step of milling is performed [14]. Figure 16 represents the paddy and rice value chain in Sri Lanka.

Small-scale farmers in villages and semi-urban areas produce sufficient rice for personal use and store it throughout the season or the year. Mid-level farmers store for their consumption and sell excess paddy. Farmers sell these quantities directly to millers, and paddy collectors act as intermediaries in this buying and selling process. After collecting a substantial amount of paddy, they will sell it to co-operative or industrial-scale mills. Large rice farmers who grow rice on an industrial scale sell their crops directly to rice processing mills or process them in their mills. In most cases, small-scale private millers process only a small amount of waste in a particular area throughout the year or during a specific period.

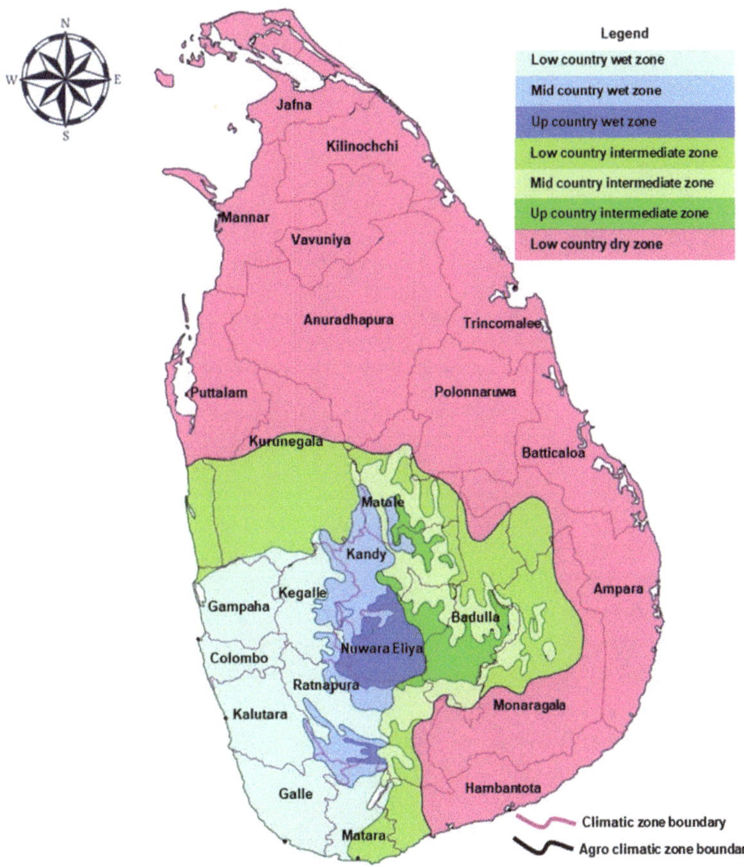

Figure 15. Map of Sri Lanka district distribution by climate zone and geographic regions adopted from [240].

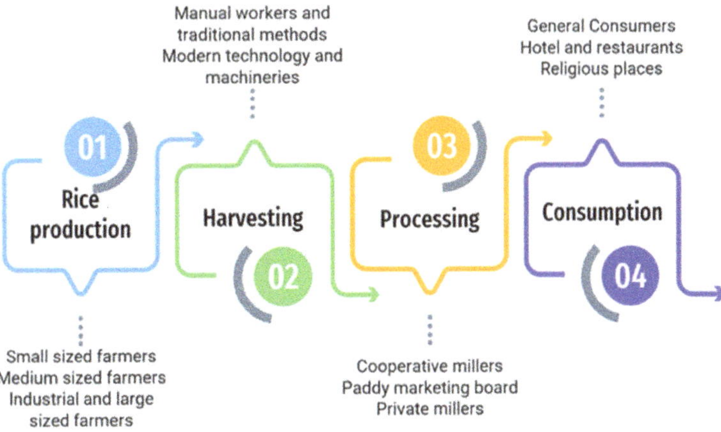

Figure 16. Paddy and rice value chain of Sri Lanka adopted from [14].

6.2. Rice Waste Availability in Sri Lanka

The vegetative portion of the rice plant is called rice straw (*Oryza sativa* L.). After the grain has been harvested or sliced, rice straw is produced. According to several researchers, the ratio of rice straw to grain is between 1.0 and 1.5 kg [2,16,39]. Previous studies have shown that 0.1 kg of rice bran is produced for every kilogram of rice [2,78]. Various scholars have revealed that approximately 20–28% wt of rice husk generates greater grain weight [2,35,37,38]. Abbas and Ansumali [79] noted that around 50% of the country's rice mill husk is burned to fulfill its energy needs for steam production. According to some literature [38], approximately 25% of rice husk ash is generated during the burning process. According to the data gathered from Sri Lanka Rice Research Center, in the 2019 Yala season, Sri Lanka produced 1,519,475 metric tons of rice, 1,899,343.75 metric tons of rice straw, 151,947.5 metric tons of rice bran, 364,674 metric tons of rice husk, and 45,584.25 metric tons of rice husk ash as shown in Figure 17. Because of varying weather conditions and several other variables, rice production differs among districts.

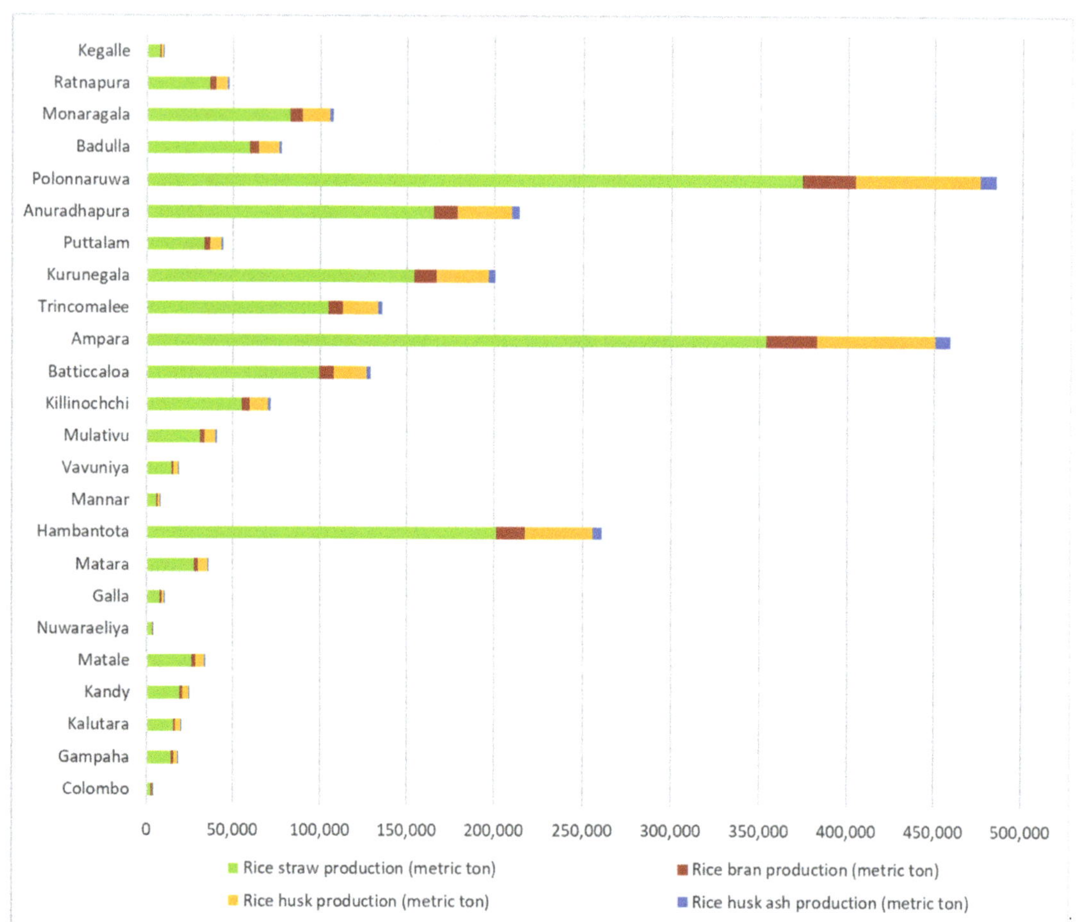

Figure 17. Rice straw, Rice bran, rice husk, and rice husk ash availability in each district of Sri Lanka (Source: Sri Lanka Rice Research Center).

6.3. Potentiality of Rice Waste Valorization in Sri Lanka

Currently, rice industry by-products are resold and recycled but have yet to be fully utilized. They are high in nutrients and have chemical–physical properties, making them useful in various sectors of the economy (food, fertilizer, building materials, energy, etc.). Therefore, sustainable and efficient agricultural waste management has emerged as a critical concern for all parties involved in the agricultural value chain in Sri Lanka. According to Figure 17, massive volumes of waste are discharged into the environment at every point along the value chain, permanently harming the air, water, and land. Proper management and implementation of a sustainable valorization system is a complex transformation for developing countries. It needs more funding, laws, and regulations to enhance the capabilities and techniques for agricultural waste management and coordinated efforts by local, regional, and global stakeholders. Therefore, any waste management system that applies to this industry should be founded on value principles and promote the circular economy throughout the process. Figure 18 displays the challenges that should be overcome to implement a sustainable valorization waste management system. The authors have found some feasible valorization techniques for rice processing by-products in the Sri Lankan context:

- Animal feed: Rice straw and rice bran can be used as animal feed, and a fitting technology can be used to produce nutritious and high-quality goods.
- Energy generation: Rice straw and rice husk can be used in grate-fired combustion boilers using steam turbine cycle technology to produce energy.
- Energy storage: activated porous rice husk and bran biochar can be used as hydrogen-storing carbon-based material with significant added value.
- Adsorbent in water treatment: Activated carbon from rice husk and rice straw can be utilized for wastewater treatment.
- Anaerobic co-digestion: A more efficient method of valorizing rice-related waste to produce green energy.
- Fertilizer production: Rice waste has a high potential for producing biochar and fertilizer to improve soil structure and organic matter content.
- Adsorption: The industrial production of adsorption by rice husks has enormous potential.
- Construction sector: Waste from the rice value chain can be effectively used in the building industry without affecting the end product negatively.
- Rice husk can be used in the ceramics sector, according to several studies.
- Rubber industry: Rice can may be used as filler in the rubber industry since Sri Lanka is one of the greatest rubber-producing nations in the world.

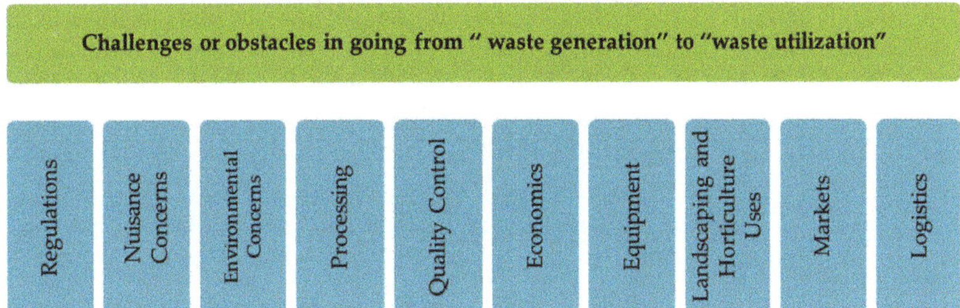

Figure 18. Challenges of valorization of rice industry by-products.

7. Management Issues for Handling Rice Industry By-Products

Lastly, the authors would like to stress the importance of taking precautions around rice husk remnants since they always include small dust particles created during processing. Breathing filters should be worn while working with these substances since they may irritate the upper respiratory tract and trigger allergic responses, including rhinitis, asthma, bronchitis, COPD, and extrinsic allergic alveolitis. Moreover, the dust from rice husk and straw may be readily ignited due to the small size of the dust particles. They can produce explosive concentrations in the air and may smolder when exposed to heat [241,242]. As a result, when working with rice wastes the same precautions should be followed as when working with other flammable dust.

8. Conclusions and Outlook

Rice husk, bran, and straw are often considered low-value waste. However, agriculture, energy generation and storage, pollution control and water treatment, construction materials, and many other vital valorization processes have already been adopted for them. Consequently, to implement these options, the legal requirements governing their disposal methods should be considered. Figures 7, 9 and 13 illustrate rice straw, bran, and husk valorization options. These lignocellulosic materials do, however, have further potential applications. As science and technology improve, identification of many more applications is anticipated as the scientific community and societies become more concerned with sustainability. According to the analysis of relevant research that has already been carried out, rice straw has considerable potential as a renewable energy source. However, there is a significant research gap in using rice bran biochar as an energy storage material. Additionally, modified rice husk biochar has a high promise as an adsorbent in the bio-based water treatment industry. Therefore, further research and development are needed to fill these gaps permanently. In the future, these by-products are expected to be used in fields such as the pharmaceutical industry, space science, etc. Most approaches are anticipated to take place in very small-scale operations, particularly if governments adopt a rural development strategy to halt urbanization, and non-scientific variables might judge the usefulness of research. Therefore, the present review mainly discusses the economics of different procedures. However, in the end, the conclusions about specific applications must be made by politics. For example, taxes or laws can be used to encourage each type of activity, while traditional activities such as burning straw and husk are intended to be discontinued. This will lead to laws that make alternative ways of doing things more fascinating.

Author Contributions: Conceptualization, W.A.M.A.N.I., C.M. and S.S.; methodology, W.A.M.A.N.I., C.M. and S.S.; software, W.A.M.A.N.I.; validation, W.A.M.A.N.I. and C.M.; formal analysis, W.A.M.A.N.I.; investigation, W.A.M.A.N.I.; data curation, W.A.M.A.N.I.; writing—original draft preparation, W.A.M.A.N.I.; writing—review and editing, W.A.M.A.N.I., C.M., M.C.C. and S.S.; supervision, C.M., M.C.C. and S.S. All authors have read and agreed to the published version of the manuscript.

Funding: This research received no external funding.

Data Availability Statement: The data presented in this study are available on request from the corresponding author.

Acknowledgments: Darusha Witharana of the Sri Lanka Rice Research Center deserves our deepest thanks for supplying important data that helped make this research a success. Hashani Ruwanthika Padmasiri is acknowledged for her invaluable assistance in making this research a success.

Conflicts of Interest: The authors declare no conflict of interest.

References

1. Bhandari, H. keynote speech Global Rice Production, Consumption and Trade: Trends and Future Directions. In *Proceedings of the Korean Society of Crop Science Conference*; The Korean Society of Crop Science: Suwon, Republic of Korea, 2019.
2. Patsios, S.I.; Plakas, K.V.; Kontogiannopoulos; Mitrouli, S.T.; Karabelas, A.J. *Characterisation of Agricultural Waste Co- and By-Products*; Chemical Process & Energy Resources Institute: Thessaloniki, Greece, 2016.

3. Muthayya, S.; Sugimoto, J.D.; Montgomery, S.; Maberly, G.F. An Overview of Global Rice Production, Supply, Trade, and Consumption. *Ann. N. Y. Acad. Sci.* **2014**, *1324*, 7–14. [CrossRef]

4. Bandumula, N. Rice Production in Asia: Key to Global Food Security. *Proc. Natl. Acad. Sci. India Sect. B Biol. Sci.* **2018**, *88*, 1323–1328. [CrossRef]

5. Toenniessen, G.; Adesina, A.; DeVries, J. Building an Alliance for a Green Revolution in Africa. *Ann. N. Y. Acad. Sci.* **2008**, *1136*, 233–242. [CrossRef]

6. Papademetriou, M.K. Rice Production in the Asia-Pacific Region: Issues and Perspectives. Available online: https://www.fao.org/3/X6905e/x6905e04.htm (accessed on 24 November 2022).

7. Yevich, R.; Logan, J.A. An Assessment of Biofuel Use and Burning of Agricultural Waste in the Developing World. *Glob. Biogeochem. Cycles* **2003**, *17*. [CrossRef]

8. Kim, J.Y. *Transport for Health: The Global Burden of Disease from Motorized Road Transport*; World Bank Group: Washington, DC, USA, 2014.

9. He, K.; Zhang, J.; Zeng, Y. Knowledge Domain and Emerging Trends of Agricultural Waste Management in the Field of Social Science: A Scientometric Review. *Sci. Total Environ.* **2019**, *670*, 236–244. [CrossRef]

10. Koul, B.; Yakoob, M.; Shah, M.P. Agricultural Waste Management Strategies for Environmental Sustainability. *Environ. Res.* **2022**, *206*, 112285. [CrossRef]

11. Crops and Livestock Products. Available online: https://www.fao.org/faostat/en/#data/QCL/visualize (accessed on 8 October 2022).

12. Kalkanis, K.; Alexakis, D.E.; Kyriakis, E.; Kiskira, K.; Lorenzo-Llanes, J.; Themelis, N.J.; Psomopoulos, C.S. Transforming Waste to Wealth, Achieving Circular Economy. *Circ. Econ. Sustain.* **2022**, *2*, 1541–1559. [CrossRef]

13. Fiore, E.; Stabellini, B.; Tamborrini, P. A Systemic Design Approach Applied to Rice and Wine Value Chains. The Case of the InnovaEcoFood Project in Piedmont (Italy). *Sustainability* **2020**, *12*, 9272. [CrossRef]

14. Senanayake, S.M.P.; Premaratne, S.P. An Analysis of the Paddy/Rice Value Chains in Sri Lanka. *Asia Pac. J. Rural Dev.* **2016**, *26*, 105–126. [CrossRef]

15. Drake, D.J.; Nader, G.; Forero, L. *Feeding Rice Straw to Cattle*; ANR Publication: Davis, CA, USA, 2002; Volume 8079.

16. Illankoon, W.A.M.A.N.; Milanese, C.; Girella, A.; Rathnasiri, P.G.; Sudesh, K.H.M.; Llamas, M.M.; Collivignarelli, M.C.; Sorlini, S. Agricultural Biomass-Based Power Generation Potential in Sri Lanka: A Techno-Economic Analysis. *Energies* **2022**, *15*, 8984. [CrossRef]

17. Bhat, S.A.; Huang, N.-F. Big Data and AI Revolution in Precision Agriculture: Survey and Challenges. *IEEE Access* **2021**, *9*, 110209–110222. [CrossRef]

18. Gupta, R.; Sharma, A.K.; Garg, O.; Modi, K.; Kasim, S.; Baharum, Z.; Mahdin, H.; Mostafa, S.A. WB-CPI: Weather Based Crop Prediction in India Using Big Data Analytics. *IEEE Access* **2021**, *9*, 137869–137885. [CrossRef]

19. Hoffman, G.J.; Martin, D.L. Engineering Systems to Enhance Irrigation Performance. *Irrig. Sci.* **1993**, *14*, 53–63. [CrossRef]

20. Aslam, M. Agricultural Productivity Current Scenario, Constraints and Future Prospects in Pakistan. *Sarhad J. Agric.* **2016**, *32*, 289–303. [CrossRef]

21. Barker, R.; Herdt, R.W.; Rose, B. *The Rice Economy of Asia*; Routledge: London, UK, 2014; ISBN 9781315060521.

22. Hobbs, P.R.; Gupta, R.K. Resource-Conserving Technologies for Wheat in the Rice-Wheat System. In *Improving Productivity and Sustainability of Rice-Wheat Systems: Issues and Impact*; American Society of Agronomy Special Publication: Madison, WI, USA, 2015; pp. 149–171.

23. Priefer, C.; Jörissen, J.; Bräutigam, K.-R. Food Waste Prevention in Europe—A Cause-Driven Approach to Identify the Most Relevant Leverage Points for Action. *Resour. Conserv. Recycl.* **2016**, *109*, 155–165. [CrossRef]

24. Mesterházy, Á.; Oláh, J.; Popp, J. Losses in the Grain Supply Chain: Causes and Solutions. *Sustainability* **2020**, *12*, 2342. [CrossRef]

25. Beitzen-Heineke, E.F.; Balta-Ozkan, N.; Reefke, H. The Prospects of Zero-Packaging Grocery Stores to Improve the Social and Environmental Impacts of the Food Supply Chain. *J. Clean. Prod.* **2017**, *140*, 1528–1541. [CrossRef]

26. Mohidem, N.A.; Hashim, N.; Shamsudin, R.; Man, H.C. Rice for Food Security: Revisiting Its Production, Diversity, Rice Milling Process and Nutrient Content. *Agriculture* **2022**, *12*, 741. [CrossRef]

27. Wilson, R.T.; Lewis, I. *The RICE Value Chain in Tanzania*; Food and Agriculture Organization of the United Nations: Rome, Italy, 2015.

28. Soullier, G.; Demont, M.; Arouna, A.; Lançon, F.; Mendez del Villar, P. The State of Rice Value Chain Upgrading in West Africa. *Glob. Food Sec.* **2020**, *25*, 100365. [CrossRef]

29. Hong, B.H.; How, B.S.; Lam, H.L. Overview of Sustainable Biomass Supply Chain: From Concept to Modelling. *Clean. Technol. Environ. Policy* **2016**, *18*, 2173–2194. [CrossRef]

30. Afzalinia, S.; Shaker, M.; Zare, E. Comparison of Different Rice Milling Methods. *Can. Biosyst. Eng.* **2004**, *46*, 363–366.

31. Dhankhar, P. Rice Milling. *IOSR J. Eng.* **2014**, *4*, 34–42. [CrossRef]

32. Kamari, S.; Ghorbani, F. Extraction of Highly Pure Silica from Rice Husk as an Agricultural By-Product and Its Application in the Production of Magnetic Mesoporous Silica MCM–41. *Biomass Convers. Biorefin.* **2021**, *11*, 3001–3009. [CrossRef]

33. Wiboonsirikul, J.; Kimura, Y.; Kanaya, Y.; Tsuno, T.; Adachi, S. Production and Characterization of Functional Substances from a By-Product of Rice Bran Oil and Protein Production by a Compressed Hot Water Treatment. *Biosci. Biotechnol. Biochem.* **2008**, *72*, 384–392. [CrossRef]
34. Migo-Sumagang, M.V.P.; van Hung, N.; Detras, M.C.M.; Alfafara, C.G.; Borines, M.G.; Capunitan, J.A.; Gummert, M. Optimization of a Downdraft Furnace for Rice Straw-Based Heat Generation. *Renew. Energy* **2020**, *148*, 953–963. [CrossRef]
35. Vaskalis, I.; Skoulou, V.; Stavropoulos, G.; Zabaniotou, A. Towards Circular Economy Solutions for The Management of Rice Processing Residues to Bioenergy via Gasification. *Sustainability* **2019**, *11*, 6433. [CrossRef]
36. Magnago, R.F.; Costa, S.C.; de Assunção-Ezirio, M.J.; de Godoy Saciloto, V.; Cremona Parma, G.O.; Gasparotto, E.S.; Gonçalves, A.C.; Tutida, A.Y.; Barcelos, R.L. Briquettes of Citrus Peel and Rice Husk. *J. Clean. Prod.* **2020**, *276*, 123820. [CrossRef]
37. Payá, J.; Monzó, J.; Borrachero, M.V.; Peris-Mora, E.; Ordóñez, L.M. Studies on Crystalline Rice Husk Ashes and the Activation of Their Pozzolanic Properties. *Waste Manag. Ser.* **2000**, *1*, 493–503.
38. Hossain, S.S.; Mathur, L.; Roy, P.K. Rice Husk/Rice Husk Ash as an Alternative Source of Silica in Ceramics: A Review. *J. Asian Ceram. Soc.* **2018**, *6*, 299–313. [CrossRef]
39. Collivignarelli, M.C.; Sorlini, S.; Milanese, C.; Illankoon, W.A.M.A.N.; Caccamo, F.M.; Calatroni, S. Rice Industry By-Products as Adsorbent Materials for Removing Fluoride and Arsenic from Drinking Water—A Review. *Appl. Sci.* **2022**, *12*, 3166. [CrossRef]
40. FAO. *OECD-FAO Agricultural Outlook 2021–2030*; FAO: Paris, France, 2021.
41. World Production Volume of Milled Rice from 2008/2009 to 2021/2022. Available online: https://www.statista.com/statistics/27 1972/world-husked-rice-production-volume-since-2008/ (accessed on 2 October 2022).
42. Belc, N.; Mustatea, G.; Apostol, L.; Iorga, S.; Vlăduţ, V.-N.; Mosoiu, C. Cereal Supply Chain Waste in the Context of Circular Economy. *E3S Web Conf.* **2019**, *112*, 03031. [CrossRef]
43. Parfitt, J.; Barthel, M.; Macnaughton, S. Food Waste within Food Supply Chains: Quantification and Potential for Change to 2050. *Philos. Trans. R. Soc. B Biol. Sci.* **2010**, *365*, 3065–3081. [CrossRef] [PubMed]
44. Thyberg, K.L.; Tonjes, D.J. Drivers of Food Waste and Their Implications for Sustainable Policy Development. *Resour. Conserv. Recycl.* **2016**, *106*, 110–123. [CrossRef]
45. Rosmiza, M.Z.; Davies, W.; Aznie, C.R.; Mazdi, M.; Jabil, M.J. Farmers' Knowledge on Potential Uses of Rice Straw: An Assessment in MADA and Sekinchan, Malaysia. *Malays. J. Soc. Space* **2014**, *10*, 30–43.
46. Singh, G.; Arya, S.K. A Review on Management of Rice Straw by Use of Cleaner Technologies: Abundant Opportunities and Expectations for Indian Farming. *J. Clean. Prod.* **2021**, *291*, 125278. [CrossRef]
47. Abraham, A.; Mathew, A.K.; Sindhu, R.; Pandey, A.; Binod, P. Potential of Rice Straw for Bio-Refining: An Overview. *Bioresour. Technol.* **2016**, *215*, 29–36. [CrossRef]
48. Van Soest, P.J. Rice Straw, the Role of Silica and Treatments to Improve Quality. *Anim. Feed Sci. Technol.* **2006**, *130*, 137–171. [CrossRef]
49. McManus, W.R.; Choung, C.C. Studies on Forage Cell Walls: 2. Conditions for Alkali Treatment of Rice Straw and Rice Hulls. *J. Agric. Sci.* **1976**, *86*, 453–470. [CrossRef]
50. Jackson, M.G. Rice Straw as Livestock Feed. Available online: https://www.fao.org/3/X6512E/X6512E07.htm (accessed on 12 October 2022).
51. Djajanegara, A.; Doyle, P.T. Urea Supplementation Compared with Pretreatment. 1. Effects on Intake, Digestion and Live-Weight Change by Sheep Fed a Rice Straw. *Anim. Feed Sci. Technol.* **1989**, *27*, 17–30. [CrossRef]
52. Kanjanapruthipong, J.; Thaboot, B. Effects of Neutral Detergent Fiber from Rice Straw on Blood Metabolites and Productivity of Dairy Cows in the Tropics Center. *Asian-Australas. J. Anim. Sci.* **2006**, *19*, 356–362. [CrossRef]
53. Prasad, C.S.; Gowda, N.K.S. *Importance of Trace Minerals and Relevance of Their Supplementation in Tropical Animal Feeding System: A Review*; Directorate of Knowledge Management in Agriculture: New Delhi, India, 2005.
54. Nishida, T.; Suzuki, T.; Phaophaisal, I.; Pholsen, P.; Narmsilee, R.; Indramanee, S.; Oshio, S. Research on Feeding Standard of Beef Cattle and Feedstuff Database in the Indochinese Peninsula. In *Education for Sustainable Development (ESD) on Relationships between Agriculture and Global Environmental Issues*; Dairy Japan Co., Ltd.: Tokyo, Japan, 2008.
55. Phyllis2, Database for (Treated) Biomass, Algae, Feedstocks for Biogas Production and Biochar. Available online: https://phyllis.nl/ (accessed on 13 November 2022).
56. Animal Feed Resources Information System. Available online: https://www.feedipedia.org/ (accessed on 13 November 2022).
57. Zahari, N.I.; Shah, U.K.M.; Asa'ari, A.Z.M.; Mohamad, R. Selection of Potential Fungi for Production of Cellulase Poor Xylanase from Rice Straw. *Bioresources* **2016**, *11*, 1162–1175. [CrossRef]
58. Rahnama, N.; Mamat, S.; Shah, U.K.M.; Ling, F.H.; Rahman, N.A.; Ariff, A.B. Effect of Alkali Pretreatment of Rice Straw on Cellulase and Xylanase Production by Local Trichoderma Harzianum SNRS3 under Solid State Fermentation. *Bioresources* **2013**, *8*, 2881–2896. [CrossRef]
59. Jung, S.-H.; Kang, B.-S.; Kim, J.-S. Production of Bio-Oil from Rice Straw and Bamboo Sawdust under Various Reaction Conditions in a Fast Pyrolysis Plant Equipped with a Fluidized Bed and a Char Separation System. *J. Anal. Appl. Pyrolysis* **2008**, *82*, 240–247. [CrossRef]

60. Sayre, R.N.; Earl, L.; Kratzer, F.H.; Saunders, R.M. Nutritional Qualities of Stabilized and Raw Rice Bran for Chicks. *Poult. Sci.* **1987**, *66*, 493–499. [CrossRef]
61. Spears, J.K.; Grieshop, C.M.; Fahey, G.C. Evaluation of Stabilized Rice Bran as an Ingredient in Dry Extruded Dog Diets. *J. Anim. Sci.* **2004**, *82*, 1122–1135. [CrossRef]
62. Nagendra Prasad, M.N.; Sanjay, K.R.; Shravya Khatokar, M.; Vismaya, M.N.; Nanjunda Swamy, S. Health Benefits of Rice Bran—A Review. *J. Nutr. Food Sci.* **2011**, *1*, 1–7. [CrossRef]
63. Bhatnagar, A.S.; Prabhakar, D.S.; Prasanth Kumar, P.K.; Raja Rajan, R.G.; Gopala Krishna, A.G. Processing of Commercial Rice Bran for the Production of Fat and Nutraceutical Rich Rice Brokens, Rice Germ and Pure Bran. *LWT Food Sci. Technol.* **2014**, *58*, 306–311. [CrossRef]
64. Rosniyana, A.; Hashifah, M.A.; Shariffah, S.A.N. Nutritional Content and Storage Stability of Stabilised Rice. Bran-MR 220. *J. Trop. Agric. Food Sci.* **2009**, *37*, 163–170.
65. Lavanya, M.N.; Venkatachalapathy, N.; Manickavasagan, A. Physicochemical Characteristics of Rice Bran. In *Brown Rice*; Springer International Publishing: Cham, Switzerland, 2017; pp. 79–90.
66. Ausman, L.M.; Rong, N.; Nicolosi, R.J. Hypocholesterolemic Effect of Physically Refined Rice Bran Oil: Studies of Cholesterol Metabolism and Early Atherosclerosis in Hypercholesterolemic Hamsters. *J. Nutr. Biochem.* **2005**, *16*, 521–529. [CrossRef]
67. Schmidt, J.E.; Ahring, B.K. Treatment of Waste Water from a Multi-Product Food Processing Company in Upflow Anaerobic Sludge Blanket (UASB) Reactors: The Effect of Seasonal Variation. *Pure Appl. Chem.* **1997**, *69*, 2447–2452. [CrossRef]
68. Prakash, J.; Ramaswamy, H.S. Rice Bran Proteins: Properties and Food Uses. *Crit. Rev. Food Sci. Nutr.* **1996**, *36*, 537–552. [CrossRef] [PubMed]
69. Hong, G.-B.; Yu, T.-J.; Lee, H.-C.; Ma, C.-M. Using Rice Bran Hydrogel Beads to Remove Dye from Aqueous Solutions. *Sustainability* **2021**, *13*, 5640. [CrossRef]
70. Qi, J.; Yokoyama, W.; Masamba, K.G.; Majeed, H.; Zhong, F.; Li, Y. Structural and Physico-Chemical Properties of Insoluble Rice Bran Fiber: Effect of Acid–Base Induced Modifications. *RSC Adv.* **2015**, *5*, 79915–79923. [CrossRef]
71. Oliveira, M.D.S.; Feddern, V.; Kupski, L.; Cipolatti, E.P.; Badiale-Furlong, E.; de Souza-Soares, L.A. Physico-Chemical Characterization of Fermented Rice Bran Biomass Caracterización Fisico-Química de La Biomasa Del Salvado de Arroz Fermentado. *CyTA-J. Food* **2010**, *8*, 229–236. [CrossRef]
72. Ameh, M.O.; Gernah, D.I.; Igbabul, B.D. Physico-Chemical and Sensory Evaluation of Wheat Bread Supplemented with Stabilized Undefatted Rice Bran. *Food Nutr. Sci.* **2013**, *4*, 43–48. [CrossRef]
73. Huang, S.C.; Shiau, C.Y.; Liu, T.E.; Chu, C.L.; Hwang, D.F. Effects of Rice Bran on Sensory and Physico-Chemical Properties of Emulsified Pork Meatballs. *Meat Sci.* **2005**, *70*, 613–619. [CrossRef]
74. Sairam, S.; Krishna, A.G.G.; Urooj, A. Physico-Chemical Characteristics of Defatted Rice Bran and Its Utilization in a Bakery Product. *J. Food Sci. Technol.* **2011**, *48*, 478–483. [CrossRef] [PubMed]
75. Li, H.; Wang, C.; Chen, X.; Xiong, L.; Guo, H.; Yao, S.; Wang, M.; Chen, X.; Huang, C. Anaerobic Digestion of Rice Straw Pretreatment Liquor without Detoxification for Continuous Biogas Production Using a 100 L Internal Circulation Reactor. *J. Clean. Prod.* **2022**, *349*, 131450. [CrossRef]
76. Nguyen, H.M.; Duong, M.H. *Rice Husk Gasification for Electricity Generation in Cambodia in December 2014*; Hanoi University of Science And Technology: Hanoi, Vietnam, 2014.
77. Waheed, Q.M.K.; Williams, P.T. Hydrogen Production from High Temperature Pyrolysis/Steam Reforming of Waste Biomass: Rice Husk, Sugar Cane Bagasse, and Wheat Straw. *Energy Fuels* **2013**, *27*, 6695–6704. [CrossRef]
78. Dagnino, E.P.; Chamorro, E.R.; Romano, S.D.; Felissia, F.E.; Area, M.C. Optimization of the Acid Pretreatment of Rice Hulls to Obtain Fermentable Sugars for Bioethanol Production. *Ind. Crops Prod.* **2013**, *42*, 363–368. [CrossRef]
79. Abbas, A.; Ansumali, S. Global Potential of Rice Husk as a Renewable Feedstock for Ethanol Biofuel Production. *Bioenergy Res.* **2010**, *3*, 328–334. [CrossRef]
80. Della, V.P.; Kühn, I.; Hotza, D. Rice Husk Ash as an Alternate Source for Active Silica Production. *Mater. Lett.* **2002**, *57*, 818–821. [CrossRef]
81. Jawad, Z.F.; Ghayyib, R.J.; Salman, A.J. Microstructural and Compressive Strength Analysis for Cement Mortar with Industrial Waste Materials. *Civ. Eng. J.* **2020**, *6*, 1007–1016. [CrossRef]
82. Kalapathy, U. A Simple Method for Production of Pure Silica from Rice Hull Ash. *Bioresour. Technol.* **2000**, *73*, 257–262. [CrossRef]
83. Bakar, R.A.; Yahya, R.; Gan, S.N. Production of High Purity Amorphous Silica from Rice Husk. *Procedia Chem.* **2016**, *19*, 189–195. [CrossRef]
84. Cavalcante, D.G.; Marques, M.G.D.S.; Filho, J.D.A.M.; de Vasconcelos, R.P. Influence of the Levels of Replacement of Portland Cement by Metakaolin and Silica Extracted from Rice Husk Ash in the Physical and Mechanical Characteristics of Cement Pastes. *Cem. Concr. Compos.* **2018**, *94*, 296–306. [CrossRef]
85. Korotkova, T.; Ksandopulo, S.; Donenko, A.; Bushumov, S.; Danilchenko, A. Physical Properties and Chemical Composition of the Rice Husk and Dust. *Orient. J. Chem.* **2016**, *32*, 3213–3219. [CrossRef]
86. Nenadis, N.; Kyriakoudi, A.; Tsimidou, M.Z. Impact of Alkaline or Acid Digestion to Antioxidant Activity, Phenolic Content and Composition of Rice Hull Extracts. *LWT Food Sci. Technol.* **2013**, *54*, 207–215. [CrossRef]

87. Das, S.K.; Mishra, J.; Mustakim, S.M.; Adesina, A.; Kaze, C.R.; Das, D. Sustainable Utilization of Ultrafine Rice Husk Ash in Alkali Activated Concrete: Characterization and Performance Evaluation. *J. Sustain. Cem.-Based Mater.* **2022**, *11*, 100–112. [CrossRef]
88. Raheem, A.A.; Kareem, M.A. Chemical Composition and Physical Characteristics of Rice Husk Ash Blended Cement. *Int. J. Eng. Res. Afr.* **2017**, *32*, 25–35. [CrossRef]
89. Rukzon, S.; Chindaprasirt, P.; Mahachai, R. Effect of Grinding on Chemical and Physical Properties of Rice Husk Ash. *Int. J. Miner. Metall. Mater.* **2009**, *16*, 242–247. [CrossRef]
90. Prasad, R.K.; Chatterjee, S.; Mazumder, P.B.; Gupta, S.K.; Sharma, S.; Vairale, M.G.; Datta, S.; Dwivedi, S.K.; Gupta, D.K. Bioethanol Production from Waste Lignocelluloses: A Review on Microbial Degradation Potential. *Chemosphere* **2019**, *231*, 588–606. [CrossRef] [PubMed]
91. Sivagurunathan, P.; Raj, T.; Chauhan, P.S.; Kumari, P.; Satlewal, A.; Gupta, R.P.; Kumar, R. High-Titer Lactic Acid Production from Pilot-Scale Pretreated Non-Detoxified Rice Straw Hydrolysate at High-Solid Loading. *Biochem. Eng. J.* **2022**, *187*, 108668. [CrossRef]
92. Oskoueian, E.; Jahromi, M.F.; Jafari, S.; Shakeri, M.; Le, H.H.; Ebrahimi, M. Manipulation of Rice Straw Silage Fermentation with Different Types of Lactic Acid Bacteria Inoculant Affects Rumen Microbial Fermentation Characteristics and Methane Production. *Vet. Sci.* **2021**, *8*, 100. [CrossRef]
93. Sahu, T.K.; Sahu, V.K.; Mondal, A.; Shukla, P.C.; Gupta, S.; Sarkar, S. Investigation of Sugar Extraction Capability from Rice Paddy Straw for Potential Use of Bioethanol Production towards Energy Security. *Energy Sources Part A Recovery Util. Environ. Eff.* **2022**, *44*, 272–286. [CrossRef]
94. Madzingira, O.; Hepute, V.; Mwenda, E.N.; Kandiwa, E.; Mushonga, B.; Mupangwa, J.F. Nutritional Assessment of Three Baled Rice Straw Varieties Intended for Use as Ruminant Feed in Namibia. *Cogent Food Agric.* **2021**, *7*, 1950402. [CrossRef]
95. Jaspreet, K.; Meenakshi, G.; Singh, L.J.; Kumar, A.R. Sugarcane Tops and Additives Influence Nutritional Quality and Fermentation Characteristics of Mixed Silage Prepared with Rice Straw. *Range Manag. Agrofor.* **2022**, *43*, 309–316.
96. Singh, L.; Brar, B.S. A Review on Rice Straw Management Strategies. *Nat. Environ. Pollut. Technol. Int. Q. Sci. J.* **2021**, *20*, 1485–1493. [CrossRef]
97. Kumar, A.; Nayak, A.K.; Sharma, S.; Senapati, A.; Mitra, D.; Mohanty, B.; Prabhukarthikeyan, S.R.; Sabarinathan, K.G.; Mani, I.; Garhwal, R.S.; et al. Recycling of Rice Straw—A Sustainable Approach for Ensuring Environmental Quality and Economic Security: A Review. *Pedosphere*, 2022; *in press*. [CrossRef]
98. Jerold, M.; Santhiagu, A.; Babu, R.S.; Korapatti, N. *Sustainable Bioprocessing for a Clean and Green Environment*; CRC Press: Boca Raton, FL, USA, 2021; ISBN 9781003035398.
99. Haque, S.; Singh, R.; Pal, D.B.; Harakeh, S.; Alghanmi, M.; Teklemariam, A.D.; Abujamel, T.S.; Srivastava, N.; Gupta, V.K. Recent Update on Anaerobic Digestion of Paddy Straw for Biogas Production: Advancement, Limitation and Recommendations. *Environ. Res.* **2022**, *215*, 114292. [CrossRef]
100. Luo, T.; Ge, Y.; Yang, Y.; Fu, Y.; Kumar Awasthi, M.; Pan, J.; Zhai, L.; Mei, Z.; Liu, H. The Impact of Immersed Liquid Circulation on Anaerobic Digestion of Rice Straw Bale and Methane Generation Improvement. *Bioresour. Technol.* **2021**, *337*, 125368. [CrossRef] [PubMed]
101. Susana, I.G.B.; Alit, I.B. Rice Husk as Sustainable Waste Energy for Small Farmers—A Review. *World J. Adv. Eng. Technol. Sci.* **2021**, *3*, 001–006. [CrossRef]
102. Dafiqurrohman, H.; Safitri, K.A.; Setyawan, M.I.B.; Surjosatyo, A.; Aziz, M. Gasification of Rice Wastes toward Green and Sustainable Energy Production: A Review. *J. Clean. Prod.* **2022**, *366*, 132926. [CrossRef]
103. Umar, H.A.; Sulaiman, S.A.; Said, M.A.; Gungor, A.; Ahmad, R.K. An Overview of Biomass Conversion Technologies in Nigeria. In *Clean Energy Opportunities in Tropical Countries*; Springer: Singapore, 2021; pp. 133–150.
104. Mohiuddin, O.; Mohiuddin, A.; Obaidullah, M.; Ahmed, H.; Asumadu-Sarkodie, S. Electricity Production Potential and Social Benefits from Rice Husk, a Case Study in Pakistan. *Cogent Eng.* **2016**, *3*, 1177156. [CrossRef]
105. Hasan, K.M.F.; Horváth, P.G.; Bak, M.; Le, D.H.A.; Mucsi, Z.M.; Alpár, T. Rice Straw and Energy Reed Fibers Reinforced Phenol Formaldehyde Resin Polymeric Biocomposites. *Cellulose* **2021**, *28*, 7859–7875. [CrossRef]
106. Basta, A.H.; Lotfy, V.F. Impact of Pulping Routes of Rice Straw on Cellulose Nanoarchitectonics and Their Behavior toward Indigo Dye. *Appl. Nanosci.* **2022**, *12*, 1–5. [CrossRef]
107. Zhu, L.; Feng, Z.; Wang, D.; Wu, J.; Qiu, J.; Zhu, P. Highly-Efficient Isolation of Cellulose Microfiber from Rice Straw via Gentle Low-Temperature Phase Transition. *Cellulose* **2021**, *28*, 7021–7031. [CrossRef]
108. Thabah, W.; Kumar Singh, A.; Bedi, R. Tensile Properties of Urea Treated Rice Straw Reinforced Recycled Polyethylene Terephthalate Composite Materials. *Mater. Today Proc.* **2022**, *56*, 2151–2157. [CrossRef]
109. Vinoth, V.; Sathiyamurthy, S.; Ananthi, N.; Elaiyarasan, U. Chemical Treatments and Mechanical Characterisation of Natural Fibre Reinforced Composite Materials—A Review. *Int. J. Mater. Eng. Innov.* **2022**, *13*, 208. [CrossRef]
110. Buzarovska, A.; Bogoeva, G.G.; Grozdanov, A.; Avella, M.; Gentile, G.; Errico, M. Potential Use of Rice Straw as Filler in Eco-Composite Materials. *Aust. J. Crop Sci.* **2008**, *1*, 37–42.

111. Jena, P.K.; Bhoi, P.; Behera, R. Mechanical and Thermal Properties of Rice/Wheat Straw Fiber Reinforced Epoxy Composites: A Comparative Study. In *Recent Advances in Mechanical Engineering. Lecture Notes in Mechanical Engineering*; Springer: Singapore, 2023; pp. 727–735.
112. Quintana-Gallardo, A.; Romero Clausell, J.; Guillén-Guillamón, I.; Mendiguchia, F.A. Waste Valorization of Rice Straw as a Building Material in Valencia and Its Implications for Local and Global Ecosystems. *J. Clean. Prod.* **2021**, *318*, 128507. [CrossRef]
113. Xie, X.; Zhang, W.; Luan, X.; Gao, W.; Geng, X.; Xue, Y. Thermal Performance Enhancement of Hollow Brick by Agricultural Wastes. *Case Stud. Constr. Mater.* **2022**, *16*, e01047. [CrossRef]
114. Tachaudomdach, S.; Hempao, S. Investigation of Compression Strength and Heat Absorption of Native Rice Straw Bricks for Environmentally Friendly Construction. *Sustainability* **2022**, *14*, 12229. [CrossRef]
115. Awoyera, P.O.; Akinrinade, A.D.; de Sousa Galdino, A.G.; Althoey, F.; Kirgiz, M.S.; Tayeh, B.A. Thermal Insulation and Mechanical Characteristics of Cement Mortar Reinforced with Mineral Wool and Rice Straw Fibers. *J. Build. Eng.* **2022**, *53*, 104568. [CrossRef]
116. El-Kassas, A.M.; Elsheikh, A.H. A New Eco-Friendly Mechanical Technique for Production of Rice Straw Fibers for Medium Density Fiberboards Manufacturing. *Int. J. Environ. Sci. Technol.* **2021**, *18*, 979–988. [CrossRef]
117. Zhu, Y.; Tang, W.; Jin, X.; Shan, B. Using Biochar Capping to Reduce Nitrogen Release from Sediments in Eutrophic Lakes. *Sci. Total Environ.* **2019**, *646*, 93–104. [CrossRef]
118. Xu, Y.; Chen, J.; Chen, R.; Yu, P.; Guo, S.; Wang, X. Adsorption and Reduction of Chromium(VI) from Aqueous Solution Using Polypyrrole/Calcium Rectorite Composite Adsorbent. *Water Res.* **2019**, *160*, 148–157. [CrossRef]
119. Okeke, E.S.; Olagbaju, O.A.; Okoye, C.O.; Addey, C.I.; Chukwudozie, K.I.; Okoro, J.O.; Deme, G.G.; Ewusi-Mensah, D.; Igun, E.; Ejeromedoghene, O.; et al. Microplastic Burden in Africa: A Review of Occurrence, Impacts, and Sustainability Potential of Bioplastics. *Chem. Eng. J. Adv.* **2022**, *12*, 100402. [CrossRef]
120. Guo, X.; Zheng, P.; Zou, X.; Chen, X.; Zhang, Q. Influence of Pyroligneous Acid on Fermentation Parameters, CO_2 Production and Bacterial Communities of Rice Straw and Stylo Silage. *Front. Microbiol.* **2021**, *12*, 701434. [CrossRef]
121. Ahmed, M.A.; Rafii, M.Y.; Ain Izzati, M.Z.N.; Khalilah, A.K.; Awad, E.A.; Kaka, U.; Chukwu, S.C.; Liang, J.B.; Sazili, A.Q. Biological Additives Improved Qualities. *Anim. Prod. Sci.* **2022**, *62*, 1414–1429. [CrossRef]
122. Chen, Y.F.; Ouyang, J.L.; Shahzad, K.; Qi, R.X.; Wang, M.Z. Effects of Facultative Heterofermentative Lab, Enzymes and Fermentable Substrates on the Fermentation Quality, Aerobic Stability and In Vitro Ruminal Fermentation of Rice Straw. *Anim. Nutr. Feed Technol.* **2022**, *22*, 67–78. [CrossRef]
123. Wang, Y.; Xu, B.; Ning, S.; Shi, S.; Tan, L. Magnetically Stimulated Azo Dye Biodegradation by a Newly Isolated Osmo-Tolerant Candida Tropicalis A1 and Transcriptomic Responses. *Ecotoxicol. Environ. Saf.* **2021**, *209*, 111791. [CrossRef] [PubMed]
124. Rusdy, M. Chemical Composition and Nutritional Value of Urea Treated Rice Straw for Ruminants. *Livest. Res. Rural. Dev.* **2022**, *34*, 10.
125. Buasri, A.; Chaiyut, N.; Tapang, K.; Jaroensin, S.; Panphrom, S. Removal of Cu^{2+} from Aqueous Solution by Biosorption on Rice Straw—An Agricultural Waste Biomass. *Int. J. Environ. Sci. Dev.* **2012**, *3*, 10. [CrossRef]
126. Rocha, C.G.; Zaia, D.A.M.; Alfaya, R.V.D.S.; Alfaya, A.A.D.S. Use of Rice Straw as Biosorbent for Removal of Cu(II), Zn(II), Cd(II) and Hg(II) Ions in Industrial Effluents. *J. Hazard. Mater.* **2009**, *166*, 383–388. [CrossRef]
127. Nawar, N.; Ebrahim, M.; Sami, E. Removal of Heavy Metals Fe^{3+}, Mn^{2+}, Zn^{2+}, Pb^{2+} and Cd^{2+} from Wastewater by Using Rice Straw as Low Cost Adsorbent. *Acad. J. Interdiscip. Stud.* **2013**, *2*, 85. [CrossRef]
128. Cao, L.; Zhang, C.; Hao, S.; Luo, G.; Zhang, S.; Chen, J. Effect of Glycerol as Co-Solvent on Yields of Bio-Oil from Rice Straw through Hydrothermal Liquefaction. *Bioresour. Technol.* **2016**, *220*, 471–478. [CrossRef] [PubMed]
129. Bishnoi, N.R.; Bajaj, M.; Sharma, N.; Gupta, A. Adsorption of Cr(VI) on Activated Rice Husk Carbon and Activated Alumina. *Bioresour. Technol.* **2004**, *91*, 305–307. [CrossRef]
130. Gao, H.; Liu, Y.; Zeng, G.; Xu, W.; Li, T.; Xia, W. Characterization of Cr(VI) Removal from Aqueous Solutions by a Surplus Agricultural Waste—Rice Straw. *J. Hazard. Mater.* **2008**, *150*, 446–452. [CrossRef]
131. Rungrodnimitchai, S. Modification of Rice Straw for Heavy Metal Ion Adsorbents by Microwave Heating. *Macromol. Symp.* **2010**, *295*, 100–106. [CrossRef]
132. Cao, W.; Dang, Z.; Zhou, X.-Q.; Yi, X.-Y.; Wu, P.-X.; Zhu, N.-W.; Lu, G.-N. Removal of Sulphate from Aqueous Solution Using Modified Rice Straw: Preparation, Characterization and Adsorption Performance. *Carbohydr. Polym.* **2011**, *85*, 571–577. [CrossRef]
133. Amin, M.N.; Mustafa, A.I.; Khalil, M.I.; Rahman, M.; Nahid, I. Adsorption of Phenol onto Rice Straw Biowaste for Water Purification. *Clean. Technol. Environ. Policy* **2012**, *14*, 837–844. [CrossRef]
134. Fathy, N.A.; El-Shafey, O.I.; Khalil, L.B. Effectiveness of Alkali-Acid Treatment in Enhancement the Adsorption Capacity for Rice Straw: The Removal of Methylene Blue Dye. *ISRN Phys. Chem.* **2013**, *2013*, 208087. [CrossRef]
135. Goodman, B.A. Utilization of Waste Straw and Husks from Rice Production: A Review. *J. Bioresour. Bioprod.* **2020**, *5*, 143–162. [CrossRef]
136. Eladel, H.; Abd-Elhay, R.; Anees, D. Effect of Rice Straw Application on Water Quality and Microalgal Flora in Fish Ponds. *Egypt. J. Bot.* **2018**, *59*, 171–184. [CrossRef]
137. Yang, W.; Guo, F.; Wan, Z. Yield and Size of Oyster Mushroom Grown on Rice/Wheat Straw Basal Substrate Supplemented with Cotton Seed Hull. *Saudi J. Biol. Sci.* **2013**, *20*, 333–338. [CrossRef] [PubMed]

138. Li, T.; Gao, J.; Bai, L.; Wang, Y.; Huang, J.; Kumar, M.; Zeng, X. Influence of Green Manure and Rice Straw Management on Soil Organic Carbon, Enzyme Activities, and Rice Yield in Red Paddy Soil. *Soil Tillage Res.* **2019**, *195*, 104428. [CrossRef]
139. Lu, K.; Yang, X.; Shen, J.; Robinson, B.; Huang, H.; Liu, D.; Bolan, N.; Pei, J.; Wang, H. Effect of Bamboo and Rice Straw Biochars on the Bioavailability of Cd, Cu, Pb and Zn to Sedum Plumbizincicola. *Agric. Ecosyst. Environ.* **2014**, *191*, 124–132. [CrossRef]
140. Cui, H.-J.; Wang, M.K.; Fu, M.-L.; Ci, E. Enhancing Phosphorus Availability in Phosphorus-Fertilized Zones by Reducing Phosphate Adsorbed on Ferrihydrite Using Rice Straw-Derived Biochar. *J. Soils Sediments* **2011**, *11*, 1135–1141. [CrossRef]
141. Wannapeera, J.; Worasuwannarak, N.; Pipatmanomai, S. Product Yields and Characteristics of Rice Husk, Rice Straw and Corncob during Fast Pyrolysis in a Drop-Tube/Fixed-Bed Reactor. *Songklanakarin J. Sci. Technol.* **2008**, *30*, 393–404.
142. Ye, S.; Zeng, G.; Wu, H.; Liang, J.; Zhang, C.; Dai, J.; Xiong, W.; Song, B.; Wu, S.; Yu, J. The Effects of Activated Biochar Addition on Remediation Efficiency of Co-Composting with Contaminated Wetland Soil. *Resour. Conserv. Recycl.* **2019**, *140*, 278–285. [CrossRef]
143. Liu, W.-J.; Jiang, H.; Yu, H.-Q. Emerging Applications of Biochar-Based Materials for Energy Storage and Conversion. *Energy Environ. Sci.* **2019**, *12*, 1751–1779. [CrossRef]
144. Jing, X.-R.; Wang, Y.-Y.; Liu, W.-J.; Wang, Y.-K.; Jiang, H. Enhanced Adsorption Performance of Tetracycline in Aqueous Solutions by Methanol-Modified Biochar. *Chem. Eng. J.* **2014**, *248*, 168–174. [CrossRef]
145. Zaccheria, F.; Mariani, M.; Ravasio, N. The Use of Rice Bran Oil within a Biorefinery Concept. *Chem. Biol. Technol. Agric.* **2015**, *2*, 23. [CrossRef]
146. Roth-Maier, D.A.; Kettler, S.I.; Kirchgessner, M. Availability of Vitamin B 6 from Different Food Sources. *Int. J. Food Sci. Nutr.* **2002**, *53*, 171–179. [CrossRef]
147. Nyström, L.; Mäkinen, M.; Lampi, A.-M.; Piironen, V. Antioxidant Activity of Steryl Ferulate Extracts from Rye and Wheat Bran. *J. Agric. Food Chem.* **2005**, *53*, 2503–2510. [CrossRef]
148. Nyström, L.; Achrenius, T.; Lampi, A.-M.; Moreau, R.A.; Piironen, V. A Comparison of the Antioxidant Properties of Steryl Ferulates with Tocopherol at High Temperatures. *Food Chem.* **2007**, *101*, 947–954. [CrossRef]
149. Bhunia, R.K.; Sinha, K.; Kaur, R.; Kaur, S.; Chawla, K. A Holistic View of the Genetic Factors Involved in Triggering Hydrolytic and Oxidative Rancidity of Rice Bran Lipids. *Food Rev. Int.* **2021**, *37*, 1–26. [CrossRef]
150. Ranjan, A.; Kumar, S.; Sahu, N.P.; Jain, K.K.; Deo, A.D. Strategies for Maximizing Utilization of De-Oiled Rice Bran (DORB) in the Fish Feed. *Aquac. Int.* **2022**, *30*, 99–114. [CrossRef]
151. Jha, B.; Chandra, R.; Vijay, V.K.; Subbarao, P.M.V.; Isha, A. Utilization of De-Oiled Rice Bran as a Feedstock for Renewable Biomethane Production. *Biomass Bioenergy* **2020**, *140*, 105674. [CrossRef]
152. Zaidel, D.N.A.; Muhamad, I.I.; Daud, N.S.M.; Muttalib, N.A.A.; Khairuddin, N.; Lazim, N.A.M. Production of Biodiesel from Rice Bran Oil. In *Biomass, Biopolymer-Based Materials, and Bioenergy*; Elsevier: Amsterdam, The Netherlands, 2019; pp. 409–447.
153. Hoang, A.T.; Tabatabaei, M.; Aghbashlo, M.; Carlucci, A.P.; Ölçer, A.I.; Le, A.T.; Ghassemi, A. Rice Bran Oil-Based Biodiesel as a Promising Renewable Fuel Alternative to Petrodiesel: A Review. *Renew. Sustain. Energy Rev.* **2021**, *135*, 110204. [CrossRef]
154. Marchetti, J.M. A Summary of the Available Technologies for Biodiesel Production Based on a Comparison of Different Feedstock's Properties. *Process Saf. Environ. Prot.* **2012**, *90*, 157–163. [CrossRef]
155. Abbaszadeh, A.; Ghobadian, B.; Najafi, G.; Yusaf, T. An Experimental Investigation of the Effective Parameters on Wet Washing of Biodiesel Purification. *Int. J. Automot. Mech. Eng.* **2014**, *9*, 1525–1537. [CrossRef]
156. Zhang, Y.; Wong, W.-T.; Yung, K.-F. One-Step Production of Biodiesel from Rice Bran Oil Catalyzed by Chlorosulfonic Acid Modified Zirconia via Simultaneous Esterification and Transesterification. *Bioresour. Technol.* **2013**, *147*, 59–64. [CrossRef] [PubMed]
157. Sutanto, S.; Go, A.W.; Chen, K.-H.; Ismadji, S.; Ju, Y.-H. Maximized Utilization of Raw Rice Bran in Microbial Oils Production and Recovery of Active Compounds: A Proof of Concept. *Waste Biomass Valorization* **2017**, *8*, 1067–1080. [CrossRef]
158. Gupte, A.P.; Basaglia, M.; Casella, S.; Favaro, L. Rice Waste Streams as a Promising Source of Biofuels: Feedstocks, Biotechnologies and Future Perspectives. *Renew. Sustain. Energy Rev.* **2022**, *167*, 112673. [CrossRef]
159. Alexandri, M.; López-Gómez, J.P.; Olszewska-Widdrat, A.; Venus, J. Valorising Agro-Industrial Wastes within the Circular Bioeconomy Concept: The Case of Defatted Rice Bran with Emphasis on Bioconversion Strategies. *Fermentation* **2020**, *6*, 42. [CrossRef]
160. Friedman, M. Rice Brans, Rice Bran Oils, and Rice Hulls: Composition, Food and Industrial Uses, and Bioactivities in Humans, Animals, and Cells. *J. Agric. Food Chem.* **2013**, *61*, 10626–10641. [CrossRef]
161. Chandel, A.K.; Narasu, M.L.; Rudravaram, R.; Pogaku, R.; Rao, L.V. Bioconversion of De-Oiled Rice Bran (DORB) Hemicellulosic Hydrolysate into Ethanol by Pichia Stipitis NCM3499 under Optimized Conditions. *Int. J. Food Eng.* **2009**, *5*. [CrossRef]
162. Beliya, E.; Tiwari, S.; Jadhav, S.K.; Tiwari, K.L. De-Oiled Rice Bran as a Source of Bioethanol. *Energy Explor. Exploit.* **2013**, *31*, 771–782. [CrossRef]
163. Alexandri, M.; Neu, A.; Schneider, R.; López-Gómez, J.P.; Venus, J. Evaluation of Various *Bacillus coagulans* Isolates for the Production of High Purity L-lactic Acid Using Defatted Rice Bran Hydrolysates. *Int. J. Food Sci. Technol.* **2019**, *54*, 1321–1329. [CrossRef]

164. Kurniasih, N.; Dinna, F.; Amalia, V.; Widiastuti, D. The effect of fortification of brands and chitosan on tempeh on fiber levels and probiotic bacteria growth. *Helium J. Sci. Appl. Chem.* **2021**, *1*, 37–41. [CrossRef]

165. Nontasan, S.; Moongngarm, A.; Deeseenthum, S. Application of Functional Colorant Prepared from Black Rice Bran in Yogurt. *APCBEE Procedia* **2012**, *2*, 62–67. [CrossRef]

166. Caplice, E. Food Fermentations: Role of Microorganisms in Food Production and Preservation. *Int. J. Food Microbiol.* **1999**, *50*, 131–149. [CrossRef] [PubMed]

167. Al-Shorgani, N.K.N.; Al-Tabib, A.I.; Kadier, A.; Zanil, M.F.; Lee, K.M.; Kalil, M.S. Continuous Butanol Fermentation of Dilute Acid-Pretreated De-Oiled Rice Bran by Clostridium Acetobutylicum YM1. *Sci. Rep.* **2019**, *9*, 4622. [CrossRef] [PubMed]

168. Ranjan, D.; Talat, M.; Hasan, S.H. Rice Polish: An Alternative to Conventional Adsorbents for Treating Arsenic Bearing Water by Up-Flow Column Method. *Ind. Eng. Chem. Res.* **2009**, *48*, 10180–10185. [CrossRef]

169. Lai, Q.D.; Doan, N.T.T.; Nguyen, H.D.; Nguyen, T.X.N. Influence of Enzyme Treatment of Rice Bran on Gamma-aminobutyric Acid Synthesis by *Lacto bacillus*. *Int. J. Food Sci. Technol.* **2021**, *56*, 4722–4729. [CrossRef]

170. Chugh, P.; Soni, R.; Soni, S.K. Deoiled Rice Bran: A Substrate for Co-Production of a Consortium of Hydrolytic Enzymes by Aspergillus Niger P-19. *Waste Biomass Valorization* **2016**, *7*, 513–525. [CrossRef]

171. Tang, S.; Hettiarachchy, N.S.; Shellhammer, T.H. Protein Extraction from Heat-Stabilized Defatted Rice Bran. 1. Physical Processing and Enzyme Treatments. *J. Agric. Food Chem.* **2002**, *50*, 7444–7448. [CrossRef]

172. Tandon, M.; Thakur, V.; Tiwari, K.L.; Jadhav, S.K. Enterobacter Ludwigii Strain IF2SW-B4 Isolated for Bio-Hydrogen Production from Rice Bran and de-Oiled Rice Bran. *Environ. Technol. Innov.* **2018**, *10*, 345–354. [CrossRef]

173. Soares, J.F.; Confortin, T.C.; Todero, I.; Mayer, F.D.; Mazutti, M.A. Dark Fermentative Biohydrogen Production from Lignocellulosic Biomass: Technological Challenges and Future Prospects. *Renew. Sustain. Energy Rev.* **2020**, *117*, 109484. [CrossRef]

174. Azman, N.F.; Abdeshahian, P.; Al-Shorgani, N.K.N.; Hamid, A.A.; Kalil, M.S. Production of Hydrogen Energy from Dilute Acid-Hydrolyzed Palm Oil Mill Effluent in Dark Fermentation Using an Empirical Model. *Int. J. Hydrog. Energy* **2016**, *41*, 16373–16384. [CrossRef]

175. Soltani, N.; Bahrami, A.; Pech-Canul, M.I.; González, L.A. Review on the Physicochemical Treatments of Rice Husk for Production of Advanced Materials. *Chem. Eng. J.* **2015**, *264*, 899–935. [CrossRef]

176. Pode, R. Potential Applications of Rice Husk Ash Waste from Rice Husk Biomass Power Plant. *Renew. Sustain. Energy Rev.* **2016**, *53*, 1468–1485. [CrossRef]

177. Ho, Y.-S.; Chiang, C.-C.; Hsu, Y.-C. Sorption kinetics for dye removal from aqueous solution using activated clay. *Sep. Sci. Technol.* **2001**, *36*, 2473–2488. [CrossRef]

178. Zhang, Y.; Zheng, R.; Zhao, J.; Ma, F.; Zhang, Y.; Meng, Q. Characterization of -Treated Rice Husk Adsorbent and Adsorption of Copper(II) from Aqueous Solution. *Biomed. Res. Int.* **2014**, *2014*, 496878. [CrossRef] [PubMed]

179. Da Rosa, M.P.; Igansi, A.V.; Lütke, S.F.; Sant'Anna Cadaval, T.R.; do Santos, A.C.R.; de Oliveira Lopes Inacio, A.P.; de Almeida Pinto, L.A.; Beck, P.H. A New Approach to Convert Rice Husk Waste in a Quick and Efficient Adsorbent to Remove Cationic Dye from Water. *J. Environ. Chem. Eng.* **2019**, *7*, 103504. [CrossRef]

180. Chandrasekhar, S.; Pramada, P.N. Rice Husk Ash as an Adsorbent for Methylene Blue—Effect of Ashing Temperature. *Adsorption* **2006**, *12*, 27–43. [CrossRef]

181. Chowdhury, S.; Mishra, R.; Saha, P.; Kushwaha, P. Adsorption Thermodynamics, Kinetics and Isosteric Heat of Adsorption of Malachite Green onto Chemically Modified Rice Husk. *Desalination* **2011**, *265*, 159–168. [CrossRef]

182. Deshmukh, W.S.; Attar, S.J.; Waghmare, M.D. Investigation on Sorption of Fluoride in Water Using Rice Husk as an Adsorbent. *Nat. Environ. Pollut. Technol. Int. Q. Sci. J.* **2009**, *8*, 217–223.

183. Ahmaruzzaman, M.; Gupta, V.K. Rice Husk and Its Ash as Low-Cost Adsorbents in Water and Wastewater Treatment. *Ind. Eng. Chem. Res.* **2011**, *50*, 13589–13613. [CrossRef]

184. Masoumi, A.; Hemmati, K.; Ghaemy, M. Low-Cost Nanoparticles Sorbent from Modified Rice Husk and a Copolymer for Efficient Removal of Pb(II) and Crystal Violet from Water. *Chemosphere* **2016**, *146*, 253–262. [CrossRef] [PubMed]

185. Hubadillah, S.K.; Othman, M.H.D.; Harun, Z.; Ismail, A.F.; Rahman, M.A.; Jaafar, J. A Novel Green Ceramic Hollow Fiber Membrane (CHFM) Derived from Rice Husk Ash as Combined Adsorbent-Separator for Efficient Heavy Metals Removal. *Ceram. Int.* **2017**, *43*, 4716–4720. [CrossRef]

186. Al-Sultani Kadhim, F.; Al-Seroury, F.A. Characterization the Removal of Phenol from Aqueous Solution in Fluidized Bed Column by Rice Husk Adsorbent. *Res. J. Recent Sci.* **2012**, *1*, 145–151.

187. Shen, Y.; Fu, Y. KOH-Activated Rice Husk Char via CO_2 Pyrolysis for Phenol Adsorption. *Mater. Today Energy* **2018**, *9*, 397–405. [CrossRef]

188. Nakbanpote, W.; Thiravetyan, P.; Kalambaheti, C. Preconcentration of Gold by Rice Husk Ash. *Miner. Eng.* **2000**, *13*, 391–400. [CrossRef]

189. Sarker, N.; Fakhruddin, A.N.M. Removal of Phenol from Aqueous Solution Using Rice Straw as Adsorbent. *Appl. Water Sci.* **2017**, *7*, 1459–1465. [CrossRef]

190. Anshar, A.M.; Taba, P.; Raya, I. Kinetic and Thermodynamics Studies on the Adsorption of Phenol on Activated Carbon from Rice Husk Activated by $ZnCl_2$. *Indones. J. Sci. Technol.* **2016**, *1*, 47–60. [CrossRef]

191. Singh, D.K.; Bhavana, S. *Basic Dyes Removal from Wastewater by Adsorption on Rice Husk Carbon*; NISCAIR-CSIR: New Delhi, India, 2001; Volume 8, pp. 133–139.
192. El-Maghraby, A.; El-Deeb, H.A. Removal of a basic dye from aqueous solution by adsorption using rice hulls. *Glob. NEST J.* **2011**, *13*, 90–98.
193. Mohamed, M.M. Acid Dye Removal: Comparison of Surfactant-Modified Mesoporous FSM-16 with Activated Carbon Derived from Rice Husk. *J. Colloid Interface Sci.* **2004**, *272*, 28–34. [CrossRef]
194. Malik, P.K. Use of Activated Carbons Prepared from Sawdust and Rice-Husk for Adsorption of Acid Dyes: A Case Study of Acid Yellow 36. *Dye. Pigment.* **2003**, *56*, 239–249. [CrossRef]
195. Bansal, M.; Garg, U.; Singh, D.; Garg, V.K. Removal of Cr(VI) from Aqueous Solutions Using Pre-Consumer Processing Agricultural Waste: A Case Study of Rice Husk. *J. Hazard. Mater.* **2009**, *162*, 312–320. [CrossRef] [PubMed]
196. Sugashini, S.; Begum, K.M.M.S. Preparation of Activated Carbon from Carbonized Rice Husk by Ozone Activation for Cr(VI) Removal. *New Carbon Mater.* **2015**, *30*, 252–261. [CrossRef]
197. Ma, J.; Li, T.; Liu, Y.; Cai, T.; Wei, Y.; Dong, W.; Chen, H. Rice Husk Derived Double Network Hydrogel as Efficient Adsorbent for Pb(II), Cu(II) and Cd(II) Removal in Individual and Multicomponent Systems. *Bioresour. Technol.* **2019**, *290*, 121793. [CrossRef] [PubMed]
198. Hagemann, N.; Spokas, K.; Schmidt, H.-P.; Kägi, R.; Böhler, M.; Bucheli, T. Activated Carbon, Biochar and Charcoal: Linkages and Synergies across Pyrogenic Carbon's ABCs. *Water* **2018**, *10*, 182. [CrossRef]
199. Jeyasubramanian, K.; Thangagiri, B.; Sakthivel, A.; Dhaveethu Raja, J.; Seenivasan, S.; Vallinayagam, P.; Madhavan, D.; Malathi Devi, S.; Rathika, B. A Complete Review on Biochar: Production, Property, Multifaceted Applications, Interaction Mechanism and Computational Approach. *Fuel* **2021**, *292*, 120243. [CrossRef]
200. Alkurdi, S.S.A.; Herath, I.; Bundschuh, J.; Al-Juboori, R.A.; Vithanage, M.; Mohan, D. Biochar versus Bone Char for a Sustainable Inorganic Arsenic Mitigation in Water: What Needs to Be Done in Future Research? *Environ. Int.* **2019**, *127*, 52–69. [CrossRef]
201. Sakhiya, A.K.; Anand, A.; Kaushal, P. Production, Activation, and Applications of Biochar in Recent Times. *Biochar* **2020**, *2*, 253–285. [CrossRef]
202. Quispe, I.; Navia, R.; Kahhat, R. Energy Potential from Rice Husk through Direct Combustion and Fast Pyrolysis: A Review. *Waste Manag.* **2017**, *59*, 200–210. [CrossRef]
203. Decker, S.R.; Brunecky, R.; Tucker, M.P.; Himmel, M.E.; Selig, M.J. High-Throughput Screening Techniques for Biomass Conversion. *Bioenergy Res.* **2009**, *2*, 179–192. [CrossRef]
204. Faaij, A. Modern Biomass Conversion Technologies. *Mitig. Adapt. Strateg. Glob. Chang.* **2006**, *11*, 343–375. [CrossRef]
205. Yahaya, D.B.; Ibrahim, T.G. Development of rice husk briquettes for use as fuel. *Res. J. Eng. Appl. Sci.* **2012**, *1*, 130–133.
206. Wu, H.-C.; Ku, Y.; Tsai, H.-H.; Kuo, Y.-L.; Tseng, Y.-H. Rice Husk as Solid Fuel for Chemical Looping Combustion in an Annular Dual-Tube Moving Bed Reactor. *Chem. Eng. J.* **2015**, *280*, 82–89. [CrossRef]
207. Islam, M.N.; Ani, F.N. Techno-Economics of Rice Husk Pyrolysis, Conversion with Catalytic Treatment to Produce Liquid Fuel. *Bioresour. Technol.* **2000**, *73*, 67–75. [CrossRef]
208. Salam, M.A.; Ahmed, K.; Hossain, T.; Habib, M.S.; Uddin, M.S.; Papri, N. Prospect of Molecular Sieves Production Using Rice Husk in Bangladesh: A Review. *Int. J. Chem. Math. Phys.* **2019**, *3*, 105–134. [CrossRef]
209. Sun, L.; Gong, K. Silicon-Based Materials from Rice Husks and Their Applications. *Ind. Eng. Chem. Res.* **2001**, *40*, 5861–5877. [CrossRef]
210. Li, D.; Ma, T.; Zhang, R.; Tian, Y.; Qiao, Y. Preparation of Porous Carbons with High Low-Pressure CO_2 Uptake by KOH Activation of Rice Husk Char. *Fuel* **2015**, *139*, 68–70. [CrossRef]
211. Illankoon, W.A.M.A.N.; Milanese, C.; Girella, A.; Medina-Llamas, M.; Magnani, G.; Pontiroli, D.; Ricco, M.; Collivignarelli, M.C.; Sorlini, S. Biochar derived from the rice industry by-products as sustainable energy storage material. In Proceedings of the 30th European Biomass Conference and Exhibition (EUBCE), Online, 9–12 May 2022; Chevet, P.-F., Scarlat, N., Grassi, A., Eds.; ETA-Florence Renewable Energies: Florence, Italy, 2022.
212. Feng, Q.; Lin, Q.; Gong, F.; Sugita, S.; Shoya, M. Adsorption of Lead and Mercury by Rice Husk Ash. *J. Colloid Interface Sci.* **2004**, *278*, 1–8. [CrossRef]
213. Naiya, T.K.; Bhattacharya, A.K.; Mandal, S.; Das, S.K. The Sorption of Lead(II) Ions on Rice Husk Ash. *J. Hazard. Mater.* **2009**, *163*, 1254–1264. [CrossRef]
214. Ganvir, V.; Das, K. Removal of Fluoride from Drinking Water Using Aluminum Hydroxide Coated Rice Husk Ash. *J. Hazard. Mater.* **2011**, *185*, 1287–1294. [CrossRef]
215. Chakraborty, S.; Chowdhury, S.; das Saha, P. Adsorption of Crystal Violet from Aqueous Solution onto NaOH-Modified Rice Husk. *Carbohydr. Polym.* **2011**, *86*, 1533–1541. [CrossRef]
216. Lawagon, C.P.; Amon, R.E.C. Magnetic Rice Husk Ash "cleanser" as Efficient Methylene Blue Adsorbent. *Environ. Eng. Res.* **2019**, *25*, 685–692. [CrossRef]
217. Sarkar, D.; Bandyopadhyay, A. Adsorptive Mass Transport of Dye on Rice Husk Ash. *J. Water Resour. Prot.* **2010**, *2*, 424–431. [CrossRef]

218. Lakshmi, U.R.; Srivastava, V.C.; Mall, I.D.; Lataye, D.H. Rice Husk Ash as an Effective Adsorbent: Evaluation of Adsorptive Characteristics for Indigo Carmine Dye. *J. Environ. Manag.* **2009**, *90*, 710–720. [CrossRef] [PubMed]
219. Manique, M.C.; Faccini, C.S.; Onorevoli, B.; Benvenutti, E.V.; Caramão, E.B. Rice Husk Ash as an Adsorbent for Purifying Biodiesel from Waste Frying Oil. *Fuel* **2012**, *92*, 56–61. [CrossRef]
220. Zou, Y.; Yang, T. Rice Husk, Rice Husk Ash and Their Applications. In *Rice Bran and Rice Bran Oil*; Elsevier: Amsterdam, The Netherlands, 2019; pp. 207–246.
221. Azadi, M.; Bahrololoom, M.E.; Heidari, F. Enhancing the Mechanical Properties of an Epoxy Coating with Rice Husk Ash, a Green Product. *J. Coat. Technol. Res.* **2011**, *8*, 117–123. [CrossRef]
222. Tipsotnaiyana, N.; Jarupan, L.; Pechyen, C. Synthesized Silica Powder from Rice Husk for Printing Raw Materials Application. *Adv. Mater. Res.* **2012**, *506*, 218–221. [CrossRef]
223. Nair, D.G.; Fraaij, A.; Klaassen, A.A.K.; Kentgens, A.P.M. A Structural Investigation Relating to the Pozzolanic Activity of Rice Husk Ashes. *Cem. Concr. Res.* **2008**, *38*, 861–869. [CrossRef]
224. Zahedi, M.; Ramezanianpour, A.A.; Ramezanianpour, A.M. Evaluation of the Mechanical Properties and Durability of Cement Mortars Containing Nanosilica and Rice Husk Ash under Chloride Ion Penetration. *Constr. Build Mater.* **2015**, *78*, 354–361. [CrossRef]
225. Ramezanianpour, A.A.; Pourbeik, P.; Moodi, F.; Sabzi, M.Z. Sulfate Resistance of Concretes Containing Rice Husk Ash. In Proceedings of the 9th International Congress on Advances in Civil Engineering, Karadeniz Technical University, Trabzon, Turkey, 30 September 2010.
226. Zareei, S.A.; Ameri, F.; Dorostkar, F.; Ahmadi, M. Rice Husk Ash as a Partial Replacement of Cement in High Strength Concrete Containing Micro Silica: Evaluating Durability and Mechanical Properties. *Case Stud. Constr. Mater.* **2017**, *7*, 73–81. [CrossRef]
227. Haxo, H.E.; Mehta, P.K. Ground Rice-Hull Ash as a Filler for Rubber. *Rubber Chem. Technol.* **1975**, *48*, 271–288. [CrossRef]
228. Nzila, C.; Oluoch, N.; Kiprop, A.; Ramkat, R.; Kosgey, I. *Advances in Phytochemistry, Textile and Renewable Energy Research for Industrial Growth*; CRC Press: London, UK, 2022; ISBN 9781003221968.
229. Wang, L.; Guo, Y.; Zhu, Y.; Li, Y.; Qu, Y.; Rong, C.; Ma, X.; Wang, Z. A New Route for Preparation of Hydrochars from Rice Husk. *Bioresour. Technol.* **2010**, *101*, 9807–9810. [CrossRef]
230. Kirchherr, J.; Piscicelli, L. Towards an Education for the Circular Economy (ECE): Five Teaching Principles and a Case Study. *Resour. Conserv. Recycl.* **2019**, *150*, 104406. [CrossRef]
231. Nassereddine, H.; Seo, K.W.; Rybkowski, Z.K.; Schranz, C.; Urban, H. Propositions for a Resilient, Post-COVID-19 Future for the AEC Industry. *Front. Built. Environ.* **2021**, *7*, 117–123. [CrossRef]
232. Okorie, O.; Salonitis, K.; Charnley, F.; Moreno, M.; Turner, C.; Tiwari, A. Digitisation and the Circular Economy: A Review of Current Research and Future Trends. *Energies* **2018**, *11*, 3009. [CrossRef]
233. Kevin van Langen, S.; Vassillo, C.; Ghisellini, P.; Restaino, D.; Passaro, R.; Ulgiati, S. Promoting Circular Economy Transition: A Study about Perceptions and Awareness by Different Stakeholders Groups. *J. Clean. Prod.* **2021**, *316*, 128166. [CrossRef]
234. Let's Build a Circular Economy. Available online: https://ellenmacarthurfoundation.org/ (accessed on 8 December 2022).
235. Cecchin, A.; Salomone, R.; Deutz, P.; Raggi, A.; Cutaia, L. What Is in a Name? The Rising Star of the Circular Economy as a Resource-Related Concept for Sustainable Development. *Circ. Econ. Sustain.* **2021**, *1*, 83–97. [CrossRef]
236. EU. Directive 2008/98/EC of the European Parliament and of the Council of 19 November 2008 on Waste and Repealing Certain Directives. Official Journal of EU L 2008, 312. Available online: https://eur-lex.europa.eu/legal-content/EN/TXT/PDF/?uri=CELEX:02008L0098-20180705&from=SV (accessed on 9 November 2022).
237. Pinjing, H.; Fan, L.; Hua, Z.; Liming, S. Recent Developments in the Area of Waste as a Resource, with Particular Reference to the Circular Economy as a Guiding Principle. *Issues Environ. Sci. Technol.* **2013**, *37*, 144–161.
238. Nikolaou, I.E.; Jones, N.; Stefanakis, A. Circular Economy and Sustainability: The Past, the Present and the Future Directions. *Circ. Econ. Sustain.* **2021**, *1*, 1–20. [CrossRef]
239. Traditional Rice. Available online: https://it.lakpura.com/pages/traditional-rice?shpxid=7f0e5626-b366-4d60-9fd0-312089d431ff (accessed on 29 November 2022).
240. Jayaneththi.Blogspot.Com. Available online: http://jayaneththi.blogspot.com/2011/03/trunk-of-rubber-tree.html (accessed on 3 January 2023).
241. Fu, F.; Wang, Q. Removal of Heavy Metal Ions from Wastewaters: A Review. *J. Environ. Manag.* **2011**, *92*, 407–418. [CrossRef]
242. Wang, L.; Templer, R.; Murphy, R.J. High-Solids Loading Enzymatic Hydrolysis of Waste Papers for Biofuel Production. *Appl. Energy* **2012**, *99*, 23–31. [CrossRef]

Review

Review on Aquatic Weeds as Potential Source for Compost Production to Meet Sustainable Plant Nutrient Management Needs

D. M. N. S. Dissanayaka, S. S. Udumann, D. K. R. P. L. Dissanayake, T. D. Nuwarapaksha and Anjana J. Atapattu *

Agronomy Division, Coconut Research Institute, Lunuwila 61150, Sri Lanka
* Correspondence: aaajatapattu@gmail.com

Abstract: As a result of the increase in agricultural production and environmental pollution, waste management and disposal are becoming vital. Proper treatments, such as converting abundant bio-mass wastes into beneficial materials, might mitigate the negative effects and convert waste into reusable resources. Aquatic weeds are a significant concern in the majority of water bodies. Their quick growth, rapid ecological adaptations, and lack of natural enemies make these plants invasive, problematic, and challenging to manage over time. Although there are many methods to manage aquatic weeds, composting has been identified as one of the easily adapted and eco-friendly methods for transferring nutrients to the cropping cycle. Their short life cycle, higher biomass yield, higher nutrient compositions, and allelopathic and phytoremediation properties confirm their suitability as raw materials for composting. Most aquatic ecosystems can be maintained in optimum conditions while facilitating maximum benefits for life by identifying and developing proper composting techniques. Studying the ecology and morphological features of aquatic weeds is essential for this purpose. This is an overview of identifying the potential of aquatic weeds as a source of composting, targeting sustainable plant nutrient management while managing weeds.

Keywords: agriculture; agricultural waste management; allelopathic effect; invasive species; phytoremediation

Citation: Dissanayaka, D.M.N.S.; Udumann, S.S.; Dissanayake, D.K.R.P.L.; Nuwarapaksha, T.D.; Atapattu, A.J. Review on Aquatic Weeds as Potential Source for Compost Production to Meet Sustainable Plant Nutrient Management Needs. *Waste* **2023**, *1*, 264–280. https://doi.org/10.3390/waste1010017

Academic Editor: Vassilis Athanasiadis

Received: 17 November 2022
Revised: 10 January 2023
Accepted: 12 January 2023
Published: 25 January 2023

1. Introduction

Modern conventional high-input agriculture has caused many problems in all agricultural ecosystems, even in nature. This happens mainly due to low soil organic matter content, unbalanced soil nutrient levels, inappropriate irrigation, and higher biomass waste generation, resulting in the depletion of arable lands and crop yield [1–3]. Therefore, effective soil nutrient management with limited resources available is essential for maximizing crop output to satisfy the growing food demand in the twenty-first century.

Composting is one of the eco-friendly approaches for recycling bio-waste productively and cost-effectively [2]. Composting is a natural biological decomposition process that converts different types of organic waste into beneficial organic products under controlled conditions [4]. End-products such as humus, biomass, carbon dioxide, and heat are produced during the composting process [5]. It is an aerobic process controlled by various decomposing agents, including bacteria, fungi, and actinomycetes, which vary with the different stages of compost formation [5]. Quality raw materials, including nutrients such as carbon (C), nitrogen (N), phosphorus (P), and potassium (K); supportive carbon/nitrogen (C/N) ratio; moisture content; natural qualities (porosity, particle size, and bulk density); adequate oxygen supply; and conditions of the composting process such as temperature and pH changes should be present for better growth of micro-communities and thus fast and high-quality composting production [5–8]. Furthermore, for a sufficient nutritional balance during composting, the C/N ratio of primary organic waste should be between 25 and 35 under optimal conditions [9].

Depending on farmers' preferences and availability, different raw materials such as municipal organic waste [10]; sewage sludge [11]; all kinds of farm wastes (garden debris, weed residues, stalks and stems, fallen leaves, pruning parts, chaff, fodder, and leftovers) [12]; livestock (buffalo, cow, goat, and swine) and poultry manure [13,14]; oil palm wastage, such as effluent of the oil palm mill; empty fruit bunches; and immature fallen fruits [4] are used for composting.

Aquatic weeds have also been used as a raw material for composting due to their rapid establishment and growth, which has caused the degradation of aquatic ecosystems [15]. Aquatic weeds are the plants that complete their life cycle at either the water–land interface or in the water environment, hence interfering with ecosystem processes [16]. Their rapid growth, fast spread, fast ecological adaptations, and lack of natural enemies will make these plants invasive, problematic, and difficult to manage over time [17,18]. Most of the aquatic weeds are exotic species and have been introduced for ornamental and landscaping purposes [15].

Depending on the mode of adaptation and nature of the harm and control, these plants can be classified into three groups: floating, submerged, and emergent (Figure 1) [19,20]. However, this is not a unique classification. Some common aquatic weeds found in the world are listed in Table 1 [19,21].

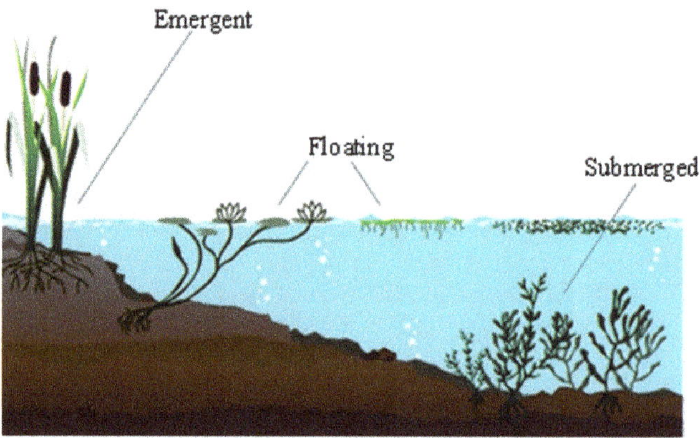

Figure 1. Three types of aquatic weeds.

Table 1. Common aquatic weeds in the world [19,21–23].

Plant Type	Botanical Name	Family
Floating	*Azolla* spp.	Azollaceae
	Eichhornia crassipes	Pontederiaceae
	Hydrilla verticillata	Hydrocharitaceae
	Lemna spp.	Lemnaceae
	Pistia stratiotes	Araceae
	Salvinia molesta	Salviniaceae
Submersed	*Cabomba caroliniana*	Cabombaceae
	Ceratophyllum demersum	Ceratophyllaceae
	Chara spp.	Characeae
	Crassula helmsii	Crassulaceae
	Egeria spp.	Hydrocharitaceae
	Lagarosiphon major	Hydrocharitaceae
	Nitella spp.	Characeae
	Potamogeton spp.	Potamogetonaceae
	Utricularia spp.	Lentibulariaceae
	Vallisneria spp.	Hydrocharitaceae

Table 1. *Cont.*

Plant Type	Botanical Name	Family
Emersed	*Brachiaria* spp.	Poaceae
	Ipomoea spp.	Convolvulaceae
	Limnocharis flava	Limnocharitaceae
	Ludwigia spp.	Onagraceae
	Lythrum salicaria	Lythraceae
	Monochoria spp.	Pontederiaceae
	Myriophyllum spp.	Haloragaceae
	Nelumbo nucifera	Nymphaeaceae
	Nuphar luteum (*Nyphaea* spp.)	Nymphaeaceae
	Nymphaea stellata	Nymphaeaceae
	Phragmites karka	Poaceae
	Polygonum spp.	Polygonaceae
	Sagittaria spp.	Alismataceae
	Scirpus spp.	Cyperaceae
	Spartina spp.	Poaceae
	Sphenoclea zeylanica	Sphenocleaceae
	Typha spp.	Typhaceae
	Vossia cuspidata	Poaceae
Wetland tree	*Melaleuca quinquenervia*	Myrtaceae
Mat-forming	*Alternanthera philoxeroides*	Amaranthaceae

Sri Lanka is an island that contains a large extent of fresh-water reservoirs. Approximately 23.3% of agricultural lands are cultivated with those resources in both of the two major cropping seasons [24]. As shown by Kariyawasam et al., the aquatic invasive alien plant community in the local tank cascade system has been spread over the past few decade creating many difficulties ecologically, environmentally, and economically [22]. Variations in environmental factors, anthropogenic activities, and ecological factors change the abundance of such aquatic plants from time to time. A few locally abundant aquatic invasive weed species have been listed in Figure 2.

Figure 2. Common aquatic weeds found in Sri Lankan aquatic ecosystems: (**a**) *Azolla* spp.; (**b**) *Eichhornia crassipes*; (**c**) *Hydrilla verticillata*; (**d**) *Lemna* spp.; (**e**) *Limnocharis flava*; (**f**) *Pistia* spp.; (**g**) *Salvinia molesta*; (**h**) *Typha* spp.; (**i**) *Vallisneria* spp. [22,24,25].

The majority of aquatic weeds pose risks to agricultural lands, aquatic habitats, and natural and forest ecosystems. A few of them can be listed as follows:

1. Interruptions to hydropower generation due to the reduction of water levels in water bodies, siltation, the inefficiency of related machines and equipment (such as turbines), and high maintenance costs [23,26].
2. Disturbance in navigation thereby influencing fisheries and water recreation [21].
3. Impact on the life of aquatic organisms such as fish, submerged plants, and phytoplankton [27]. That is mainly due to interrupting the photosynthesis of aquatic flora (by limiting light penetration into deep water layers) and changing the oxygen: carbon dioxide gas balance [23,28].
4. Blocking irrigation and water channels leading to flooding and bank erosion incidences [19].
5. Noxious aquatic plants in water bodies reducing irrigation water for agriculture as it increases water loss by evapotranspiration [27].
6. Influence on drinking and other household consumption, health-related issues such as insect-borne (primarily mosquito-infested) and poisonous incidences (plants, snakes, insects, and fish) of water bodies on the surrounding human community [29].
7. Reduction of aesthetic appearance of water bodies as a result of eutrophication and foul smell [30].
8. Economic losses related to social, ecological, and policy issues occurring mainly due to the eutrophication phenomenon in water bodies. This may cause the increasing purifying cost of polluted water and negative impacts on tourism [31].
9. Challenges with the sustainability of aquatic ecosystems.

The probability of the successful control of aquatic weeds is inversely related to the size of the aquatic ecosystem, and adapting intergraded weed management will be more effective than a single weed control method [32]. Controlling these plants needs more attention due to their higher growth rate, dispersion ability, and capacity for the formation of resistance structures for survival. The commonly used control methods are mentioned below.

1.1. Preventive Management

The spread and further introduction of aquatic weeds can be avoided by following rules and regulations on plant trades, plant and animal transportation, quarantine, and proper construction and maintenance of water bodies [33].

1.2. Manual and Mechanical Management

This method is more labor-intensive and time-consuming compared to other methods [34]. Reduction in water levels in water bodies may influence the growth of submerged and floating aquatic weeds due to changing in the nutrient composition of the water [35]. Introducing barriers like booms and cables in the water channels, manual removal of weeds using rakes and fine-meshed nets, and mechanical removal using tractors and excavators are the common manual and mechanical methods used to manage aquatic weeds [36]. With advanced technology, types of equipment like autonomous rotary-wing unmanned air vehicles have been developed for easy handling [37]. Manual and mechanical removal of submerged and floating aquatic weeds such as *Hydrilla verticillata* and *Egeria* spp. can lead to canopy fragmentation resulting in subsequent dispersal and an increase in weed population [38].

1.3. Ecological Aquatic Weed Management

These plants are essential to the aquatic ecosystem as they offer the majority of the food, nutrients, and habitat for aquatic life while regulating the concentrations of dissolved gases in the water. As a result, the unexpected removal of a significant amount of these could disrupt ecosystem services. Therefore, understanding the importance of aquatic weeds in a particular aquatic environment, and studying the ecological strategies, adaptations of the plants, and environmental consequences after removing them from that environment are important before adapting weed management strategies [39].

1.4. Chemical Management

Selecting the most suitable herbicides which are specific to aquatic weeds and applying them at the correct time with the recommended dose will control the aquatic weeds effectively without harming their ecosystems [16]. Some components (ingredients and/or surfactants) in such herbicides can be toxic to beneficial aquatic organisms and terrestrial biological controlling agents [40]. The toxicity of diquat, glyphosate, and glyphosate-trimesium chemicals (which is used for controlling *Eichhornia crassipes*) on aquatic insects such as *Eccritotarsus catarinensis* and *Neochetina eichhorniae* had been proven by previous literature [40].

1.5. Biological Management

Since 99% of aquatic weeds are new entries to the ecosystems, natural biological controlling agents cannot be found in existing environments [41]. Therefore, specific controlling agents for particular weeds should be identified and transported after analyzing their negative effects on the ecosystems as well [32]. In some areas, this method is considered as a slow and insufficient method for short-term benefits [42].

Potential biological control agents are fungal species such as *Fusarium* spp. for *Egeria densa*, *Cercospora pistiae* for *Pistia stratiotes*, *Phaeoramularia* spp. and *Phoma* spp. for *Echinochloa polystachya* and *Paspalum repens* [43], insects such as *Agasicles hygrophila* and *Vogtia malloi* for *Alternanthera philoxeroides*, *Bagous affinis* Hustache and *Hydrellia pakistanae* Deonier for *Hydrilla verticillata*, *Stenopelmus rufinasus* for *Azolla* spp., *Cornops aquaticum* and *Eccritotarsus catarinensis* for *Eichhornia crassipes* [34,44], and fish species such as *Ctenopharyngodon idellus* and *Cyprinus carpio* for controlling a variety of weeds, especially Filamentous algae [45].

1.6. Aquatic Weed Management through Utilization

Despite the fact that aquatic plants have become invasive in some regions of the world, certain weeds have directly led to socioeconomic livelihood in other areas. Some aquatic weeds have a greater potential to be used as a bio-fertilizer due to their allelopathic behaviors [46] and a mulching material for supplying plant nutrients and organic matter, retaining soil moisture, and increasing soil microbial population [47]. Its higher cellulose, hemicellulose, and low lignin content can easily be used for making low-cost bags, paper plates, paper boards, and decorative paper [48]. Moreover, aquatic weeds are potential raw materials for producing biogas and thereby generating electricity [49]. *Azolla* spp. is considered as a feed supplement for livestock, poultry, and aquaculture farming due to its high nutritional value, especially with favorable amino acids and higher protein content [50,51]. Other than the ornamental value, free radical scavenging pigments in *Nymphaea pubescens* extract have a medicinal value that can be used for treating melanoma skin cancers [52]. *Eichhornia crassipes* is an ornamental plant and is also popular as a phytoremediation plant, a source of biomass energy, and a source of raw materials for animal feed, construction, handicraft, paper, and board making [53]. *Pistia stratiotes* oil extract is good medicine, especially for worm infections, asthma, and skin disease, while leaves and roots are excellent sources of antioxidants [54]. Other than that, every part of *Nelumbo nucifera*, including leaves, rhizomes, seeds, and flowers, have been involved with traditional human livelihood as a part of the human diet, ayurvedic medicine, pharmaceuticals, and also landscaping [55].

Controlling aquatic weeds is very difficult in developing countries such as Sri Lanka due to limited resources. As a result of that, several types of aquatic weed species have grown at an alarming rate, causing a disturbance in nature and agriculture in Sri Lanka. The ultimate aim of this work is to evaluate the feasibility of aquatic weeds to be used as a raw material for compost production to meet the requirements in sustainable plant nutrient management in the local context. Aquatic weed management will thereby be more effective in terms of the agriculture and ecosystem level.

2. Aquatic Weeds as a Potential Raw Material in Composting

According to the literature, composting is one of the best methods for transferring nutrients in aquatic weeds into crop production [23,56]. Its short life cycle with higher biomass yield confirms its continuous availability as raw materials for composting [57]. Aquatic weeds are known as excellent sources of minerals, vitamins, proteins, and carbohydrates that are beneficial for plant health (Table 2). In addition, its allelopathic action, phytoremediation properties, and lower technology involvement also point out the suitability of aquatic weeds for composting. Even though most aquatic plants can be applied as green manure, with the process of composting, more nutrients can be available for plants [16].

Composting agents use carbon as a source of energy, which causes them to break down complex carbon compounds such as cellulose, hemicellulose, and lignin into simpler forms, leaving higher organic matter losses [58,59]. As shown by Gusain et al., *Pistia* spp. reduces its total organic carbon content from 461 g/kg to 449.87 g/kg through composting. The losses happen mainly as carbon dioxide gas. When *Salvinia molesta*, *Eichhornia crassipes*, and *Lagenandra toxicaria* are vermicomposted with cow dung, the resulting organic carbon contents (at 70% of moisture content) were 8.37%, 8.24%, and 13.21%, respectively, whereas the *Gliricidia* and grass mixture produced 9.54% in 70% of moistened samples [57]. Concentrations of the primary nutrients can be increased by composting (Figure 3). The reasons for increasing total kjeldahl nitrogen content (TKN) in prepared compost may include: (1) the nourished activities of earthworms, with the excreting nitrogen sources (nitrogenous excretory products, mucus, body fluids, and various enzymes) and their feeding mechanisms; (2) the conversion of ammonium nitrogen into nitrates in the vermicomposting process; (3) the mineralization of organic materials; and (4) the conversion of proteins in the aquatic weeds [56]. These fluctuations lead to a decrease in the C/N ratio, demonstrating the compost's suitability in terms of stability and maturity [59,60]. Prior research has found that compost materials with a C/N ratio of 20 to 25 produce the greatest results because a greater C/N ratio delays the degradation of raw materials [59]. Higher phosphorous content (TP) in prepared compost than weed sediments may result due to earthworm gut reactions at the production process. Phosphorus-solubilizing micro-organisms will help mineralize and increase the availability of phosphorus as different phosphate components. Mainly, the group of bacteria known as "diazotrophic bacteria" (for example, *Pseudomonas* spp., *Burkholderia* spp., *Agrobacterium* spp., *Azotobacter* spp., and *Erwinia* spp.) have been identified as phosphorus-solubilizing micro-organisms [61]. It not only increases the end phosphorus content, but it also activates the phosphorus recycling process. The least amount of phosphorus can be leached down the compost pile [58]. A slight increase in potassium concentration (TK) in the end product could be observed due to releasing acids that can convert insoluble potassium into a plant-available form during composting by micro-organisms [56].

Most aquatic weeds absorb toxic metals from the surrounding environment [62]. Therefore, it is important to monitor the potentially toxic metal contents to avoid the bioavailability risks. Standards and guidelines for that are promulgated by a variety of agencies in worldwide [63]. As mentioned in the previous literature, the majority of potentially toxic metal concentrations are well within accepted standards which are stated by the Colombian Technical Norms (NTC) 5167/04 and the standard of the Fertilizer Control Order 1985 [56,64]. In the local context of the Sri Lankan Standards Institution, Sri Lanka had placed important specifications on toxic elements for compost made from raw materials of organic origin to maintain the quality. A comparison of maximum permissible limits of potentially toxic metal elements in compost has been given in Table 2.

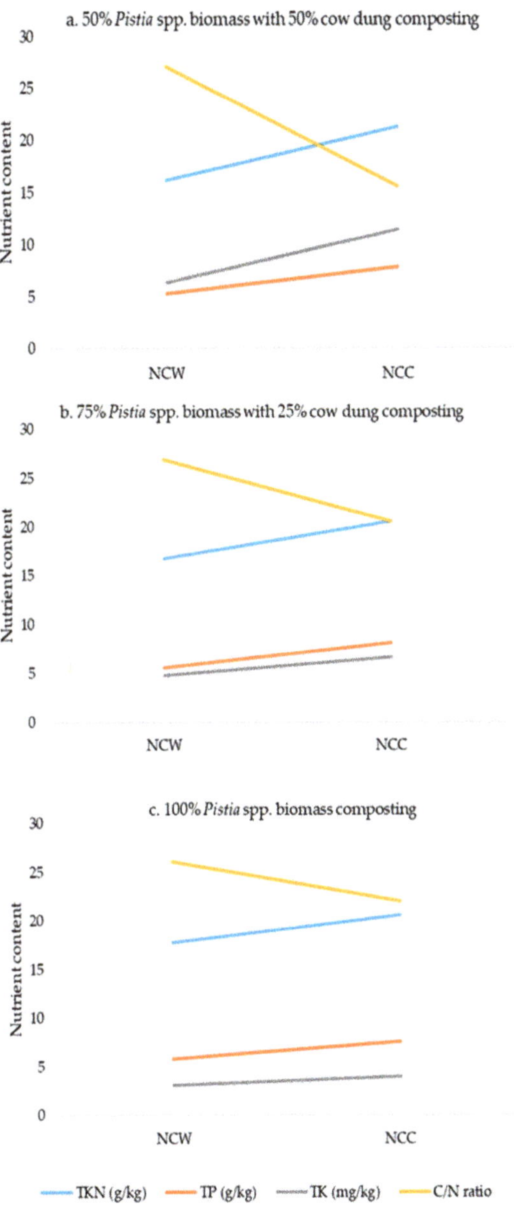

Figure 3. Chemical analysis of initial and composted waste mixtures produced by (**a**) 50% *Pistia* spp. biomass and 50% cow dung composting; (**b**) 75% *Pistia* spp. biomass and 25% cow dung composting; and (**c**) 100% *Pistia* spp. biomass composting [59]. Note: NCW—Nutrient content of weed sediment, NCC—Nutrient content of prepared compost. Methods which are used to measure TKN are the micro-kjeldahl digestion method, the method of ascolbic acid-reduced molybdophosphoric acid method for TP, and the flame photometer (Systronic 128) method for TK in Dehradun, India.

Table 2. The maximum limits for potentially toxic elements promulgated by different agencies [63,65].

Potentially Toxic Element	EU Range (mg/kg)	USA Biosolids (mg/kg)	Sri Lankan Standards (SLS) (mg/kg Dry Basis, AOAC Testing Method)
Cadmium (Cd)	0.7–10	39	1.5
Chromium (Cr)	70–200	1200	50
Mercury (Hg)	0.7–10	17	0.5
Nickel (Ni)	20–200	420	40
Lead (Pb)	70–1000	300	30

Most aquatic weeds have a considerable positive chance to be a good source of nutrients in the local context and this had been confirmed by several works in the literature [57,63,66]. The dynamic interaction between bacteria and earthworms affects the fate of heavy metals and alters their concentrations. Microbes present in the gut of the earthworms have good potential to detoxify, absorb, and accumulate the heavy metals from the raw material leaving a low concentration in compost and high concentration in earthworm bodies [56]. Later, those earthworms can be removed from the content, especially in vermicomposting. In contrast, fungi-like organisms and physiochemical processes such as oxidation, formation of organometallic complexes, and solubilization, as well as naturally occurring zeolites/hydrated alum-inosilicate minerals, would contribute to the changing bioavailability of potentially toxic metal concentrations during composting [67]. The addition of lime, sodium sulfide, bamboo charcoal, bamboo vinegar, or red-mud also has good potential to alter the forms of heavy metals and their bioavailability. One can see a slight fluctuation of concentrations of nickel, cadmium, molybdenum, cobalt, barium, and chromium while increasing the concentrations of zinc, lead, manganese, copper, and lithium in prepared vermicompost compared to initial substrates [56]. Milojkovic et al. introduced a "low-cost biosorbent" method with *Myriophyllum spicatum* L., which showed a lower metal absorption potential for lead(+2), copper(+2), cadmium(+2), and nickel(+2) ions compared to normal composting techniques [68]. Additionally, blending various source materials can alter the risk of bioavailability associated with aquatic weeds (Table 3). It changes the nutrient content, especially potentially toxic metal concentrations as well. Furthermore, considering previous research, we are able to confirm that the performance of aquatic weeds is same as other types of organic waste (Table 4).

Table 3. Variation of potentially toxic metal concentration with different raw material mixing ratios [66].

Potentially Toxic Metals (mg/kg Dry Basis)	Different Compost Mixing Methods						
	50% WH + 25% DLL + 25% CM	50% WH + 45% DLL + 5% WA	50% WH + 45% DLL + 5% ERP	50% WH + 50% DLL	50% WH + 25% DLL + 5% ERP + 5% WA + 15% SLP	50% WH + 25% DLL + 15% CM + 5% ERP + 5% WA	100% WH
Cu	18.5 ± 0.14	13.83 ± 0.18	8.28 ± 0.08	6.44 ± 0.14	5.71 ± 0.08	17.24 ± 0.38	14.6 ± 0.0
Cd	-	-	-	-	-	-	-
Pb	19.59 ± 0.93	6.74 ± 0.26	10.58 ± 0.30	5.78 ± 0.51	5.10 ± 0.48	18.34 ± 0.13	-
Zn	25.16 ± 0.16	31.43 ± 0.70	5.77 ± 0.02	15.97 ± 0.25	26.51 ± 0.32	21.93 ± 0.19	32.47 ± 0.29
Ni	-	-	-	-	-	-	-
As	1.24 ± 0.03	0.28 ± 0.01	1.0 ± 0.05	0.28 ± 0.02	0.21 ± 0.03	1.35 ± 0.04	0.79 ± 0.01

Note: WH—water hyacinth, DLL—dry leaf litter, CM—cattle manure, WA—wood ash, ERP—Eppawala rock phosphate, SPL—spent poultry litter.

Table 4. Comparison of different chemical properties in different compost materials from various raw materials [65,66,69].

Organic Waste Category	Main Ingredients	pH	Electrical Conductivity (dS m^{-1})
Vegetal Residue	Cucumber and zucchini crop residues	9.18	8.48
	Cucumber and zucchini crop residues	8.08	17.36
	Pepper crop residues	9.68	9.97
Municipal Solid Waste	Different sources	8.66	4.97
		7.50	5.58
		6.00	10.29
Agri-food Waste	Citric sludge and palm tree pruning (1:3 v/v)	6.64	7.24
	Cull tomatoes and tomato plant (stalks and leaves)	7.83	5.1
	Citric sludge, pig slurry, and pruning wastes (mainly palm tree) (3:1:1.5 v/v)	6.67	2.72
Water hyacinth	50% water hyacinth, 25% cattle manure, 25% dry leaf litter	8.50	4.78
	50% water hyacinth, 45% cattle manure, 5% Eppawala rock phosphate	7.30	1.59
	50% water hyacinth, 25% cattle manure, 15% dry leaf litter, 5% Eppawala rock phosphate, 5% wood ash	7.15	3.49
	100% water hyacinth	7.60	5.28
Requirement set by Sri Lankan Standards Institution		6.5–8.5	4.0

Note: pH and EC of vegetal residue, municipal solid waste, and agri-food waste are measured in a 1:10 (w/v) (compost: water extract) solution. pH of water hyacinth is determined in a 1:2.5 w/v solution, and 1:5 w/v is for determining EC.

Additionally, aquatic weed compost has a greater capacity of making nutrients available for plants and soil rehabilitation. As an example, the potential of *Eichhornia crassipes* compost to rehabilitate the salt-affected soil had been identified by Ahmed et al. [70]. According to these results, part of the chemical requirement (gypsum fertilizer) could be easily cut off with these nutrient-rich compost materials, saving the farmers' wealth as well as environmental health (Table 5).

Table 5. Effect of *Eichhornia crassipes* compost and gypsum on rice and wheat yield [70].

Treatment	Rice Yield (t ha^{-1})	Wheat Yield (t ha^{-1})
Control (no fertilizer application)	1.85	1.52
Application of 100% Gypsum requirement	3.64	3.53
Application of 50% Gypsum requirement with 10 t ha^{-1} *Eichhornia crassipes* compost	3.71	3.58
Application of 10 t ha^{-1} *Eichhornia crassipes* compost	2.44	2.68

Dhadse et al. experimented with investigating the effect of different vermicompost mixtures prepared from different aquatic weeds (*Hydrilla verticilata* (L.f.) Royle., *Ceratophyllum demursum* L., *Nelumbo nucifera* Gaerth., *Ludwigia palustris* L., *Pistia stratioles* L., and *Eicchornia crassipis* Mart.) on the plant growth [56]. All the vermicompost mixtures had shown better plant height, initiation of new leaves, and maturing of exiting leaves compared to soil, indicating their nutrient richness for plant growth and development. This will indicate the suitability of aquatic weed composting as a soil amendment.

3. Methods for Compost Production with Aquatic Weeds

Different composting techniques can effectively transfer many aquatic weeds into valuable manure. Diverse composting processes for different source materials have been developed in various regions of the world, employing newer technology to suit crop requirements while considering local concerns and needs [12,71] as follows:

1. Wind-row methods (turned wind-rows, passively aerated wind-rows, and aerated static pile)
2. In-vessel composting methods (bin composting, passively aerated bin composting, rectangular agitated beds, silos, rotating drums, transportable containers)
3. Traditional methods (anaerobic composting, aerobic composting through passive aeration/static composting)
4. Rapid methods (aerobic high-temperature composting, aerobic high-temperature composting with inoculation, IBS rapid composting)
5. Vermicomposting methods

All the above-mentioned composting methods may not be suitable for composting with aquatic weeds. As examples, biodynamic fortified composting and microbe mediated composting are suited Dal weeds [72], and vermicomposting is suitable for *Salvinia molesta*, *Eichhornia crassipes* and *Lagenandra toxicaria* [57] can be shown. Figures 4 and 5 explain the effect of the different composting methods on *Limnocharis flava* (L.) Buchenau for its recovery percentage and number of days taken for quality compost production. It explains the importance of selecting the proper composting method, as it changes the final yield in terms of quality and quantity. As described in Figure 4, the maximum recovery % or weight of compost gained from unit biomass weight is higher in vermicomposting compared with other composting methods. Considerably, *Limnocharis flava* (L.) Buchenau can be converted into a nutrient-rich amendment with a shorter period than normal composting techniques (Figure 5).

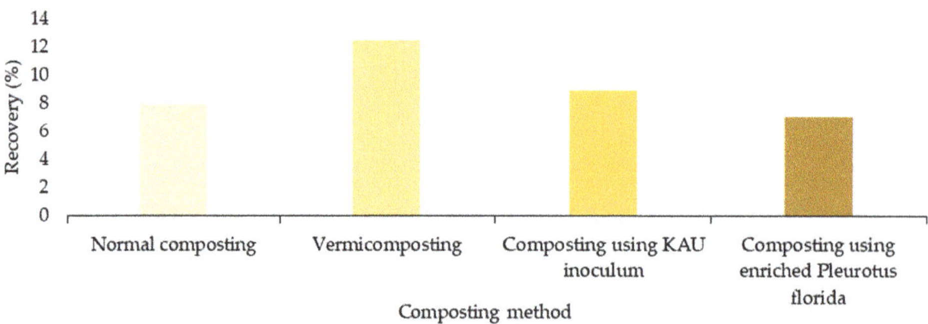

Figure 4. Effect of different composting methods on recovery % [73].

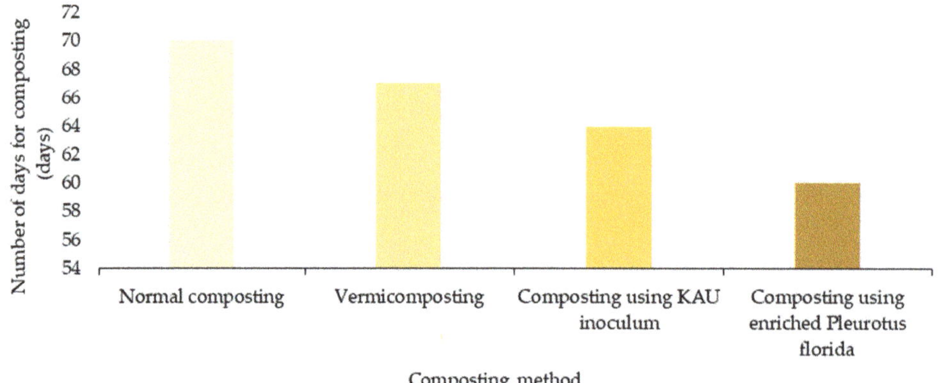

Figure 5. Effect of different composting methods on time taken for composting [73].

Different production techniques lead to different processing conditions, composting times, and micro-organism corporations with various stability, maturity, and sanitation potencies. That will result in different final products with different qualities and compositions as different nutrient contents (Figure 6).

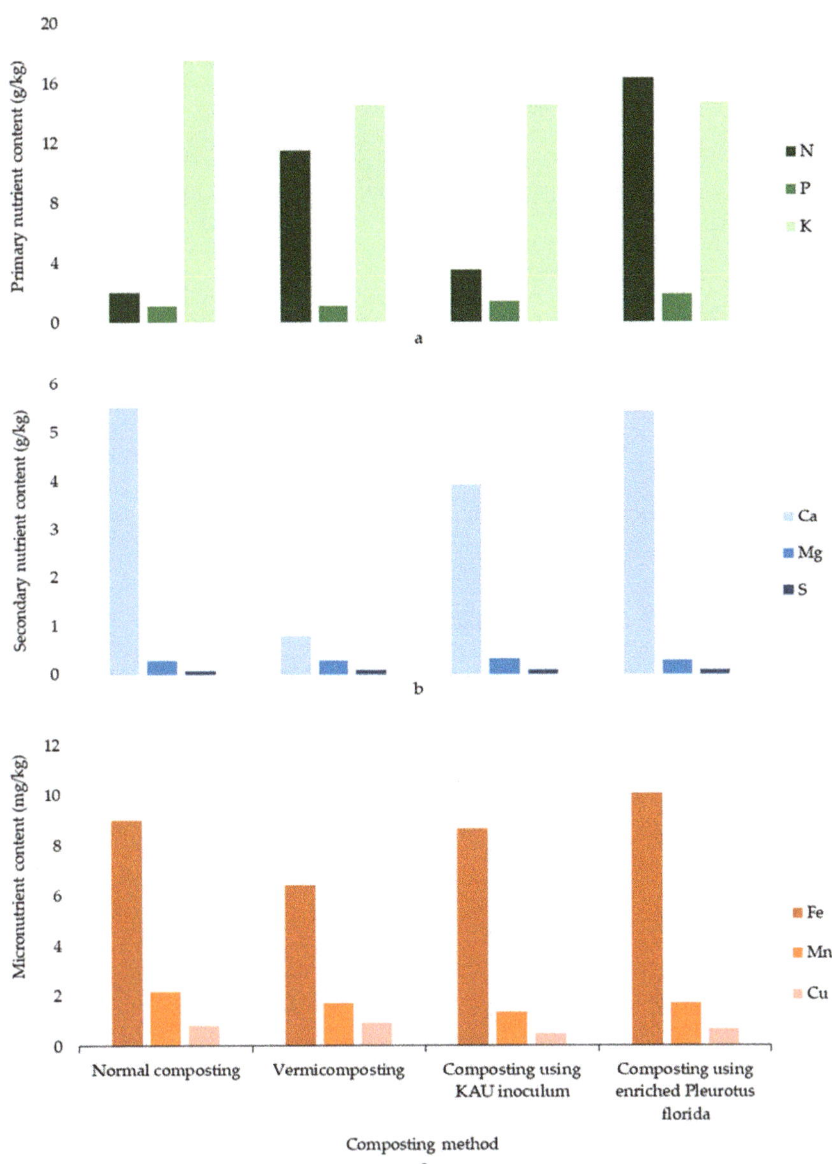

Figure 6. Effect of different composting methods on (**a**) primary, (**b**) secondary, and (**c**) micronutrient content of compost from *Limnocharis flava* (L.) Buchenau [73].

4. Steps for Producing Compost with Aquatic Weeds

Aquatic weeds should be collected at the particular stage that shows optimum plant growth. The location chosen for compost production should have shade without standing water. A foundation layer with bricks and stones, followed by a coarse sand layer (about 6–7.5 cm), can be created to maintain good drainage. Rather than applying the weed mass as is, it is better to chop the weeds into 5–10 cm long small pieces to increase the aeration and surface area. Thereby, microbial activity and decomposition rate can be enhanced. Then, the chopped weed mass can be spread on the foundation as a 30 cm thick layer. Ash, soil, organic matter, animal manure, or other materials such as municipal waste, lime, Eppawala rock phosphate, etc. can be mixed with weeds to achieve better qualities. For example, fresh cattle manure as a 10–15 L of diluted solution, and 1 L of molasses can be applied as a layer on the top of the weed pile. Followed by that, another 30 cm thick layer of weed mass can be added. This process should be revised up to receiving a $15 \times 5 \times 5^{(1/2)}$ ft compost pile. Then, the pile should be covered with black polythene. Appropriate turning and mixing steps with frequent application of water should be practised. The correct moisture content can be identified by a simple field method such as squeezing a handful of compost. Finally, the end product can be applied to the plants as a mulch or incorporated with the soil [53,63,69].

A few experiments done on composting with aquatic weeds are summarized in Table 6. Most of these experiments had been concluded with the improvement in the growth of targeted plants.

Table 6. Experiments on composting with aquatic weeds.

Name of the Aquatic Weed	Reference	Remarks
Phragmites australis, Typha angustata, Azolla sp., *Nymphoides peltatum, Nelumbo nucifera, Nymphaea* sp., and *Cerutophyllum demersum, Myriophyllum spicatum*	[74]	Production of Vermipellet is more effective than the production of vermicomposting from these mentioned aquatic weeds due to less disease transmission potential, lower heavy metal concentrations, minimum weed growth ability, enhanced dispersal nature, high C/N ratio, and maximum nutrient composition in vermipellets.
Ceratophyllum demursum, Nelumbo nucifera, Ludwigia palustris	[56]	Even though the parameters of compost produced with these weeds are varied, they are generally recognized as good sources for nutrient-rich vermicomposting.
Limnocharis flava	[73]	According to this research, the potential of *Limnocharis flava* for vermicomposting compared to normal composting and KAU inoculum composting was confirmed with the highest recovery percentage (the quantity of compost given by the unit amount of biomass) within 60–70 days.
Myriophyllum spp.	[68,75]	Compost made by *Myriophyllum* spp. (*M. spicatum*) had significantly higher bio-sorption capacity and ability, and thereby can be used to purify heavy metals from waste. Furthermore, herbicide application is not advisable after applying compost produced with *Myriophyllum* spp.
Eicchornia crassipis	[76]	*Eicchornia crassipis* can be transferred into nutrient-rich vermicompost materials within 60 days with the help of probiotics *Lactobacillus sporogens*.
Hydrilla verticilata	[15]	Since the whole plant is decomposable, shredding is not a must, and it can easily supply enough moisture (more than 60%) and plant growth nutrients such as P, K, Mg, and Ca.
Lagenandra toxicaria	[57]	Vermicomposting with *L. toxicaria* gives a better-quality, nutrient-rich end product than normal composting techniques. Since the end product has 6.75 dS m^{-1} of electrical conductivity, 13.21% organic carbon, 3.61% P content, 5.03% K content, and 6.12% Ca content with good microbial activities, it can be an excellent organic fertilizer source for coconut.

Table 6. *Cont.*

Name of the Aquatic Weed	Reference	Remarks
Pistia spp.	[60]	After inoculating 60–80 % of cow dung and *Eisenia fetida*, *Pistia* spp. can be transferred into odor-free, nutrient-rich (N, P, K, Cu, Zn, Fe) vermicompost.
Azolla filiculoides and *Typha latifolia*	[64]	This research justified the suitability of compost produced from these plants for agronomic usage depending on its nutritional and physical properties. Here, an autonomous, self-powered fixed bed gasifier with a gyrating cylinder bioreactor was used to reduce the time of the composting process.
Salvinia molesta	[77]	*S. molesta* has a greater potential to be an excellent vermicomposting material resulting in an ideal reduction in C: N ratio, humification index, allelopathy, and other kinds of toxicity.
Egeria densa	[78]	Composting of *Egeria densa* can be induced by inoculating microbial bio-preparations with bacteria, fungi, and actinomycetes.
Alternanthera philoxeroides	[79]	Because of the higher survival rate of these seeds, the mixing should be done properly, and the whole mixture should be uniformly subjected to a temperature higher than 55 °C over three consecutive days to avoid spreading aquatic weeds on the terrestrial composing.

5. Potential Environmental Risks Associated with the Application of Aquatic Weed Composts

Dorahy et al. summarized the environmental risk associated with applying compost prepared from aquatic weeds to terrestrial plants using different matrixes [79]. According to that, high risk was associated with the ability to survive and the spread of weeds on the land. Other than that, eutrophication of waterbodies, heavy metal accumulation, and phytotoxicity could also happen with composting. Nevertheless, the associated risks with those three were low. This will highlight the importance of having a proper composting method and ongoing management of the application. To avoid these risks, it is advisable to (1) turn and mix the compost mixture as recommended; (2) reduce the time spent stockpiling before composting, (3) increase the size of the compost pile to achieve higher internal temperature to destroy weed seeds and harmful substances [80]); (4) field application of well-processed, good-quality compost (Immature and unstable compost has a negative impact on soil fertility and seed germination) [81]); (5) avoid application of aquatic weed compost to the lands located close to water bodies; (6) maintain good site sanitary conditions; and (7) clean all machineries and tools used to handle compost, as directed [79].

6. Conclusions

Invasive aquatic weeds have been one of the most predominant threats that have many adverse effects on aquatic habitats creating negative impacts on the economy, ecology, and environment in the world. It is fast-growing with a wider range of adaptation mechanisms causing difficulties in the utilization and management of most of the water bodies. Composting is one of the best eco-friendly and sustainable ways of transferring nutrients in aquatic weeds into crop production. Short life cycle, higher biomass yield, higher plant nutrient compositions, allelopathic behaviours, and phytoremediation properties confirm their suitability as raw materials for composting. Following proper composting techniques and parameters after studying ecology and morphological features of particular aquatic weeds, most aquatic environments can be saved from soil and water pollution while enhancing crop production. Further research studies should focus more on cost–benefit analysis, legal and regulation activities, health risks, and pollutant removal of those weeds before including those weeds in the cropping cycle.

Author Contributions: Conceptualization, D.M.N.S.D. and A.J.A.; Methodology, S.S.U.; Validation, D.K.R.P.L.D. and T.D.N.; Writing—Original Draft Preparation, D.M.N.S.D. and S.S.U.; Writing—Review and Editing, A.J.A., T.D.N. and D.K.R.P.L.D.; Supervision, A.J.A.; Visualization, D.M.N.S.D. and S.S.U.; Project Administration, A.J.A. All authors have read and agreed to the published version of the manuscript.

Funding: This research received no external funding.

Institutional Review Board Statement: Not applicable.

Informed Consent Statement: Not applicable.

Data Availability Statement: Data sharing not applicable—no new data generated.

Acknowledgments: We would like to express our appreciation to the technical staff of the Agronomy Division of the Coconut Research Institute. We would like to express our deep gratitude to the editor and three anonymous reviewers for their valuable comments and critical evaluation.

Conflicts of Interest: The authors declare no conflict of interest.

References

1. Dissanayaka, D.; Nuwarapaksha, T.; Udumann, S.; Dissanayake, D.; Atapattu, A.J. A Sustainable Way of Increasing Productivity of Coconut Cultivation Using Cover Crops: A Review. *Circ. Agric. Syst.* **2022**, *2*, 7. [CrossRef]
2. Mengqi, Z.; Shi, A.; Ajmal, M.; Ye, L.; Awais, M. Comprehensive Review on Agricultural Waste Utilization and High-Temperature Fermentation and Composting. *Biomass Convers. Biorefin.* **2021**, 1–25. [CrossRef]
3. Zahra, M.B.; Fayyaz, B.; Aftab, Z.E.H.; Haider, M.S. Mitigation of Degraded Soils by Using Biochar and Compost: A Systematic Review. *J. Soil Sci. Plant Nutr.* **2021**, *21*, 2718–2738. [CrossRef]
4. Mohammad, N.; Alam, M.Z.; Kabbashi, N.A.; Ahsan, A. Effective Composting of Oil Palm Industrial Waste by Filamentous Fungi: A Review. *Resour. Conserv. Recycl.* **2012**, *58*, 69–78. [CrossRef]
5. Tuomela, M.; Vikman, M.; Hatakka, A.; Itavaara, M. Biodegradation of Lignin in a Compost Environment: A review. *Bioresour. Technol.* **2000**, *72*, 169–183. [CrossRef]
6. Richard, T.L.; Hamelers, H.V.M.B.; Veeken, A.; Silva, T. Moisture Relationships in Composting Processes. *Compos. Sci. Util.* **2002**, *10*, 286–302. [CrossRef]
7. Akyol, Ç.; Ince, O.; Ince, B. Crop-based Composting of Lignocellulosic Digestates: Focus on Bacterial and Fungal Diversity. *Bioresour. Technol.* **2019**, *288*, 121549. [CrossRef] [PubMed]
8. Agnew, J.M.; Leonard, J.J. The Physical Properties of Compost. *Compos. Sci. Util.* **2003**, *11*, 238–264. [CrossRef]
9. Bernal, M.P.; Alburquerque, J.A.; Moral, R. Composting of Animal Manures and Chemical Criteria for Compost Maturity Assessment. A Review. *Bioresour. Technol.* **2009**, *100*, 5444–5453. [CrossRef]
10. Mia, S.; Uddin, M.E.; Kader, M.A.; Ahsan, A.; Mannan, M.A.; Hossain, M.M.; Solaiman, Z.M. Pyrolysis and Co-composting of Municipal Organic Waste in Bangladesh: A Quantitative Estimate of Recyclable Nutrients, Greenhouse Gas Emissions, and Economic Benefits. *Waste Manag.* **2018**, *75*, 503–513. [CrossRef] [PubMed]
11. Ma, C.; Hu, B.; Wei, M.B.; Zhao, J.H.; Zhang, H.Z. Influence of Matured Compost Inoculation on Sewage Sludge Composting: Enzyme Activity, Bacterial and Fungal Community Succession. *Bioresour. Technol.* **2019**, *294*, 122165. [CrossRef]
12. Ayilara, M.; Olanrewaju, O.; Babalola, O.; Odeyemi, O. Waste Management through Composting: Challenges and Potentials. *Sustainability* **2020**, *12*, 4456. [CrossRef]
13. Irshad, M.; Eneji, A.E.; Hussain, Z.; Ashraf, M. Chemical Characterization of Fresh and Composted Livestock Manures. *J. Soil Sci. Plant Nutr.* **2013**, *13*, 115–121. [CrossRef]
14. Wang, J.; Gu, J.; Wang, X.; Song, Z.; Dai, X.; Guo, H.; Yu, J.; Zhao, W.; Lei, L. Enhanced Removal of Antibiotic Resistance Genes and Mobile Genetic Elements during Swine Manure Composting Inoculated with Mature Compost. *J. Hazard. Mater.* **2021**, *411*, 125135. [CrossRef]
15. Jain, M.S.; Kalamdhad, A.S. A Review on Management of *Hydrilla verticillata* and its Utilization as Potential Nitrogen-rich Biomass for Compost or Biogas Production. *Bioresour. Technol. Rep.* **2018**, *1*, 69–78. [CrossRef]
16. Sarkar, S.; Sarkar, U.K.; Ali, S.; Kumari, S.; Puthiyotti, M. Status, ecological services and management of aquatic weeds of floodplain wetlands in India: An overview. *Lakes Reserv. Sci. Policy Manag. Sustain. Use* **2021**, *26*, 76–91. [CrossRef]
17. McFadyen, R.E.C. Biological Control of Weeds. *Annu. Rev. Entomol.* **1998**, *43*, 369–393. [CrossRef] [PubMed]
18. Ayanda, O.I.; Ajayi, T.; Asuwaju, F.P. *Eichhornia crassipes* (Mart.) Solms: Uses, Challenges, Threats, and Prospects. *Sci. World J.* **2020**, *2020*, 3452172. [CrossRef] [PubMed]
19. Jayan, P.R.; Sathyanathan, N. Aquatic Weed Classification, Environmental Effects and the Management Technologies for its Effective Control in Kerala, India. *Int. J. Agric. Biol. Eng.* **2012**, *5*, 76–91. [CrossRef]

20. Raney, F.C. Geobotany. *Ecology* **1966**, *47*, 173–174. [CrossRef]
21. Holm, L.G.; Weldon, L.W.; Blackburn, R.D. Aquatic Weeds. *Pest Artic. News Summ. (PANS)* **1970**, *16*, 576–589. [CrossRef]
22. Kariyawasam, C.S.; Kumar, L.; Kogo, B.K.; Ratnayake, S.S. Long-Term Changes of Aquatic Invasive Plants and Implications for Future Distribution: A Case Study Using a Tank Cascade System in Sri Lanka. *Climate* **2021**, *9*, 31. [CrossRef]
23. Kaur, M.; Kumar, M.; Sachdeva, S.; Puri, S.K. Aquatic Weeds as the Next Generation Feedstock for Sustainable Bioenergy Production. *Bioresour. Technol.* **2018**, *251*, 390–402. [CrossRef]
24. Ratnayake, S.S.; Khan, A.; Reid, M.; Dharmasena, P.B.; Hunter, D.; Kumar, L.; Herath, K.; Kogo, B.; Kadupitiya, H.K.; Dammalage, T.; et al. Land Use-Based Participatory Assessment of Ecosystem Services for Ecological Restoration in Village Tank Cascade Systems of Sri Lanka. *Sustainability* **2022**, *14*, 10180. [CrossRef]
25. Perera, P.C.D.; Dahanayake, N. Review of Major Abundant Weeds of Cultivation in Sri Lanka. *Int. J. Sci. Res. Publ.* **2015**, *5*, 2250–3153.
26. Mzuza, M.K.; Chapola, L.; Kapute, F.; Chikopa, I.; Gondwe, J. Analysis of the Impact of Aquatic Weeds in the Shire River on Generation of Electricity in Malawi: A Case of Nkula Falls Hydro-Electric Power Station in Mwanza District, Southern Malawi. *Int. J. Geosci.* **2015**, *6*, 636–643. [CrossRef]
27. Howard, G.W.; Harley, K.L.S. How do Floating Aquatic Weeds Affect Wetland Conservation and Development? How Can These Effects be Minimised? *Wetl. Ecol. Manag.* **1997**, *5*, 215–225. [CrossRef]
28. Samiei, J.; Mobaraki, R. Investigation of Effects of Control Method on Immersed Aquatic Weeds. *Medbiotech J.* **2019**, *3*, 88–92. [CrossRef]
29. Honlah, E.; Yao Segbefia, A.; Odame Appiah, D.; Mensah, M.; Atakora, P.O. Effects of Water Hyacinth Invasion on the Health of the Communities, and the Education of Children along River Tano and Abby-Tano Lagoon in Ghana. *Cogent Soc. Sci.* **2019**, *5*, 1619652. [CrossRef]
30. Ekwealor, K.U.; Echereme, C.B.; Ofobeze, T.N.; Okereke, C.N. Economic Importance of Weeds: A Review. *Asian Plant Res. J.* **2019**, *December*, 1–11. [CrossRef]
31. Pretty, J.N.; Mason, C.F.; Nedwell, D.B.; Hine, R.E.; Leaf, S.; Dils, R. Environmental Costs of Freshwater Eutrophication in England and Wales. *Environ. Sci. Technol.* **2003**, *37*, 201–208. [CrossRef]
32. Sousa, W.T.Z. *Hydrilla verticillata* (Hydrocharitaceae), a Recent Invader Threatening Brazil's Freshwater Environments: A Review of the Extent of the Problem. *Hydrobiologia* **2011**, *669*, 1–20. [CrossRef]
33. Madsen, J.D. Methods for Management of Nonindigenous Aquatic Plants. In *Assessment and Management of Plant Invasions*; Luken, J.O., Thieret, J.W., Eds.; Springer Series on Environmental Management; Springer: New York, NY, USA, 1997; p. 27. [CrossRef]
34. Hill, M.P.; Coetzee, J. The Biological Control of Aquatic Weeds in South Africa: Current Status and Future Challenges. *Bothalia* **2017**, *47*, 1–12. [CrossRef]
35. Thomaz, S.M.; Pagioro, T.A.; Bini, L.M.; Murphy, K.J. Effect of Reservoir Drawdown on Biomass of Three Species of Aquatic Macrophytes in a Large Sub-Tropical Reservoir (Itaipu, Brazil). *Hydrobiologia* **2006**, *570*, 53–59. [CrossRef]
36. Cilliers, C.J.; Hill, M.P.; Ogwang, J.A.; Ajuonu, O. Aquatic Weeds in Africa and Their Control. In *Biological Control in IPM Systems in Africa*; Neuenschwander, P., Borgemeister, C., Langewald, J., Eds.; CABI Publishing: Wallingford, UK, 2003; Volume 1991, pp. 161–178. [CrossRef]
37. Goktogan, A.H.; Sukkarieh, S.; Bryson, M.; Randle, J.; Lupton, T.; Hung, C. A Rotary-wing Unmanned Air Vehicle for Aquatic Weed Surveillance and Management. *J. Intell. Robot. Syst.* **2010**, *57*, 467–484. [CrossRef]
38. Dayan, F.E.; Netherland, M.D. Hydrilla, the Perfect Aquatic Weed, Becomes More Noxious Than Ever. *Outlooks Pest Manag.* **2005**, *16*, 277–282. [CrossRef]
39. Mitchell, D.S. African Aquatic Weeds and Their Management. In *The Ecology and Management of African Wetland Vegetation*; Denny, P., Ed.; Springer: Dordrecht, The Netherlands, 1985; Volume 6, pp. 177–202. [CrossRef]
40. Hill, M.P.; Coetzee, J.A.; Ueckermann, C. Toxic Effect of Herbicides Used for Water Hyacinth Control on Two Insects Released for Its Biological Control in South Africa. *Biocontrol. Sci. Technol.* **2012**, *22*, 1321–1333. [CrossRef]
41. Lovell, S.J.; Stone, S.F.; Fernandez, L. The Economic Impacts of Aquatic Invasive Species: A Review of the Literature. *Agric. Resour. Econ. Rev.* **2006**, *35*, 195–208. [CrossRef]
42. Coetzee, J.A.; Byrne, M.J.; Hill, M.P. Impact of Nutrients and Herbivory by *Eccritotarsus catarinensis* on the Biological Control of Water Hyacinth, *Eichhornia crassipes*. *Aquat. Bot.* **2007**, *86*, 179–186. [CrossRef]
43. Barreto, R.; Charudattan, R.; Pomella, A.; Hanada, R. Biological Control of Neotropical Aquatic Weeds with Fungi. *Crop Prot.* **2000**, *19*, 697–703. [CrossRef]
44. Purcell, M.; Harms, N.; Grodowitz, M.; Zhang, J.; Ding, J.; Wheeler, G.; Zonneveld, R.; de Chenon, R.D. Exploration for Candidate Biological Control Agents of the Submerged Aquatic Weed *Hydrilla verticillata*, in Asia and Australia 1996–2013. *BioControl* **2019**, *64*, 233–247. [CrossRef]
45. Avault, J.W. Preliminary studies with grass carp for aquatic weed control. *Progress. Fish-Cult.* **1965**, *27*, 207–209. [CrossRef]
46. Fu, Y.; Bhadha, J.H.; Rott, P.; Beuzelin, J.M.; Kanissery, R. Investigating the Use of Aquatic Weeds as Biopesticides towards Promoting Sustainable Agriculture. *PLoS ONE* **2020**, *15*, e0237258. [CrossRef] [PubMed]
47. Tate, R.L.; Riemer, D.N. Aquatic Weed Biomass Disposal: Effect on Soil Organic Matter. *J. Environ. Qual.* **1988**, *17*, 163–168. [CrossRef]
48. Nawaj Alam, S.; Singh, B.; Guldhe, A. Aquatic Weed as a Biorefinery Resource for Biofuels and Value-Added Products: Challenges and Recent Advancements. *Clean. Eng. Technol.* **2021**, *4*, 100235. [CrossRef]
49. Bote, M.A.; Naik, V.R.; Jagdeeshgouda, K.B. Production of Biogas with Aquatic Weed Water Hyacinth and Development of Briquette Making Machine. *Mater. Sci. Energy Technol.* **2020**, *3*, 64–71. [CrossRef]

50. Brouwer, P.; Schluepmann, H.; Nierop, K.G.; Elderson, J.; Bijl, P.K.; van der Meer, I.; de Visser, W.; Reichart, G.J.; Smeekens, S.; van der Werf, A. Growing Azolla to Produce Sustainable Protein Feed: The Effect of Differing Species and CO_2 Concentrations on Biomass Productivity and Chemical Composition. *J. Sci. Food Agric.* **2018**, *98*, 4759–4768. [CrossRef]
51. Das, M.; Rahim, F.; Hossain, M. Evaluation of Fresh *Azolla pinnata* as a Low-Cost Supplemental Feed for Thai Silver Barb *Barbonymus gonionotus*. *Fishes* **2018**, *3*, 15. [CrossRef]
52. Aimvijarn, P.; Rodboon, T.; Payuhakrit, W.; Suwannalert, P. *Nymphaea pubescens* Induces Apoptosis, Suppresses Cellular Oxidants-Related Cell Invasion in B16 Melanoma Cells. *Pharm. Sci.* **2018**, *24*, 199–206. [CrossRef]
53. Jafari, N. Ecological and Socio-Economic Utilization of Water Hyacinth (*Eichhornia crassipes* Mart Solms). *J. Appl. Sci. Environ. Manag.* **2010**, *14*, 43–49. [CrossRef]
54. Wasagu, R.S.; Lawal, M.; Shehu, S.; Alfa, H.H.; Muhammad, C. Nutritive Values, Mineral and Antioxidant Properties Of *Pistia stratiotes* (Water lettuce). *Niger. J. Basic Appl. Sci.* **2014**, *21*, 253. [CrossRef]
55. Chen, G.; Zhu, M.; Guo, M. Research Advances in Traditional and Modern Use of *Nelumbo nucifera* : Phytochemicals, Health Promoting Activities and Beyond. *Crit. Rev. Food Sci. Nutr.* **2019**, *59* (Suppl. 1), S189–S209. [CrossRef] [PubMed]
56. Dhadse, S.; Alam, S.N.; Mallikarjuna Rao, M. Development of Nutrient Rich Biofertilizer by Co-vermistabilization of Aquatic Weeds Using Herbal Pharmaceutical Wastewater along with Sediment of Lake. *Bioresour. Technol. Rep.* **2021**, *13*, 100633. [CrossRef]
57. Senarathne, S.H.S.; Dayananda, H.N.; Atapattu, A.A.A.J.; Raveendra, S.A.S.T. Feasibility of Using Problematic Aquatic Weeds in Productive Manner by Generating Vermicompost in Coconut Triangle Area of Sri Lanka. *CORD* **2017**, *33*, 15. [CrossRef]
58. Larney, F.J.; Sullivan, D.M.; Buckley, K.E.; Eghball, B. The Role of Composting in Recycling Manure Nutrients. *Can. J. Soil Sci.* **2006**, *86*, 597–611. [CrossRef]
59. Gusain, R.; Pandey, B.; Suthar, S. Composting as a Sustainable Option for Managing Biomass of Aquatic Weed *Pistia*: A Biological Hazard to Aquatic System. *J. Clean. Prod.* **2018**, *177*, 803–812. [CrossRef]
60. Suthar, S.; Pandey, B.; Gusain, R.; Gaur, R.Z.; Kumar, K. Nutrient Changes and Biodynamics of *Eisenia fetida* during Vermicomposting of Water Lettuce (*Pistia* sp.) Biomass: A Noxious Weed of Aquatic System. *Environ. Sci. Pollut. Res.* **2017**, *24*, 199–207. [CrossRef] [PubMed]
61. Busato, J.G.; Lima, L.S.; Aguiar, N.O.; Canellas, L.P.; Olivares, F.L. Changes in Labile Phosphorus Forms during Maturation of Vermicompost Enriched with Phosphorus-solubilizing and Diazotrophic Bacteria. *Bioresour. Technol.* **2012**, *110*, 390–395. [CrossRef]
62. Abhayawardhana, M.L.D.D.; Bandara, N.J.G.J.; Rupasinge, S.K.L.S. Removal of Heavy Metals and Nutrients from Municipal Wastewater using *Salvinia molesta* and *Lemna gibba*. *J. Trop. For. Environ.* **2019**, *9*, 65–77. [CrossRef]
63. Brinton, W.F. Compost Quality Standards and Guidelines. 2000, p. 42. Available online: https://compost.css.cornell.edu/Brinton.pdf (accessed on 9 January 2023).
64. Mesa, F.; Torres, J.; Sierra, O.; Escobedo, F.J. Enhanced Production of Compost from Andean Wetland Biomass Using a Bioreactor and Photovoltaic System. *Biomass Bioenergy* **2017**, *106*, 21–28. [CrossRef]
65. *SLS 1704:2021*; Specification for Compost from Municipal Solid Waste and Agricultural Waste. Sri Lanka Standards Institute: Colombo, Sri Lanka, 2021. Available online: https://mfa.gov.lk/wp-content/uploads/2021/06/sls-1704_2021.pdf (accessed on 9 January 2023).
66. Amarasinghe, S.R. Use of Invasive Water Hyacinth for Composting of Ordinary Leaf Litter. *Sri Lankan J. Agric. Ecosyst.* **2021**, *3*, 5. [CrossRef]
67. Singh, J.; Kalamdhad, A.S. Reduction of Heavy Metals during Composting—A Review. *Int. J. Environ. Prot.* **2012**, *2*, 36–43.
68. Milojkovic, J.V.; Stojanovic, M.D.; Mihajlovic, M.L.; Lopicic, Z.R.; Petrovic, M.S.; Sostaric, T.D.; Ristic, M.D. Compost of Aquatic Weed *Myriophyllum spicatum* as Low-Cost Biosorbent for Selected Heavy Metal Ions. *Water Air Soil Pollut.* **2014**, *225*, 1927. [CrossRef]
69. Siles-Castellano, A.B.; Lopez, M.J.; Lopez-Gonzalez, J.A.; Suarez-Estrella, F.; Jurado, M.M.; Estrella-Gonzalez, M.J.; Moreno, J. Comparative Analysis of Phytotoxicity and Compost Quality in Industrial Composting Facilities Processing Different Organic Wastes. *J. Clean. Prod.* **2020**, *252*, 119820. [CrossRef]
70. Ahmed, K.; Saqib, A.I.; Naseem, A.R.; Qadir, G.; Nawaz, M.Q.; Khalid, M.; Warraich, I.A.; Arif, M. Use of Hyacinth Compost in Salt-Affected Soils. *Pak. J. Agric. Res.* **2020**, *33*, 274. [CrossRef]
71. Misra, R.V.; Roy, R.N.; Hiraoka, H. *On-Farm Composting Methods*; Land and Water Discussion Paper; UN-FAO: Rome, Italy, 2003; Volume 2, p. 51.
72. Narayan, S.; Nabi, A.; Hussain, K.; Khan, F. Practical Aspects of Utilizing Aquatic Weeds in Compost Preparation. March 2017. Available online: https://www.researchgate.net/profile/Dr-Khan-11/publication/315100323_Practical_aspects_of_utilizing_aquatic_weeds_in_compost_preparation/links/58ca44eeaca27286b3b19644/Practical-aspects-of-utilizing-aquatic-weeds-in-compost-preparation.pdf (accessed on 9 January 2023). [CrossRef]
73. Jayapal, A.; Mini, V.; Resmi, A.R.; Lovely, B. Composting *Limnocharis flava* Buchenau: A Comparative Analysis. *J. Krishi Vigyan* **2021**, *10*, 28–32. [CrossRef]
74. Sharma, A.; Kumar, S.; Ahmed, N.; Nabi, S.U.; Singh, D.B.; Akbar, S.A. Managing Problematic Aquatic Macrophytes through Vermitechnology—Composting and Pelleting. *Waste Biomass Valorization* **2021**, *12*, 5561–5571. [CrossRef]
75. Nsenga Kumwimba, M.; Dzakpasu, M.; Li, X. Potential of Invasive Watermilfoil (*Myriophyllum* spp.) to Remediate Eutrophic Waterbodies with Organic and Inorganic Pollutants. *J. Environ. Manag.* **2020**, *270*, 110919. [CrossRef]
76. Sakthika, T.; Sornalaksmi, V. Nutrients Analysis of Vermicompost of Water Hyacinth Supplemented with Probiotics. *Acta Sci. Agric.* **2019**, *3*, 10–13. [CrossRef]

77. Hussain, N.; Abbasi, T.; Abbasi, S.A. Vermiremediation of an Invasive and Pernicious Weed Salvinia (*Salvinia molesta*). *Ecol. Eng.* **2016**, *91*, 432–440. [CrossRef]

78. Martinez-Nieto, P.; Bernal-Castillo, J.; Calixto-Díaz, M.; Del Basto-Riaño, M.A. Chaparro-Rico, B. Biofertilizers and Composting Accelerators of Polluting Macrophytes of a Colombian Lake. *J. Soil Sci. Plant Nutr.* **2011**, *11*, 47–61. [CrossRef]

79. Dorahy, C.G.; Pirie, A.D.; McMaster, I.; Muirhead, L.; Pengelly, P.; Chan, K.Y.; Jackson, M.; Barchia, I.M. Environmental Risk Assessment of Compost Prepared from Salvinia, *Egeria densa*, and Alligator Weed. *J. Environ. Qual.* **2009**, *38*, 1483–1492. [CrossRef] [PubMed]

80. Petric, I.; Selimbasic, V. Composting of Poultry Manure and Wheat Straw in a Closed Reactor: Optimum Mixture Ratio and Evolution of Parameters. *Biodegradation* **2008**, *19*, 53–63. [CrossRef] [PubMed]

81. Alkarimiah, R. Effects of Technical Factors towards Achieving the Thermophilic Temperature Stage in Composting Process and the Benefits of Closed Rector System Compared to Conventional Method—A Mini Review. *Appl. Ecol. Environ. Res.* **2019**, *17*, 9979–9996. [CrossRef]

Article

A Detailed Database of the Chemical Properties and Methane Potential of Biomasses Covering a Large Range of Common Agricultural Biogas Plant Feedstocks

Audrey Lallement [1], Christine Peyrelasse [1], Camille Lagnet [1], Abdellatif Barakat [2], Blandine Schraauwers [1], Samuel Maunas [1] and Florian Monlau [1,*]

[1] APESA, Plateau Technique, Cap Ecologia, Avenue Fréderic Joliot Curie, 64230 Lescar, France
[2] INRAE, UMR IATE, Place Pierre Viala, CEDEX 02, 34060 Montpellier, France
* Correspondence: florian.monlau@apesa.fr

Abstract: Agricultural biogas plants are increasingly being used in Europe as an alternative source of energy. To optimize the sizing and operation of existing or future biogas plants, a better knowledge of different feedstocks is needed. Our aim is to characterize 132 common agricultural feedstocks in terms of their chemical composition (proteins, fibers, elemental analysis, etc.) and biochemical methane potential shared in five families: agro-industrial products, silage and energy crops, lignocellulosic biomass, manure, and slurries. Among the families investigated, manures and slurries exhibited the highest ash and protein contents (10.3–13.7% DM). High variabilities in C/N were observed among the various families (19.5% DM for slurries and 131.7% DM for lignocellulosic biomass). Methane potentials have been reported to range from 63 Nm^3 CH$_4$/t VS (green waste) to 551 Nm3 CH$_4$/t VS (duck slurry), with a mean value of 284 Nm^3 CH$_4$/t VS. In terms of biodegradability, lower values of 52% and 57% were reported for lignocelluloses biomasses and manures, respectively, due to their high fiber content, especially lignin. By contrast, animal slurries, silage, and energy crops exhibited a higher biodegradability of 70%. This database will be useful for project owners during the pre-study phases and during the operation of future agricultural biogas plants.

Keywords: anaerobic digestion; agricultural inputs; biochemical methane potential; biodegradability; lignocellulosic biomasses; manures

Citation: Lallement, A.; Peyrelasse, C.; Lagnet, C.; Barakat, A.; Schraauwers, B.; Maunas, S.; Monlau, F. A Detailed Database of the Chemical Properties and Methane Potential of Biomasses Covering a Large Range of Common Agricultural Biogas Plant Feedstocks. *Waste* **2023**, *1*, 195–227. https://doi.org/10.3390/waste1010014

Academic Editors: Dimitris P. Makris and Vassilis Athanasiadis

Received: 6 November 2022
Revised: 20 December 2022
Accepted: 22 December 2022
Published: 10 January 2023

1. Introduction

Biogas production has increased in the European Union, encouraged by the European "Green Deal" and the renewable energy policies [1,2]. Between 2000 and 2017, global biogas production quadrupled, from 78 to 364 TW h, which corresponds to a global yearly volume of 61 billion m^3 biogas; it is shared mainly among Europe (54%), Asia (31%), and the Americas (14%) [1]. Anaerobic digestion (AD) unit numbers are increasing in Europe, supported by the need to improve green energy supplies. Among the typologies of biogas plants, agricultural biogas plants are gaining increasing interest as a valuable technology to treat agricultural residues and co-products, thereby generating energy and fertilizers and improving farmers' incomes. In 2021, France had approximately 401 AD on farms and 285 centralized or territorial AD (Source: SINOE). In parallel, in 2018, 1555 and 9500 biogas plants were reported in Italy and Germany, respectively [1]. Nonetheless, it appears that the biogas sector is facing a shift in its development paradigm [1]. At the European level, the biogas sector is still dominated mainly by a model based on energy crops, high feed-in tariffs, and local electrical production via combined heat and power units. However, the biogas sector is now moving towards a different model, where organic wastes, agricultural by-products, as well as sequential crops are used mainly as feedstocks, and biogas is upgraded to biomethane for various applications (transportation, chemical production, heat, etc.) [1].

As the number of biogas plants has increased, securing deposits and the need for alternative feedstocks are growing. The main families of inputs for agricultural biogas plants are animal wastes (manures and slurries), lignocellulosic biomasses, energy and sequential crops and silages, and agricultural co-products. To help industrial and biogas operators, a better knowledge of the main chemical properties (organic matter, fibers, proteins, elemental analysis, C/N, COD, etc.) along with biochemical methane potential tests are needed. The C/N ratio of feedstock is another important parameter, and for a good anaerobic digestion process, the C/N ratio must be between 20 and 30 [3,4]. Indeed, if a biogas reactor has a low C/N ratio, there is potential inhibition from ammonia [3,5]. Among the chemical parameters, the content of fibers (cellulose, hemicelluloses, and lignin) and proteins is another important issue that can affect the final biodegradability of substrates [6]. Finally, the information of the elemental analysis (C, H, N, S, and O) is of prime importance, as it will allow determination of both the theoretical chemical oxygen demand (COD) and the theoretical methane potential according to the Buswell equation [7].

Aside from chemical properties, the determination of the methane potential through BMP (biochemical methane potential) tests is important. BMPs allow laboratory-scale measurement of the maximum production of methane generated by the digestion of a single substrate, and in recent decades, several national and international inter-laboratory studies have been carried out to optimize the protocol and define good practices [8–10]. BMP tests are a popular technique to determine the methane potential and biodegradability of organic substrates [11]. Currently, the BMP test is used for the technical and economic analysis of a project, for the design of agricultural biogas plants, and for evaluation of the process performance [8]. BMP tests can also be useful when the biogas plant unit is operating and new biomasses are to be introduced. Table 1 lists recent studies that provided detailed BMP data of different organic wastes along with the ISR (inoculum-to-substrate ratio) applied. Indeed, the ISR of the BMP is one of the crucial parameters, and the generally recommended values are between 2 and 4 [9,11]. In parallel with classical BMP tests, a theoretical one can also be estimated according to the elemental composition and the Buswell equation [12] or the COD [13], the chemical composition (lipids, carbohydrates, and proteins) [13], or using McCarty's method [14], allowing determination of the biodegradation rate of a selected substrate.

It is of interest to note that a few publications have provided detailed methane potentials per substrate categories, and they generally provide only min., max., and mean values. Among these publications, Allen et al. (2016) reported methane potentials for 83 organic substrates covering different categories from first-, second-, and third-generation biomasses with agricultural wastes, agro-industrial wastes, food residues, and seaweeds [5]. In parallel, Garcia et al. (2019) reported a detailed methane potential database of more than 50 agricultural and food processing substrates [15]. Similarly, Godin et al. (2015) referenced the methane potential of 569 plant biomasses [16]. In parallel, other studies reported exhaustive lists of the methane potentials of 56 agricultural wastes [17], 48 maize sample silages [18], 43 crop species [19], 12 lignocellulosic biomasses [6], and 30 organic wastes [14].

To date, there is clearly a lack of information in the literature regarding data about the chemical and methane potentials of a large spectrum of agricultural biogas plant feedstocks. This publication aims to highlight the characterization of 132 substrates shared by five different families: cereal and residue (CER), energy crop and silage (ENSI), lignocellulosic matter (LCM), manure (MAN), and slurry (SLU). The selection of substrates was based on their frequency of inclusion in agricultural biogas plants. First, various chemical properties (organic matter, fibers, proteins, elemental analysis, C/N, COD, etc.) were analyzed for the 131 substrates. Then, methane potentials were assessed on these substrates and biodegradability rates (defined as the ratio of the BMP assay yield to the theoretical Buswell yield) were calculated.

Table 1. Literature data on large sets of BMP references for organic substrates. N.: number of samples, ISR: inoculum–substrate ratio, MSW: municipal solid wastes, and WWTP: wastewater treatment plant. Description of the samples can be found in the Appendix B.

Reference	Data Access	Sample Description	N. (Total Number)	ISR	Min BMP	Mean BMP (Nm^3 CH$_4$/t VS)	Max BMP
[14]	Yes	30 organic substrates, including 2 raw manures, 9 food residues, 5 invasive aquatic plants, and 6 other organic wastes	22	1	122	341	649
[20]	Yes	Reed canary grasses	14	0.8	283	348	417
[21]	Yes	11 crops	41	2	177	311	401
[22]	Yes	4 grasses and 2 legume species	61	2	265	338	422
[5]	Yes	Biomasses from first-, second-, and third-generation: 6 cereal crops, 3 oilseed rapes, 7 root crops, 5 grass silages, 2 baled silages, 8 other grass substrates, 7 dairy slurries, 4 other agricultural wastes, 4 milk processing wastes, 4 abattoir wastes, 7 miscellaneous wastes, 10 domestic and commercial food wastes, 3 alternative waste substrates, and 12 seaweeds	83	2	99	328	805
[23]	Yes	20 sludge samples	20	2–2.5	58	181	318
[17]	Yes 51/57	18 plants, 12 grasses, 5 bushes, 16 trees, 4 cereals, and 1 straw	57	3	104	219	479
[15]	Yes	5 energy crops, 8 lignocellulosic biomasses, 7 herbaceous and vegetable by-products, 7 fruit by-products, 6 livestock effluents, and 18 food by-products	50	2	71	325	729
This study	Yes	46 energy crops and silages, 5 slurries, 31 manures, 17 cereal and agro-industrial residues, and 32 lignocellulosic biomasses	131	3	63	283	551
[24]	Yes Appendix *	3 animal manures, 3 crop straws, 5 food and green wastes, 2 processing organic wastes, 1 energy grass, and 2 lignocellulosic biomasses	16	2	49	317	811
[18]	Yes Appendix *	48 maize genotypes selected for diverse maturity and biomass production	204	-	295	329	355
[16]	Yes Appendix *	17 Miscanthus, 16 switch grasses, 36 spelt straws, 37 fiber sorghums, 369 tall fescues, 21 immature ryes, and 73 fiber corns	569 (588)	2	147	389	589
[19]	Yes Appendix *	405 silages from 43 crop species	43	2	143	304	425
[25]	No	68 municipal solid wastes, 7 MSW mix, 9 raw substrates, and 18 lignocellulosic wastes	20 (102)	0.5	87	257	226
[26]	No	95 meadow grasses	95	-	51	288	406

Table 1. *Cont.*

Reference	Data Access	Sample Description	N. (Total Number)	ISR	Min BMP	Mean BMP (Nm^3 CH$_4$/t VS)	Max BMP
[27]	No	57 agro-industrial biomasses, 1 macroalgae, 20 biowastes, 4 energy crops, 11 fatty wastes, 14 meat wastes, 2 co-digestion mixtures, 66 WWTP, 42 plants and vegetables, 18 agro-industrial sludges, 30 sewage sludge WWTP, and 31 municipal solid wastes	296	2–5	0	291	1344
[28]	No	33 energy crops, 15 lawn grasses, 19 hedge trimmings, and 21 wild plants	88	3	104	251	502
[29]	No	23 anaerobic sludges, 30 standard compounds, 50 household wastes, 10 agriculture wastes, 19 sewage sludges, and 6 lipid-rich wastes	138	2	39	361	943
[30]	No	48% agricultural residues, 29% animal beading wastes, 6% AD feedstock, AD digestates, lipid wastes, algae, MSW, and agro-industrial wastes	289	2.8	56	287	879
[6]	No	12 lignocellulosic biomasses	12	2	155	225	300

* data are not provided directly in the publication but in an appendix of the authors publications.

2. Materials and Methods

2.1. Sampling

Feedstocks were collected in thirty agricultural biogas plant units operating with agricultural feedstocks on the national level. Of these, 75% were operating in wet AD and 25% in dry AD. These 132 inputs are regrouped into five main families: cereal and agro-industrial co-products (CER), energy crop and silage (ENSI), lignocellulosic matter (LCM), manure (MAN), and slurry (SLU). A description of the dataset is available in the Appendix A (Tables A1 and A2).

2.2. Elemental Composition and Fiber Analysis

The elemental composition of each feedstock was assessed by an elemental apparatus (varioMicro V4.0.2, Elementar®, Langenselbold, Germany), after being dried at 60 °C until constant weight and ground into 1 mm particles using a centrifuge mill (SR 200, Retsch, Haan, Germany). Each COD was then calculated on the basis of this analysis using Equation (1) [31]:

$$COD \left(\frac{gCOD}{gCxHyOz} \right) = 8 \times \frac{4x + y - 2z}{12x + 4 + 16z} \tag{1}$$

The protein content was estimated on the basis of the nitrogen elemental composition multiplied by 6.25 [32].

For fiber analysis (e.g., cellulose, hemicelluloses, and lignin-like), 80 mg of sample was hydrolyzed with 0.85 mL of H$_2$SO$_4$ acid (72%) for 1 h at 30 °C in continuously shaken tubes for thorough mixing (450 rpm) using closed vessels to prevent evaporation. Then, 23.8 mL of deionized water was added, and the vessels were heated to 121 °C for one hour under magnetic agitation (450 rpm). After cooling, the insoluble residue was separated by filtration through 1 µm glass fiber paper (GFF, WHATMAN®, Maldstone, UK) into a soluble phase (structural carbohydrates) and a solid phase (lignin and ash). The filtrate was further filtered using nylon filters (0.2 µm) and analyzed for glucose, xylose, and arabinose by high-performance liquid chromatography (1260 infinity II technology, Agilent, Santa Clara, CA, USA) equipped with a Hi.Plex H coupled to a UV detector. The crucible and the fiberglass paper were dried at 105 °C for 24 h to determine the content of Klason lignin-like material by weighing. The cellulose-like and hemicelluloses-like contents were determined using the following equations:

$$\text{Cellulose} - \text{like} \ (\% \ \text{DM}) = \frac{\text{Glucose} \ (\% \ \text{DM})}{1.11} \tag{2}$$

$$\text{Hemicelluloses} - \text{like} \ (\% \ \text{DM}) = \frac{\text{Xylose} \ (\% \ \text{DM}) + \text{Arabinose} \ (\% \ \text{DM})}{1.13} \tag{3}$$

where 1.11 is the conversion factor of polymers based on glucose-to-glucose monomers, and 1.13 is the factor for converting polymers based on xylose (arabinose and xylose) into monomers [33].

2.3. Biochemical Methane Potential Measurement (BMPexp)

The procedure for BMP tests has been well-documented in a previous study [30] and followed the inter-laboratory study recommendations [8,34]. Feedstocks were stored at 5 °C if the storage period was less than or equal to three days or at −20 °C if the storage period exceeded three days and thawed at 6 °C before testing. Used inoculum was agitated, maintained at 38 ± 1 °C, and fed regularly with green grass and wastewater sludge at the laboratory of APESA facility. Regular checks were performed by measuring the pH, dry matter, and volatile solids. DM and vs. were obtained by loss on ignition (same as for feedstocks), and the pH was assessed using a 340i pH meter fitted with Sentix® electrodes (WTW, Weilheim, Germany). The main properties of the inoculum were TS (% fresh mass): $3.8 \pm 0.3\%$; vs. (% TS): $64.4 \pm 1.5\%$; pH: 8.3 ± 0.2; volatile fatty acids (VFAs): 300 mg eq. acetate L^{-1}; and ammonium content: 2.1 g $N\text{-}NH_4^+ \ L^{-1}$. The inocula complied with the quality criteria proposed by [10].

The BMP tests were carried out under mesophilic conditions in duplicate, and 500 mL reactors were filled with 300 mL of an inoculum/substrate ratio of 3 g VS/g VS. After filling, each bottle was flushed with N_2 gas for 30 s, incubated at 39 °C, and degassed after 1 h. Each day, manual homogenization was performed, and biogas production followed using an electronic manometer device (Digitron 2023P, Digital Instrumentation Ltd., London, UK) and expressed in normal liters (at 0 °C, 1.013 hPa). Once a week, the gas composition was analyzed by gas chromatography (Varian GC-CP4900, Agilent, Santa Clara, CA, USA) equipped with two columns. For O_2, N_2, and CH_4, a Molsieve 5A PLOT column at 110 °C was used, and for CO_2 analysis, a HayeSep A set at 70 °C was used. The injector and detector temperatures were set at 110 °C and 55 °C, respectively. Two standard gases for calibration were used: one composed of 9.5% CO_2, 0.5% O_2, 81% N_2, and 10% CH_4, and the other composed of 35% CO_2, 5% O_2, 20% N_2, and 40% CH_4 (special gas from Air Liquide®, Paris, France). The BMP tests concluded when the biogas production reached a stationary state and did not vary for more than 0.5% during three consecutive days. Blank (inoculum only) and positive controls (cellulose, Tembec®, Montréal, QC, Canada) were run in parallel in duplicate.

The theoretical BMP was calculated on the basis of the elemental characterization (CxHyOzNnSs) using Equation (4) (Achinas and Euverink, 2016):

$$BMPth \ (LCH_4/kg \ VS) = \frac{22.4 \times \left(\frac{x}{2} + \frac{y}{8} - \frac{z}{4} - \frac{3n}{8} - \frac{s}{4}\right)}{12x + y + 16z + 14n + 32s} \tag{4}$$

where 22.4 is the molar volume of an ideal gas.

Finally, the percentage of biodegradation is the ratio between the experimental BMP and the theoretical BMP.

$$Biodegradation \ (\%) = \frac{BMPexp}{BMPth} \tag{5}$$

3. Results

3.1. Chemical Composition of the Various Biomasses

The feedstock compositions are described in Figure 1 (overall and for each family, more data are available in SD Table 1). Among the five families, the distribution was as follows: 47% energy crops and silages, 32% lignocellulosic biomasses, 31% manures, 17% cereal co-products and residues, and 5% slurries. The minimum, maximum, and average values of the different chemical properties (DM, VS, C/N, fibers, proteins, and COD) are shown for all the families in Tables A1 and A2. In order to have a better sense of the inter-family variability, the most important parameters (i.e., VS/DM, COD, C/N, and protein content) are presented as boxplots (Figure 1) and the fiber compositions as radar graphs (Figure 2).

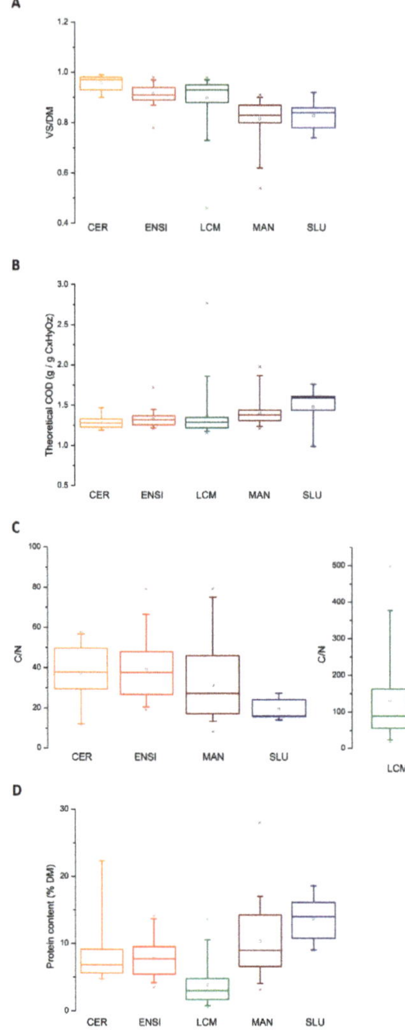

Figure 1. (**A–D**) Boxplots of chemical composition variabilities: VS/DM, COD, C/N, and protein content. Medians are the horizontal lines and means are represented by squares. Families: cereal and residue (CER), energy crop and silage (ENSI), lignocellulosic matter (LCM), manure (MAN), and slurry (SLU).

First of all, higher ash contents were reported for manures and slurries compared with the other families investigated. In terms of proteins, higher contents were also reported for manures and slurries. Indeed, mean protein contents of 10.4 and 13.7% DM were reported for animal manures and slurries, respectively. By contrast, lignocellulosic biomasses exhibited the lowest protein content, at 3.9% DM. Allen et al. (2016) reported protein contents varying from 12.3% DM to 18.5% DM for different animal slurries [5]. Similarly, Li et al. (2013) estimated protein contents of 13.7% DM, 17.5% DM, and 21% DM for swine, dairy, and chicken manures, respectively. Li et al. (2013), on the other hand, reported lower values ranging from 2.5% DM to 5.6% DM for lignocellulosic biomasses [24]. The chemical oxygen demand is another important parameter in anaerobic digestion monitoring, as it can allow determination of mass balances and the theoretical methane potential [13]. Little variability in the main COD was observed for the five families, with values ranging from 1.3 to 1.5 g/g VS. Scarce information is available in the literature regarding these parameters, as only Labatut et al. (2011) have reported it for a range of 30 substrates (mono- and co-digestion). For manure, they found a COD ranging from 0.7 to 1.3 g/g, with a mean of 1.0, which is considerably lower than our values, and higher values for biowaste substrates, with a mean of 6.4, ranging from 0.9 to 28.8 g/g [14]. The fiber content (i.e., cellulose, hemicelluloses, and lignin) was also reported for the five families, and higher contents were observed for lignocellulosic biomasses and manures, similar to the values previously reported in the literature [6,14,24].

Finally, the C/N ratio was also reported for the five families. The C/N ratio is a very important parameter for the long-term continuous digestibility of a substrate. Ideally, it should be between 25:1 and 30:1 to facilitate optimal growth of micro-organisms [5]. For this parameter, high variabilities were observed with higher values of C/N for lignocellulosic biomasses, with a median of approximately 90 and an average of 132. All the other groups have means or averages between 19 and 40. Yet, the C/N ratio is based on the elemental analysis, requiring dry samples. Volatilization of ammoniacal nitrogen or volatile compounds can differ depending on the substrate. A comparison of these results with C/N ratios in the literature points out that an overestimation occurred for slurry and manure families [15,35,36]; similar results are obtained for CER and LCM [5,15], whereas ENSI family C/N ratios are underestimated [5,15,37,38]. Extrapolations cannot be readily performed, as they can depend on the feedstock composition, type, harvest, storage, etc. As an example, manure C/N ratio means have been found to be approximately 16 for cattle manure, 9 for poultry manure, and they are higher for horse manure (between 15 to 150, depending on the type and proportion of litter) [35,36,39].

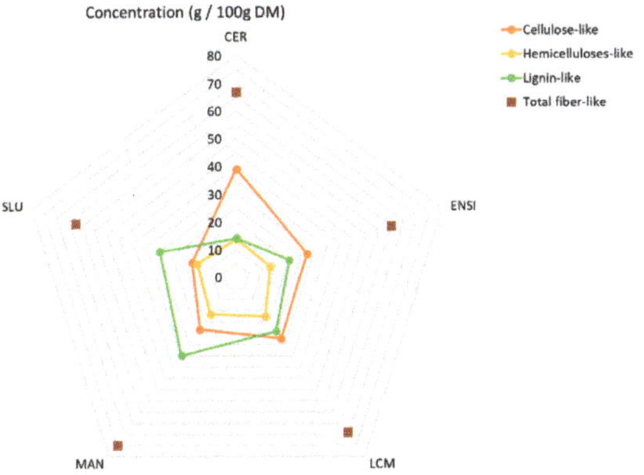

Figure 2. Means of the different family composition of fibers. Families: cereal and residue (CER), energy crop and silage (ENSI), lignocellulosic matter (LCM), manure (MAN), and slurry (SLU).

3.2. Biochemical Methane Potential of Feedstock

Another important parameter in the monitoring and optimization of agricultural biogas plants is the value of the methane potential. Methane potentials were assessed in this study by BMP tests performed on the 132 agricultural substrates shared in five families: cereals and agro-industrial co-products, lignocellulosic biomass, energy crops and silages, animal manures, and slurries (Figure 3).

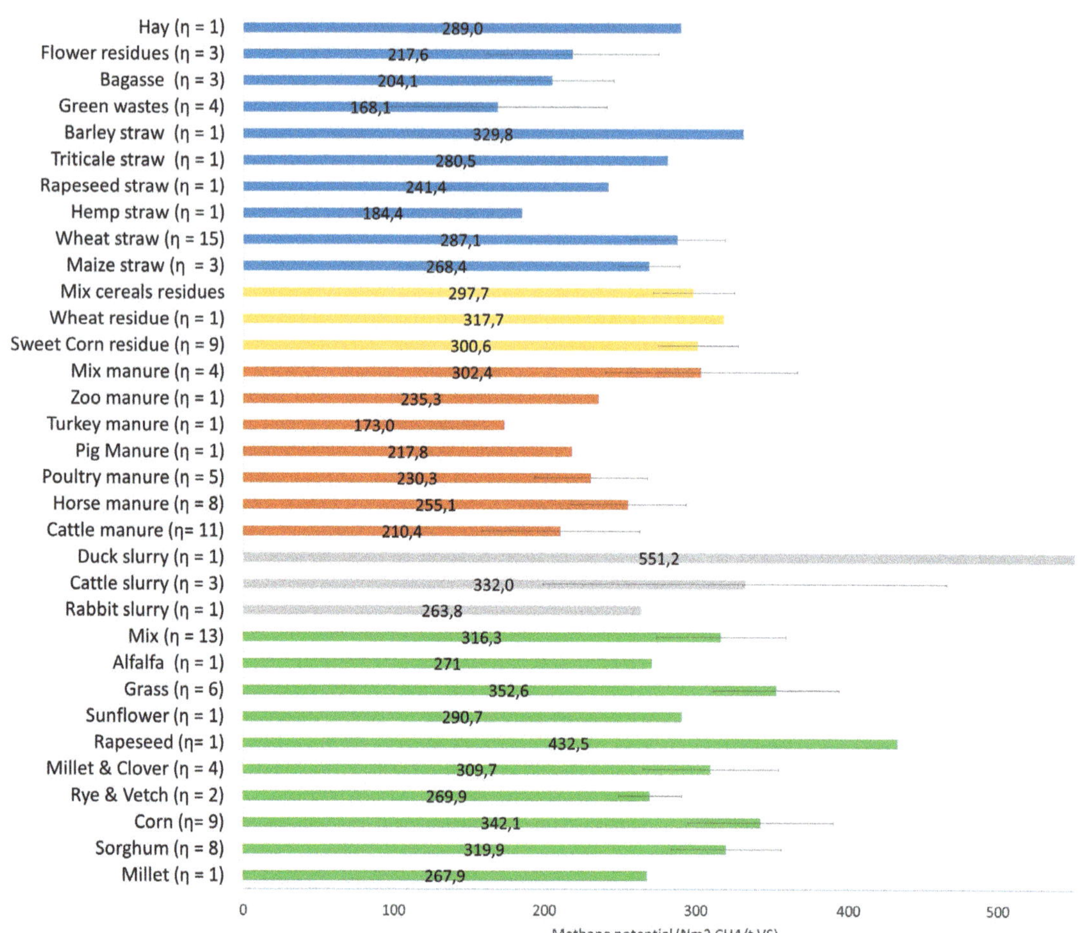

Figure 3. Methane potentials of the different categories of the five families. Families: cereal and agro-industrial residue in grey (CER), energy crop and silage in blue (ENSI), lignocellulosic matter in green (LCM), manure in red (MAN), and slurry in yellow (SLU).

As shown in Table 1, a large variability in methane potentials was observed among the different families, with methane potentials ranging from 63 Nm^3 CH_4/t VS (green waste) to 551 Nm^3 CH_4/t VS (duck slurry), with a mean value for the 132 organic samples of 284 Nm^3 CH_4/t VS.

3.2.1. Cereal and Agro-Industrial Residues (CER)

The first family investigated was cereal and agro-industrial residues ($n = 17$). The cereals were obtained from the cereal agro-industry and silos, whereas the maize was from the sweet corn industry. Methane potentials of 298, 301, and 318 Nm^3 CH_4/t VS were reported for cereal residues, sweet corn residues, and wheat residues, respectively. Garcia et al. (2019) reported a similar methane potential, with values of 345 Nm^3 CH_4/t VS for a mix of cereals [15]. Luna DeRisco et al. (2011) also investigated the methane potentials of grain mill residues, and methane potentials of 274–386 Nm^3 CH_4/t VS were reported [40]. In parallel, Garcia et al. (2019) also reported methane potentials ranging from 204 to 345 Nm^3 CH_4/t VS for ten agro-industrial co-products (from the vegetables and fruits industry) [15].

3.2.2. Manures (MAN)

The methane potential of various animal manures was investigated. Manures are organic matter, derived mostly from animal feces and urine but also normally containing plant materials (generally wheat straw) that have been used as bedding for animals. Methane potentials of 173, 210, 217, 230, 235, and 250 Nm^3 CH_4/t VS were determined for turkey, cattle, pig, poultry, zoo, and horse manures. Such data are in the same range as the values reported in the literature [24,41,42]. Kafle and Chen (2016) investigated the methane potential of five different livestock manures (dairy manure (DM), horse manure (HM), goat manure (GM), chicken manure (CM), and swine manure (SM)). The BMPs of DM, HM, GM, CM, and SM were determined to be 204, 155, 159, 259, and 323 Nm^3 CH_4/t VS, respectively [41]. Similarly, Cu et al. (2015) also reported methane potentials of various animal manures, and the highest BMP in this study was from piglet manure at 443.6 Nm^3 CH_4/t VS, followed by cow, sow, chicken, rabbit, buffalo, and sheep manures at 222, 177.7, 173, 172.8, 153, and 150.5 Nm^3 CH_4/t VS, respectively [42]. Similarly, Garcia et al. (2019) reported methane potentials of 97, 128, 200, and 208 Nm^3 CH_4/t VS for bovine, pig, rabbit, and poultry manures, respectively [15]. Yang et al., 2021 also reported methane potentials of 160 Nm^3 CH_4/t VS for dairy manure, 200 Nm^3 CH_4/t VS for goat manure, and 325 Nm^3 CH_4/t VS for swine manure [43]. It can be observed that the methane potentials of our studies are in the same range as the literature data, although some differences can be observed for the same manure families, as the methane potential can be influenced by the type of farm, the duration of storage, and the storage method. Finally, Carabeo-Perez et al. (2021) also investigated the methane potential from various herbivorous animal manures. Methane yield potentials of 245, 326, and 112 Nm^3 CH_4/t VS were obtained for horse, rabbit, and goat manures, respectively, influenced by the difference in their digestive systems to digest the grass feedstock [44]. Finally, Li et al. (2013) determined methane potentials of 51, 295, and 321 Nm^3 CH_4/t VS for dairy, chicken, and swine manure, respectively [24].

3.2.3. Animal Slurries (SLU)

Animal slurries are manure in liquid form, i.e., a mixture of excrements and urine of domestic animals, including water and/or small amounts of litter. Slurry methane potentials were also investigated in this study, with methane potentials ranging from 263 to 551 Nm^3 CH_4/t VS. As shown in Figure 3, a high variability was observed for cattle slurries, which can be explained by differences in the storage type and duration. In terms of liquid manures, little information is available in the literature [5,14]. Labatut et al. (2011) reported a methane potential of 261 Nm^3 CH_4/t VS for liquid dairy manure. Allen et al. (2016) investigated the methane potentials of different slurries (dairy, pig, and beef). Methane potentials of 99 and 311 Nm^3 CH_4/t VS were reported for pig and beef slurries, respectively. In terms of dairy slurries, methane potentials ranging from 136 to 239 have been reported [5]. Garcia et al. (2019) also reported methane potentials of 35 and 137 Nm^3 CH_4/t VS for bovine and pig slurries, respectively [15].

3.2.4. Silages and Energy Crops (ENSI)

Silages and energy crops are another type of substrate generally found in agricultural biogas plants. In our study, of the 46 organic substrates investigated, the methane potentials ranged from 187 Nm^3 CH_4/t VS to 461 Nm^3 CH_4/t VS. For instance, average methane potentials of 320, 342, and 352 Nm^3 CH$_4$/t VS were reported for sorghum, corn, and grass samples, respectively. The methane potentials of silages and energy crops have been widely investigated in the literature in recent decades, and the values obtained in this study are in the same order [5,15,18,19]. For instance, Garcia et al. (2019) investigated the methane potential of five energy crops and reported methane potentials ranging from 253 Nm^3 CH$_4$/t VS (millet, *Panicum milliaceum* L.) to 351 Nm^3 CH$_4$/t VS (triticale, *Triticum aestivum* L.). Similarly, Allen et al. (2016) reported the methane potential of 18 energy crops, and the methane potentials ranged from 281 Nm^3 CH$_4$/t VS (winter oats) to 398 Nm^3 CH$_4$/t VS (turnips). Similarly, Allen et al. (2016) also investigated the methane potentials of different silages and reported methane potentials varying from 311 Nm^3 CH$_4$/t VS (Savazi grass silage) to 433 Nm^3 CH$_4$/t VS (silage bales). Finally, Hermann et al. also investigated the methane potentials of 43 crops, including main and secondary crops, catch crops, annual grass, and perennial crops [19].

3.2.5. Lignocellulosic Biomasses (LCM)

The methane potentials of 33 lignocellulosic biomasses were also investigated. The methane potentials ranged from 63 Nm^3 CH$_4$/t VS (green waste) to 330 Nm^3 CH$_4$/t VS (barley straw). Lower methane potentials were observed for green waste residues, likely due to their high content in fibers, and especially in lignin, which has been shown to be poorly degraded in the anaerobic digestion process [19,45]. Similar methane potentials on lignocellulosic biomasses have been reported previously in the literature [6,15]. Indeed, Monlau et al. (2012) reported the methane potentials of twelve lignocellulosic biomasses ranging from 155 Nm^3 CH$_4$/t VS (sunflower stalks) to 300 Nm^3 CH$_4$/t VS (Jerusalem artichoke tubers). Similarly, Garcia et al. (2019) reported methane potentials ranging from 282 Nm^3 CH$_4$/t VS (coconut fibers) to 425 Nm^3 CH$_4$/t VS (corn, *Zea mays* L.). Similarly, Dinuccio et al. (2010) reported methane potentials ranging from 225 to 424 Nm^3 CH$_4$/t VS [46]. Perennial crops exhibited the lowest methane potentials, with values ranging from 203 Nm^3 CH$_4$/t VS (cup plant) to 260 Nm^3 CH$_4$/t VS (tall wheatgrass). The highest methane potential of the various crops investigated was reported for forage triticale, with a methane potential of 371 Nm^3 CH$_4$/t VS.

3.3. Practical Implementation of this Database

To assist the reader and user in exploiting this publication, a summary table is provided in Table 2 with the main physicochemical parameter and methane potential values for the various substrate families investigated in this study. As previously discussed, the methane potentials ranged from 63 Nm^3 CH$_4$/t VS (green waste) to 551 Nm^3 CH$_4$/t VS (duck slurry), with a mean value for the 132 organic samples of 284 Nm^3 CH$_4$/t VS.

To better understand the ability of the various organic wastes that were tested to be degraded in the AD process, a biodegradation yield (based on the ratio of the experimental and theoretical BMP) was calculated using the Buswell formula. The family biodegradation yields are presented in Figure 4.

Table 2. Chemical composition of the families (FM: fresh matter; DM: dry matter; and VS: volatile solids). Families: cereals and residues (CER), energy crops and silage (ENSI), lignocellulosic matter (LCM), manures (MAN), and slurries (SLU).

Family	CER	ENSI	LCM	MAN	SLU
Sample number	17	46	33	31	5
DM	21.9–89.4	15.5–69.0	28.4–90.6	8.0–81.6	4.7–26.3
(% FM)	57.2	27.3	71.2	39.2	13.2
VS	21.4–85.4	13.5–62.4	26.4–86.0	5.3–69.1	3.5–24.2
(% FM)	54.3	25.0	64.3	31.8	11.4
C	40.2–44.6	38.7–46.7	34.5–45.0	28.7–43.5	35.9–42.1
(% DM)	42.8	42.6	42.2	38.7	39.9
H	5.8–6.9	5.2–6.6	4.5–6.3	3.8–6.3	5.1–6.0
(% DM)	6.5	5.9	5.8	5.4	5.6
N	0.6–3.8	0.5–2.3	0.1–2.3	0.4–4.6	1.6–2.8
(% DM)	1.4	1.2	0.6	1.6	2.2
S	0.1–0.7	0.1–0.9	0.1–1.0	0.2–1.6	0.4–0.7
(% DM)	0.2	0.2	0.3	0.5	0.5
C/N	11.4–57.8	19.1–79.4	17.6–497.8	8.2–79.1	14.2–27.3
	37.3	39.2	131.7	31.4	19.5
Cellulose-like	25.8–60.9	11.2–52.3	15.4–33.6	13.6–35.0	8.7–26.4
(% VS)	39.0	27.1	27.4	23.4	17.3
Hemicellulose-like	6.0–21.3	6.4–20.5	7.8–26.0	8.8–21.5	7.7–23.5
(% VS)	13.7	12.6	17.5	15.7	15.5
Lignin-like	5.5–21.4	11.4–28.9	14.7–50.2	20.0–56.5	19.1–38.6
(% VS)	14.1	20.3	24.2	34.9	29.8
Proteins	4.5–22.5	3.5–14.1	0.6–13.6	3.1–28.1	9.0–18.6
(% DM)	9.1	7.7	3.9	10.4	13.7
COD	1.2–1.5	1.2–1.7	1.2–2.8	1.2–2.0	1.0–1.8
(g/g (CxHyOz))	1.3	1.3	1.4	1.4	1.5
Family	CER	ENSI	LCM	MAN	SLU
BMP_{th}	407–469	410–582	400–920	397–659	320–568
(Nm^3 CH_4/t VS)	434	449	466	466	483
BMP	250–336	187–461	63–330	132–366	224–551
(Nm^3 CH_4/t VS)	300	324	251	237	362
BMP	56–278	41–169	23–254	13–178	10–54
(Nm^3 CH_4/t FM)	164	78	167	75	35

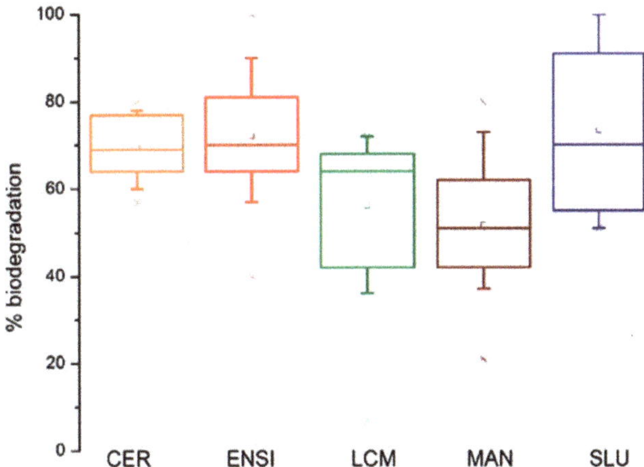

Figure 4. Boxplots of biodegradation yields of the five families. Medians are the horizontal lines and means are represented by squares. Families: cereal and residue (CER), energy crops and silage (ENSI), lignocellulosic matter (LCM), manure (MAN), and slurry (SLU).

A majority of families presented a good biodegradation rate, with means between 52 and 73%. Lower degradation rates of only 52 and 56% were reported for manure and lignocellulosic matter, respectively. As manure is a mixture of feces and bedding material, depending on the bedding material used and its concentration, it is not surprising to find similar results between these two families [47]. The biodegradability of organic substrates has been well-documented in the literature for various organic substrates [5,14,15,17,24]. Regarding lignocellulosic biomasses, Triolo et al. (2012) reported biodegradability indices of 32.7%, 39.9%, 44.9%, and 66.6% for wood cuttings, hedge cuttings, wild plants, and lawn cuttings, respectively. Similarly, Li et al. (2013) reported biodegradability indices of 51%, 54%, and 62% for corn stover, wheat straw, and rice straw, respectively. Similarly, Li et al. (2013) reported biodegradation rates of 10%, 63%, and 68% for dairy manure, chicken manure, and swine manure, respectively. Garcia et al. (2019) also reported biodegradability indices varying from 30% to 70% for different animal manures samples. Such lower biodegradation rates for LCM and MAN families can be explained by the higher fiber contents in such biomasses, especially lignin content, which is poorly degraded in the AD process [6,45]. The high nitrogen concentration in animal manures can also be a limiting factor of the expression of the methane potential [42].

In parallel, other families investigated in this study exhibited higher biodegradability rates of 69%, 72%, and 73% for cereal and agro-industrial residues, energy crops and silages, and slurries, respectively. Allen et al. (2016) reported biodegradability indices for sixteen silages from second-generation crops, and three-quarters of the samples exhibited biodegradation rates higher than 75%. Similarly, Garcia et al. (2019) reported biodegradabilities varying from 80% to 100% for various energy crops (i.e., millet, barley, maize, sorghum, and triticale). Garcia et al. (2019) also reported high biodegradabilities of 80% and 90% for flour and cereals. In terms of slurry samples, the results in the literature are more contrasted [5,15]. Indeed, biodegradabilities varying from 20% to 60% have been reported. Such variation can be explained by the difference in the origins of animal slurries as well as the storage duration and typology.

4. Discussion

At the end of 2018, annual production of biomethane from AD in the EU corresponded to 2.3 billion m^3, with 18,202 biogas plants in operation [1]. Europe is the world leader in biogas electricity production, far ahead of the USA (2.4 GW) and China (0.6 GW) [1]. At the European level, the methanization sector will greatly develop in the years to come with projections up to 64.2 billion m^3 in the EU by 2050; this would represent an energetic potential of approximately 640 TW h/year and would require a 30-fold growth of the current biomethane sector [1].

AD will continue to grow in the future, but it is clear that the sector should have better control of not only the management and the use of the deposits but also the identification of new sources of deposit. The BMP test remains an essential tool for characterizing new deposits and determining their pricing.

This publication and the results (Table 1) are intended to contribute to providing data to the scientific community and biogas developers regarding the values of methane potentials and biodegradability indices of different organic substrates and complete previous studies on the subject (Table A3 in Appendix B). In parallel, this study is intended to be a tool for the sizing, optimization, and operation of the biogas sector. All the data obtained for the different feedstocks are available in the Appendix A.

It could be interesting in the future to extend this work and to generate an overall synthesis of all the BMP values listed in the literature by taking into account the studies using a protocol based on the recommendations of interlaboratory guidelines carried out at the international level [10,34]. In parallel, the growing development of the biogas sector requires the mobilization of new resources and organic biomasses, and it will be interesting in the future to focus studies on the evaluation of the methanogenic potential of atypical biomasses (i.e., algae, paper sludges, biodegradable plastics, insect excrements, etc.). An extended open-source BMP database (based on BMP values validated by experts) could be very useful in the future in order to improve the biogas development as well as the monitoring of the energetic performances of biogas plants. Indeed, Holliger et al. (2017) compared methane production from BMPs with biogas production from the same organic materials in full-scale installations [48]. Holliger et al. (2017) highlighted that the measured weekly methane production accounted for 94.0 ± 6.8 and $89.3 \pm 5.7\%$ of the calculated weekly methane production for two biogas plants, respectively [48].

Short-term (i.e., 1–2 months), batch-mode anaerobic digestion tests, such as the bio-chemical methane potential (BMP) assay, are intended primarily to determine methane yields and the biodegradability of substrates [14]. Nonetheless, such testing may fail to truly predict the performance of full-scale anaerobic reactors. For this purpose, semi-continuous laboratory-scale experimental methods are complementary to chemical and BMP analysis. Semi-continuous flow reactors are designed to emulate the conditions of commercial-scale digesters and study their overall performance over time, taking into account co-digestion benefits and potential inhibition.

5. Conclusions

In this study, a characterization of 132 common agricultural feedstocks (shared in five families) was carried out in terms of physical properties and methane potentials. Of the various families investigated, manures and slurries exhibited the highest ash and protein contents (10.3–13.7% DM). A high degree of variability in terms of the C/N ratio was observed among the various families, with values ranging from 19.5% DM (slurries) to 131.7% DM (lignocellulosic biomass). In terms of biodegradability, lower values of 52% and 57% were reported for lignocelluloses biomasses, and manures due to their high content in fibers, especially lignin. The AD sector will continue to grow in the future, and such studies can be used as a reference for any operator/manager of units or public authority/financial provider in the future.

Author Contributions: A.L., validation, investigation, and writing—original draft; C.P., investigation and writing—original draft; C.L., supervision and investigation; A.B., investigation and writing—review and editing; B.S., methodology, analysis, and investigation; S.M., methodology, analysis, and investigation; F.M., financial support, supervision, conceptualization, and writing—review and editing. All authors have read and agreed to the published version of the manuscript.

Funding: This work is incorporated in the SPIRALE project funded by the ADEME (GRAINE 2018; 1806C0002).

Informed Consent Statement: Not applicable.

Data Availability Statement: All the data are described in Figures and Tables or in the appendix.

Acknowledgments: This work is incorporated in the SPIRALE project funded by the ADEME (GRAINE 2018; 1806C0002), whom we thank for their support along with the two other partner projects: Green Tropism and INSA Toulouse. The APESA also thanks the various operators who provided the biomasses used in this study.

Conflicts of Interest: The authors declare no conflict of interest.

Appendix A

Table A1. Description of the substrates analyzed within the families, where SD is standard deviation, DM is dry matter, FM is fresh matter, VS is volatile solids, BMP exp is the BMP measured, and BMP is the maximum methane potential based on CHNS composition.

Family	Type	Sub Type	DM	VS	VS/DM	BMP exp			Biodegradation	C/N	Carbon		Hydrogen		Nitrogen		Sulfur		Oxygen
			Mean	Mean	Mean	Mean	SD	CV(%)			Mean	SD	Mean	SD	Mean	SD	Mean	SD	Mean
			(% FM)			(Nm^3 CH$_4$/t VS)			%		(% DM)		(% DM)		(% DM)		(% DM)		(% DM)
ENSI	Millet	—	22.9	20.8	0.91	267.9	11.9	4%	61	54.1	42.7	0.1	5.3	0.1	0.8	0.1	0.2	0.1	41.9
ENSI	Sorghum	—	18.4	16.2	0.88	361.8	4.1	1%	79	28.2	42.3	0.0	5.6	0.2	1.5	0.1	0.3	0.2	38.5
ENSI	Mix	Sorghum, Millet, and Sunflower mix	16.5	14.9	0.90	406	10.9	3%	87	19.1	43.1	0.1	6.0	0.0	2.3	0.0	0.3	0.0	38.8
ENSI	Sorghum	Sucro variety	15.5	14.3	0.92	351.3	8.5	2%	79	25.1	42.6	0.3	6.0	0.0	1.7	0.0	0.2	0.0	41.8
ENSI	Sorghum	Vega variety	16.1	14.8	0.92	360.4	6.4	2%	79	19.5	43.4	0.1	5.9	0.0	2.2	0.1	0.2	0.0	39.9
ENSI	Mix	Vega sorghum variety and San Lucas sunflower variety	21.5	19.1	0.89	286.9	17.1	6%	67	25.3	41.5	0.1	5.3	0.1	1.6	0.2	0.3	0.1	40.4
ENSI	Mix	Sunflower, Millet, and Guizotia abyssinica	16.7	14.5	0.87	312.7	4.6	1%	68	27.6	42.0	0.1	5.4	0.2	1.5	0.1	0.4	0.3	37.6
ENSI	Mix	Sunflower, Millet, and Guizotia abyssinica	17.6	15.7	0.89	318.4	0.9	0%	69	30.4	42.8	0.2	5.5	0.0	1.4	0.1	0.2	0.1	39.1
ENSI	Mix	—	18.6	16.5	0.89	337.1	32.1	10%	70	32.8	42.5	0.1	6.1	0.1	1.3	0.2	0.2	0.0	38.6
ENSI	Millet and Clover	—	15.9	14.1	0.89	374.1	1.7	0%	85	26.6	40.6	0.0	5.8	0.2	1.5	0.3	0.2	0.0	40.6
ENSI	Maize	—	35.2	33.9	0.96	272.1	5	2%	63	47.5	42.6	0.3	6.5	0.1	0.9	0.2	0.1	0.0	46.2
ENSI	Mix	Residue	23.3	22.9	0.98	371.5	16.9	5%	81	62.4	45.5	0.2	6.6	0.1	0.7	0.1	0.1	0.0	45.1
ENSI	Sorghum	Sucro variety	31.3	29.9	0.96	332.7	21.5	6%	79	73.8	43.5	0.3	5.8	0.0	0.6	0.0	0.2	0.0	45.6
ENSI	Mix	Sorghum (Pacific graze), Millet (Robusta), Vetch (Bingo and Massa), and Clover (Tabor)	26.6	23.9	0.90	269.4	3.3	1%	64	27.1	40.3	0.0	5.7	0.1	1.5	0.1	0.2	0.0	42.4
ENSI	Millet and Clover	—	29.7	26.0	0.88	275.8	4.1	1%	61	21.9	41.4	0.0	5.5	0.2	1.9	0.2	0.6	0.0	38.3

Table A1. *Cont.*

Family	Type	Sub Type	DM	VS	VS/DM	BMP exp		CV(%)	Biodegradation	C/N	Carbon		Hydrogen		Nitrogen		Sulfur		Oxygen
			Mean	Mean	Mean	Mean	SD				Mean	SD	Mean	SD	Mean	SD	Mean	SD	Mean
			(% FM)			(Nm³ CH₄/t VS)			%		(% DM)		(% DM)		(% DM)		(% DM)		(% DM)
ENSI	Millet and Clover	—	27.1	24.9	0.92	303.7	17.3	6%	69	23.3	42.6	0.2	5.8	0.2	1.8	0.2	0.3	0.2	41.3
ENSI	Millet and Clover	—	27.1	24.9	0.92	285.2	4	1%	65	31.5	42.9	0.1	5.6	0.0	1.4	0.1	0.2	0.0	42.1
ENSI	Sorghum	—	19.8	18.0	0.91	285.6	2.1	1%	64	54.5	41.8	0.0	6.0	0.1	0.8	0.1	0.1	0.0	42.2
ENSI	Maize	—	36.5	35.3	0.97	336	20.4	6%	82	47.9	41.8	0.2	6.4	0.2	0.9	0.2	0.2	0.1	47.4
ENSI	Rye and Vetch	—	49.4	45.9	0.93	255.1	10.9	4%	62	40.2	42.3	0.3	5.3	0.2	1.1	0.1	0.2	0.0	44.1
ENSI	Rye and Vetch	—	28.3	26.7	0.94	284.6	14.7	5%	68	58.8	42.8	0.2	5.6	0.3	0.7	0.1	0.2	0.1	45.0
ENSI	Mix	—	28.9	25.8	0.89	368.7	4.3	1%	81	57.0	41.8	0.2	5.8	0.1	0.7	0.1	0.2	0.0	40.7
ENSI	Mix	Faba bean, Rye, and Radish	20.3	18.8	0.92	300.1	0.7	0%	72	29.4	42.0	0.2	5.5	0.0	1.4	0.1	0.2	0.0	43.2
ENSI	Mix	Faba bean, Triticale, and Radish	17.5	16.1	0.91	294.3	6.2	2%	70	30.2	41.0	0.3	5.7	0.1	1.4	0.1	0.3	0.0	43.1
ENSI	Mix	Grass	54.6	46.6	0.85	253.9	14.4	6%	55	25.7	39.7	0.2	5.8	0.1	1.5	0.2	0.3	0.0	37.9
ENSI	Mix	Sorghum and Maize	32.1	28.4	0.89	315.9	1.8	1%	74	51.5	38.7	0.2	6.1	0.2	0.8	0.2	0.3	0.2	42.8
ENSI	Mix	Peas, Vetch, Oats, and Beans	27.0	24.5	0.91	331	15	5%	67	20.6	45.0	0.0	6.1	0.0	2.2	0.2	0.2	0.0	37.4
ENSI	Maize	—	33.0	32.1	0.97	319.1	10.7	3%	73	44.4	44.4	0.2	6.4	0.1	1.0	0.0	0.1	0.0	45.4
ENSI	Sorghum	—	35.9	33.1	0.92	272.6	3.7	1%	62	51.7	43.9	0.3	5.3	0.1	0.8	0.0	0.2	0.2	42.0
ENSI	Mix	Moha and Clover	50.3	44.8	0.89	282.5	11.2	4%	58	34.9	43.9	0.2	6.0	0.1	1.3	0.1	0.2	0.0	37.7
ENSI	Rapeseed	—	18.2	17.2	0.94	432.5	5.6	1%	93	79.4	44.4	0.3	6.4	0.2	0.6	0.1	0.5	0.5	42.4
ENSI	Grass	—	27.8	24.7	0.89	264.3	15	6%	57	25.2	41.5	0.2	6.1	0.1	1.6	0.2	0.2	0.0	39.3
ENSI	Sorghum	—	23.6	21.2	0.90	305.3	3	1%	69	45.0	40.8	0.0	6.0	0.0	0.9	0.1	0.2	0.1	41.8
ENSI	Sunflower	—	15.6	14.1	0.90	290.7	0.9	0%	63	44.1	41.7	0.2	6.3	0.0	0.9	0.1	0.2	0.0	41.2
ENSI	Grass	—	17.2	13.5	0.78	406.3	4.8	1%	70	34.6	42.8	0.1	5.8	0.2	1.2	0.0	0.2	0.0	28.2
ENSI	Maize	—	28.5	27.2	0.96	376.3	31.1	8%	89	43.3	43.1	0.1	5.9	0.3	1.0	0.1	0.1	0.0	45.4
ENSI	Alfalfa	—	69.0	62.4	0.90	271	0.1	0%	60	26.7	42.9	0.2	5.7	0.0	1.6	0.1	0.2	0.0	40.0
ENSI	Sorghum	—	24.7	23.3	0.94	289.8	10.9	4%	67	43.3	43.6	0.1	5.8	0.1	1.0	0.1	0.2	0.1	43.7
ENSI	Grass	Ray-grass	17.6	16.7	0.95	393.3	26.5	7%	87	66.5	44.5	0.0	6.0	0.0	0.7	0.1	0.2	0.1	43.1
ENSI	Maize	—	19.8	19.4	0.98	418	4.2	1%	90	42.1	46.8	0.1	6.3	0.1	1.1	0.1	0.2	0.1	43.4

Table A1. *Cont.*

Family	Type	Sub Type	DM Mean	VS Mean	VS/DM Mean	BMP exp Mean	SD	CV(%)	Biodegradation %	C/N	Carbon Mean	SD	Hydrogen Mean	SD	Nitrogen Mean	SD	Sulfur Mean	SD	Oxygen Mean
			(% FM)			(Nm³ CH$_4$/t VS)			%		(% DM)		(% DM)		(% DM)		(% DM)		(% DM)
ENSI	Grass	—	21.3	18.9	0.89	461.2	3.7	1%	100	23.8	41.4	0.2	5.9	0.0	1.7	0.0	0.4	0.1	39.2
ENSI	Maize	—	33.6	32.4	0.96	383.9	1.5	0%	85	45.6	44.6	0.2	6.4	0.2	1.0	0.1	0.1	0.0	44.4
ENSI	Maize	—	30.2	28.8	0.95	335.1	10	3%	75	42.2	43.6	0.2	6.5	0.1	1.0	0.1	0.1	0.0	43.9
ENSI	Grass	—	23.4	21.3	0.91	403.2	14.2	4%	90	29.3	42.9	0.2	5.7	0.1	1.5	0.2	0.3	0.0	40.8
ENSI	Grass	—	39.9	36.1	0.91	187.2	11.5	6%	40	49.5	42.4	0.3	6.4	0.1	0.9	0.0	0.3	0.0	40.6
ENSI	Maize	—	32.0	30.3	0.95	301.1	7.4	2%	71	37.7	42.2	0.1	6.2	0.1	1.1	0.2	0.3	0.0	44.9
MAN	Mix	Manure and Spates	32.1	17.3	0.54	253.4	14.9	6%	38	27.2	31.7	0.1	4.4	0.2	1.2	0.1	0.4	0.0	16.2
MAN	Cattle	After phase separation, Straw	15.3	13.8	0.90	321.9	16.8	5%	80	15.8	39.8	0.0	5.4	0.0	2.5	0.1	0.5	0.0	41.8
MAN	Horse	—	8.0	5.3	0.66	237.6	14.3	6%	42	22.4	35.8	0.1	5.0	0.2	1.6	0.1	0.3	0.0	23.8
MAN	Cattle	Straw	27.7	25.1	0.91	257.7	14.9	6%	63	45.9	40.3	0.2	5.5	0.2	0.9	0.1	0.2	0.0	43.7
MAN	Cattle	Straw	16.6	14.3	0.86	236	21.7	9%	51	30.7	41.8	0.1	5.3	0.3	1.4	0.0	0.2	0.1	37.8
MAN	Cattle	Fern	27.6	24.1	0.87	162	5.7	4%	33	32.6	43.2	0.3	5.6	0.0	1.3	0.2	0.2	0.0	36.9
MAN	Cattle	Straw	43.1	34.5	0.80	191.3	8.4	4%	39	54.2	40.6	0.1	4.8	0.2	0.7	0.1	1.4	0.3	32.7
MAN	Horse	—	81.6	69.1	0.85	258.2	11.9	5%	57	49.6	40.8	0.1	5.1	0.2	0.8	0.1	0.8	0.1	37.1
MAN	Poultry	—	57.2	35.3	0.62	216.8	6.9	3%	49	12.0	28.7	0.3	4.0	0.3	2.4	0.0	0.4	0.1	26.2
MAN	Poultry	—	75.9	58.0	0.76	263.8	1.9	1%	63	15.3	34.7	0.3	4.7	0.2	2.3	0.0	0.6	0.1	34.3
MAN	Pig	—	31.6	27.0	0.85	217.8	13.5	6%	50	19.5	38.8	0.1	5.6	0.1	2.0	0.3	0.4	0.0	38.7
MAN	Cattle	Straw, after 1 month conservation	17.0	13.0	0.76	198.6	17.3	9%	38	29.2	39.5	0.1	5.3	0.0	1.4	0.3	0.7	0.2	29.7
MAN	Turkey	—	68.3	52.3	0.77	173	1.1	1%	37	13.5	36.6	0.2	5.1	0.1	2.7	0.3	0.5	0.0	31.6
MAN	Mix	—	15.3	13.8	0.90	243.3	1	0%	56	21.4	40.9	0.1	5.9	0.0	1.9	0.0	0.3	0.0	40.9
MAN	Poultry	—	71.2	58.3	0.82	211	16.1	8%	49	8.2	36.7	0.2	5.6	0.1	4.5	0.1	0.6	0.0	34.4
MAN	Poultry	—	60.3	50.4	0.84	274.1	13	5%	69	15.5	35.3	0.0	5.6	0.1	2.3	0.3	1.0	0.2	39.5
MAN	Cattle	Straw	45.6	34.2	0.75	131.9	1.4	1%	21	51.3	43.5	0.3	5.8	0.0	0.8	0.3	0.3	0.0	24.6
MAN	Horse	—	20.2	16.4	0.81	182.6	9.9	5%	41	27.5	39.5	0.0	4.5	0.2	1.4	0.0	0.5	0.0	35.2
MAN	Cattle	Straw	69.8	61.7	0.88	233.2	15.3	7%	55	34.0	40.6	0.3	5.2	0.0	1.2	0.2	0.6	0.1	40.9
MAN	Poultry	—	49.6	40.5	0.82	185.6	8.8	5%	42	15.3	37.7	0.2	5.4	0.1	2.5	0.2	0.7	0.0	35.4
MAN	Horse	—	34.4	29.1	0.85	308.4	15.3	5%	65	55.3	40.9	0.0	5.5	0.0	0.7	0.1	0.4	0.1	37.1
MAN	Horse	—	33.2	27.0	0.81	281.2	16.2	6%	57	79.1	39.6	0.3	5.6	0.2	0.5	0.0	0.3	0.1	35.3
MAN	Zoo	—	31.3	25.5	0.81	235.3	10.7	5%	51	30.8	38.7	0.3	5.3	0.2	1.3	0.3	0.3	0.1	35.6
MAN	Cattle	Straw	22.8	20.1	0.88	166.9	7.3	4%	38	39.2	41.2	0.2	5.5	0.1	1.1	0.1	0.3	0.1	40.1

Table A1. *Cont.*

Family	Type	Sub Type	DM Mean (% FM)	VS Mean (% FM)	VS/DM Mean	BMP exp Mean (Nm³ CH₄/t VS)	BMP exp SD	CV(%)	Biodegradation %	C/N	Carbon Mean (% DM)	Carbon SD	Hydrogen Mean (% DM)	Hydrogen SD	Nitrogen Mean (% DM)	Nitrogen SD	Sulfur Mean (% DM)	Sulfur SD	Oxygen Mean (% DM)
MAN	Mix	Straw	22.3	19.4	0.87	346.9	15.9	5%	73	18.6	41.1	0.1	6.1	0.2	2.2	0.1	0.3	0.1	37.2
MAN	Mix	Straw	20.9	18.1	0.86	366	7.1	2%	72	20.1	42.8	0.0	6.2	0.1	2.1	0.3	0.3	0.1	35.0
MAN	Horse	—	57.6	50.3	0.87	281.1	10	4%	62	62.4	40.7	0.1	5.7	0.2	0.7	0.1	0.5	0.0	39.8
MAN	Horse	—	51.0	44.8	0.88	240.9	13.6	6%	57	75.1	39.3	0.2	5.6	0.2	0.5	0.1	0.6	0.2	41.8
MAN	Cattle	Straw	43.1	34.5	0.80	191.4	8.3	4%	46	17.2	35.0	0.2	5.3	0.0	2.0	0.3	0.5	0.0	37.3
MAN	Cattle	Straw	33.7	27.9	0.83	223.8	9	4%	49	21.2	38.4	0.1	5.8	0.0	1.8	0.1	0.4	0.0	36.6
MAN	Horse	—	31.5	26.0	0.83	250.6	18.6	7%	60	13.5	36.2	0.0	5.6	0.1	2.7	0.3	0.6	0.0	37.3
CER	Maize Residues	Follicle	73.9	71.7	0.97	279.2	19	7%	64	33.1	43.1	0.2	6.8	0.1	1.3	0.0	0.2	0.0	45.6
CER	Wheat	Contaminated culture	86.2	84.0	0.97	317.7	2.1	1%	78	22.9	41.2	0.1	6.8	0.0	1.8	0.1	0.2	0.0	47.4
CER	Mix	Cereals	75.6	72.5	0.96	285.9	15.8	6%	64	29.4	43.0	0.3	6.7	0.1	1.5	0.0	0.1	0.0	44.5
CER	Mix	Cereal dust	89.4	80.1	0.90	300.1	5.6	2%	66	42.6	41.9	0.0	5.8	0.1	1.0	0.1	0.1	0.0	40.8
CER	Mix	Cereal residue	72.0	64.7	0.90	250.2	9.3	4%	57	11.4	40.6	0.3	6.2	0.1	3.6	0.2	0.3	0.0	39.2
CER	Maize	Fresh residue from sweet corn	24.6	24.2	0.98	314	20	6%	77	53.1	43.5	0.2	6.0	0.1	0.8	0.0	0.1	0.0	47.9
CER	Maize	Fresh residue from sweet corn	21.9	21.4	0.98	262.7	0.1	0%	63	43.3	43.5	0.2	6.2	0.1	1.0	0.0	0.2	0.1	46.6
CER	Maize	Fresh residue from sweet corn	23.9	23.3	0.97	306	5.4	2%	71	43.7	43.8	0.0	6.4	0.1	1.0	0.1	0.1	0.0	46.3
CER	Maize	Fresh residue from sweet corn	26.7	26.3	0.99	335.5	5.4	2%	78	49.6	44.6	0.2	6.3	0.0	0.9	0.1	0.1	0.0	46.5
CER	Maize	Fresh residue from sweet corn	23.6	23.1	0.98	312.3	14	4%	69	36.9	44.6	0.2	6.7	0.1	1.2	0.0	0.6	0.2	44.7
CER	Maize	Fresh residue from sweet corn	21.9	21.4	0.98	263.7	24.5	9%	60	57.8	44.0	0.3	6.7	0.0	0.8	0.2	0.04	0.1	46.4
CER	Maize	Fresh residue from sweet corn	26.5	26.0	0.98	306.8	5.2	2%	71	56.7	43.7	0.3	6.7	0.1	0.8	0.2	0.1	0.1	46.9
CER	Mix	Cereals	87.0	80.6	0.93	313.3	7.8	2%	67	12.1	43.3	0.0	6.7	0.0	3.6	0.3	0.2	0.0	38.9
CER	Maize	Flour	87.0	85.4	0.98	325.4	28.8	9%	80	37.8	41.4	0.2	6.8	0.0	1.1	0.1	0.1	0.0	48.8
CER	Mix	Cereals	77.0	71.9	0.93	328.4	7.2	2%	73	18.3	43.2	0.1	6.4	0.1	2.4	0.2	0.2	0.0	41.3
CER	Mix	Silo's lose	79.8	74.1	0.93	320.4	14.7	5%	77	56.2	40.2	0.0	6.2	0.0	0.7	0.1	0.3	0.0	45.4

Table A1. *Cont.*

Family	Type	Sub Type	DM	VS	VS/DM	BMP exp			Biodegradation	C/N	Carbon		Hydrogen		Nitrogen		Sulfur		Oxygen
			Mean	Mean	Mean	Mean	SD	CV(%)			Mean	SD	Mean	SD	Mean	SD	Mean	SD	Mean
			(% FM)			(Nm³ CH₄/t VS)			%		(% DM)		(% DM)		(% DM)		(% DM)		(% DM)
CER	Mix	Cereals	75.6	72.5	0.96	285.9	15.8	6%	66	29.8	42.4	0.1	6.7	0.1	1.4	0.1	0.3	0.0	44.9
SLU	Cattle	—	4.7	3.5	0.74	291	3	1%	51	15.7	40.5	0.2	5.4	0.1	2.6	0.1	0.5	0.0	25.6
SLU	Rabbit	—	18.4	15.9	0.86	263.8	4.7	2%	55	24.1	41.6	0.2	5.9	0.0	1.7	0.1	0.5	0.1	36.7
SLU	Cattle	—	26.3	24.2	0.92	224.1	3.7	2%	70	16.0	35.9	0.2	5.1	0.1	2.2	0.2	0.6	0.0	48.0
SLU	Duck	—	6.2	5.2	0.84	551.2	26.3	5%	100	14.2	42.1	0.2	6.1	0.1	3.0	0.2	0.4	0.1	31.9
SLU	Cattle	—	10.4	8.1	0.78	481	10.2	2%	91	27.3	39.5	0.1	5.8	0.0	1.4	0.3	0.5	0.2	30.6
LCM	Maize Residue	Cob	28.4	27.7	0.98	272.2	2.1	1%	63	497.8	44.1	0.2	6.3	0.1	0.1	0.0	0.1	0.1	47.2
LCM	Hemp	Dust	88.1	69.0	0.78	184.4	4.5	2%	36	48.5	39.6	0.1	5.3	0.0	0.8	0.0	0.2	0.0	32.5
LCM	Straw	Plant residues	88.0	83.9	0.95	277.6	7.6	3%	68	60.0	42.2	0.0	5.8	0.1	0.7	0.1	0.3	0.0	46.4
LCM	Straw	—	87.6	84.6	0.97	274.1	2.2	1%	67	77.8	42.7	0.1	6.0	0.0	0.5	0.0	0.3	0.1	47.1
LCM	Maize	Beans	42.0	36.4	0.87	246.8	13.3	5%	51	165.7	42.8	0.1	5.5	0.3	0.3	0.1	0.8	0.1	37.3
LCM	Bagasse and Straw	—	52.2	48.3	0.92	188.1	9.1	5%	42	245.9	43.3	0.1	5.7	0.2	0.2	0.1	0.3	0.1	43.0
LCM	Bagasse	—	43.2	40.9	0.95	173.7	7.2	4%	39	376.8	44.6	0.2	5.9	0.2	0.1	0.0	0.2	0.1	44.0
LCM	Straw	—	54.0	47.5	0.88	199.8	8.8	4%	42	193.6	43.4	0.1	5.4	0.2	0.2	0.0	0.2	0.0	38.7
LCM	Bagasse	—	56.7	41.2	0.73	250.6	16.5	7%	39	297.1	42.8	0.2	5.5	0.2	0.1	0.0	0.1	0.0	24.1
LCM	Straw	Waste	79.2	71.8	0.91	329.8	0.8	0%	72	79.1	42.8	0.1	5.8	0.0	0.5	0.1	0.2	0.1	41.4
LCM	Green waste	—	37.9	35.2	0.93	212.1	1.8	1%	47	136.8	43.4	0.0	5.8	0.1	0.3	0.1	0.6	0.1	42.7
LCM	Straw	—	86.5	82.2	0.95	277.6	21.8	8%	67	64.0	43.0	0.3	5.7	0.1	0.7	0.1	0.1	0.0	45.5
LCM	Hay	Meadow	86.0	80.4	0.93	289	7.1	2%	65	51.4	42.6	0.1	6.3	0.0	0.8	0.1	0.1	0.0	43.7
LCM	Straw	Plant residues	89.0	85.1	0.96	292.7	27.4	9%	70	89.0	42.5	0.1	6.1	0.3	0.5	0.0	0.2	0.2	46.4
LCM	Straw	Plant residues	87.2	83.2	0.95	298.9	2.3	1%	69	89.3	42.9	0.1	6.2	0.1	0.5	0.0	0.1	0.1	45.6
LCM	Straw	—	88.3	84.0	0.95	302.1	5.9	2%	72	73.6	42.3	0.2	6.0	0.3	0.6	0.1	0.3	0.3	46.0
LCM	Straw	—	88.8	86.0	0.97	290.6	7.1	2%	69	132.0	43.0	0.1	6.3	0.0	0.3	0.0	0.1	0.0	47.1
LCM	Straw	—	75.9	70.4	0.93	305.5	6.4	2%	70	102.2	42.9	0.0	5.7	0.1	0.4	0.0	0.2	0.0	43.5
LCM	Straw	Waste	84.9	81.3	0.96	293.7	1	0%	67	126.8	44.2	0.2	5.9	0.0	0.3	0.0	0.1	0.0	45.2
LCM	Flower residue	Lavender	88.7	81.2	0.92	200.5	9.6	5%	42	41.5	45.0	0.1	6.0	0.0	1.1	0.2	0.3	0.1	39.3
LCM	Maize	Leaf	37.8	35.0	0.93	286.3	23.3	8%	65	56.5	43.2	0.1	5.8	0.0	0.8	0.0	0.1	0.0	42.8
LCM	Straw	Plant residues	86.3	82.3	0.95	280.5	2.9	1%	69	68.3	41.5	0.3	6.0	0.1	0.6	0.1	0.2	0.1	47.1

Table A1. *Cont.*

Family	Type	Sub Type	DM Mean (% FM)	VS Mean (% FM)	VS/DM Mean	BMP exp Mean (Nm³ CH₄/t VS)	BMP exp SD	CV(%)	Biodegradation %	C/N	Carbon Mean (% DM)	Carbon SD	Hydrogen Mean (% DM)	Hydrogen SD	Nitrogen Mean (% DM)	Nitrogen SD	Sulfur Mean (% DM)	Sulfur SD	Oxygen Mean (% DM)
LCM	Straw	Waste	84.5	77.3	0.92	306	23.1	8%	66	128.8	43.5	0.3	5.9	0.0	0.3	0.0	0.1	0.0	41.7
LCM	Straw	Rapeseed waste	71.9	62.9	0.87	241.4	3	1%	54	31.1	40.0	0.2	5.8	0.2	1.3	0.0	0.4	0.2	40.0
LCM	Straw	—	85.0	81.2	0.96	309.7	10.1	3%	72	162.8	43.8	0.3	5.9	0.0	0.3	0.1	0.1	0.0	45.6
LCM	Straw	Waste	83.9	78.2	0.93	283.8	8.8	3%	62	119.4	43.6	0.1	6.3	0.0	0.4	0.2	0.2	0.1	42.8
LCM	Green waste	—	44.6	34.5	0.77	178.2	0.1	0%	33	23.9	40.3	0.1	5.5	0.1	1.7	0.2	0.1	0.0	29.7
LCM	Mix	Green waste	34.4	26.4	0.77	218.9	11.6	5%	45	21.9	37.5	0.3	5.3	0.2	1.7	0.1	0.2	0.0	31.9
LCM	Green waste	—	80.6	36.9	0.46	63.1	3.4	5%	7	29.2	34.5	0.0	4.5	0.0	1.2	0.2	0.3	0.0	5.3
LCM	Flower residue	Pomace	47.1	39.6	0.84	281.1	1.7	1%	64	17.6	38.3	0.1	5.5	0.0	2.2	0.1	0.6	0.0	37.3
LCM	Straw	—	90.6	85.8	0.95	240	6.3	3%	60	406.3	41.2	0.3	5.7	0.3	0.1	0.0	0.8	0.3	46.9
LCM	Flower residue	Lavender	82.8	78.2	0.94	171.3	12.4	7%	37	68.8	44.3	0.0	6.3	0.0	0.6	0.3	0.6	0.1	42.6
LCM	Straw	Waste	88.8	82.3	0.93	267	7.4	3%	64	263.7	40.3	0.1	6.1	0.3	0.2	0.1	0.8	0.1	45.3

Table A2. Description of the substrates analyzed within the families: fibers, protein content, and COD, where SD is standard deviation, DM is dry matter, VS is volatile solids, and COD is the chemical oxygen demand.

Family	Type	Sub Type	Cellulose		Hemicelluloses		Lignin		Protein	COD
			Mean	SD	Mean	SD	Mean	SD	Calculated	Calculated
			(g/100g DM)		(g/100g DM)		(g/100g DM)		(% DM)	(g COD/g CxHyOz)
ENSI	Millet	—	28.6	1.4	17.7	1.1	22.3	0.3	4.9	1.3
ENSI	Sorghum	—	21.7	0.0	10.8	0.2	21.1	0.3	9.4	1.4
ENSI	Mix	Sorghum, Millet, and Sunflower mix	22.7	0.3	12.7	0.2	19.4	0.1	14.1	1.4
ENSI	Sorghum	Sucro variety	25.1	0.5	12.1	0.6	18.1	0.1	10.6	1.3
ENSI	Sorghum	Vega variety	22.8	0.7	13.7	0.4	20.0	0.9	13.9	1.4
ENSI	Mix	Vega sorghum variety and San Lucas sunflower variety	24.7	0.2	10.1	0.1	26.7	0.7	10.3	1.3
ENSI	Mix	Sunflower, Millet, and Guizotia abyssinica	17.3	0.3	7.1	0.1	26.6	1.3	9.5	1.4
ENSI	Mix	Sunflower, Millet, and Guizotia abyssinica	17.9	0.1	8.6	0.2	25.6	0.3	8.8	1.4
ENSI	Mix	—	22.6	0.8	9.2	0.2	25.1	0.3	8.1	1.4
ENSI	Millet and Clover	—	29.6	0.2	19.1	0.3	19.6	0.5	9.5	1.3
ENSI	Maize	—	47.7	1.5	11.5	0.5	11.4	0.2	5.6	1.3
ENSI	Mix	Residue	13.5	0.3	9.0	0.2	14.0	0.1	4.6	1.3
ENSI	Sorghum	Sucro variety	31.4	0.4	18.0	0.5	17.8	0.1	3.7	1.2
ENSI	Mix	Sorghum (Pacific graze), Millet (Robusta), Vetch (Bingo and Massa), and Clover (Tabor)	30.3	2.5	14.2	0.5	19.7	0.8	9.3	1.3
ENSI	Millet and Clover	—	22.4	0.4	15.4	0.3	22.2	1.0	11.8	1.4
ENSI	Millet and Clover	—	12.9	0.4	6.6	0.1	18.6	1.8	11.4	1.3
ENSI	Millet and Clover	—	11.2	0.1	6.4	0.1	21.0	0.2	8.5	1.3
ENSI	Sorghum	—	28.3	0.6	17.2	0.4	18.7	0.1	4.8	1.3
ENSI	Maize	—	49.9	0.6	12.6	0.4	11.4	0.2	5.5	1.2
ENSI	Rye and Vetch	—	25.8	0.0	15.1	0.0	22.1	0.1	6.6	1.2
ENSI	Rye and Vetch	—	26.7	0.0	17.4	0.0	20.6	0.1	4.5	1.2
ENSI	Mix	—	22.3	0.4	13.3	0.5	19.5	0.4	4.6	1.3
ENSI	Mix	Faba bean, Rye, and Radish	19.6	0.6	10.6	0.1	22.4	0.1	8.9	1.2
ENSI	Mix	Faba bean, Triticale, and Radish	17.5	0.0	8.8	0.0	18.1	0.2	8.5	1.2
ENSI	Mix	Grass	21.7	0.0	12.3	0.0	28.9	0.5	9.6	1.4
ENSI	Mix	Sorghum and Maize	22.0	0.5	15.4	0.3	17.8	0.3	4.7	1.2
ENSI	Mix	Peas, Vetch, Oats, and Beans	21.9	0.3	10.7	0.3	20.2	0.8	13.7	1.5
ENSI	Maize	—	45.5	1.0	10.5	0.3	13.3	0.4	6.2	1.3
ENSI	Sorghum	—	30.5	0.6	20.5	0.2	24.5	0.5	5.3	1.3
ENSI	Mix	Moha and Clover	28.4	0.1	13.5	0.5	19.2	1.0	7.9	1.5

Table A2. *Cont.*

Family	Type	Sub Type	Cellulose		Hemicelluloses		Lignin		Protein	COD
			Mean	SD	Mean	SD	Mean	SD	Calculated	Calculated
			(g/100g DM)		(g/100g DM)		(g/100g DM)		(% DM)	(g COD/g CxHyOz)
ENSI	Rapeseed	—	25.7	1.2	10.2	1.0	27.4	0.3	3.5	1.4
ENSI	Grass	—	21.5	0.5	11.7	0.2	27.1	0.2	10.3	1.4
ENSI	Sorghum	—	31.6	0.2	14.0	0.1	22.9	0.3	5.7	1.3
ENSI	Sunflower	—	20.9	0.4	9.1	0.2	22.8	0.8	5.9	1.4
ENSI	Grass	—	29.9	0.2	16.9	0.2	25.7	1.6	7.7	1.7
ENSI	Maize	—	41.1	0.4	11.7	0.8	16.5	0.8	6.2	1.2
ENSI	Alfalfa	—	23.6	0.7	9.2	0.1	21.5	0.6	10.1	1.4
ENSI	Sorghum	—	29.8	0.3	13.1	0.8	16.8	0.4	6.3	1.3
ENSI	Grass	Ray-grass	31.9	1.6	14.1	0.9	21.9	1.6	4.2	1.3
ENSI	Maize	—	27.3	0.2	18.4	0.1	20.0	0.7	6.9	1.4
ENSI	Grass	—	24.9	0.2	14.8	0.5	19.5	0.3	10.9	1.4
ENSI	Maize	—	52.3	3.5	11.2	0.9	15.4	0.6	6.1	1.3
ENSI	Maize	—	41.3	1.2	9.6	0.4	13.1	1.3	6.5	1.3
ENSI	Grass	—	21.1	0.7	9.7	0.2	18.1	0.7	9.1	1.3
ENSI	Grass	—	26.0	0.4	13.7	0.3	18.4	1.1	5.3	1.4
ENSI	Maize	—	37.1	0.5	11.6	0.0	19.9	1.3	7.0	1.3
MAN	Mix	Manure and Spates	28.7	0.8	14.8	0.4	52.3	2.9	7.3	2.0
MAN	Cattle	After phase separation, Straw	24.5	0.6	17.5	0.5	26.0	0.7	15.7	1.2
MAN	Horse	—	25.3	2.5	14.2	0.6	48.7	2.2	10.0	1.7
MAN	Cattle	Straw	25.6	0.1	18.2	0.2	29.4	1.0	5.5	1.2
MAN	Cattle	Straw	22.5	1.8	15.3	0.0	34.9	0.6	8.5	1.4
MAN	Cattle	Fern	20.0	0.7	14.5	1.1	36.3	1.2	8.3	1.4
MAN	Cattle	Straw	29.2	1.4	16.3	0.6	34.9	0.4	4.7	1.5
MAN	Horse	—	29.1	0.9	17.1	0.5	32.3	0.3	5.1	1.4
MAN	Poultry	—	16.6	0.4	13.6	0.2	35.4	2.9	14.9	1.4
MAN	Poultry	—	16.8	0.1	13.5	0.2	36.3	0.5	14.2	1.3
MAN	Pig	—	19.6	1.7	11.6	0.8	36.8	0.1	12.4	1.3
MAN	Cattle	Straw, after 1 month conservation	18.8	0.8	13.6	0.3	48.5	2.6	8.5	1.6
MAN	Turkey	—	21.7	0.7	21.5	0.2	22.8	0.5	17.0	1.5
MAN	Mix	—	19.6	0.6	12.8	0.0	31.2	2.3	12.0	1.3
MAN	Poultry	—	20.1	0.2	14.4	0.1	23.5	0.6	28.1	1.4
MAN	Poultry	—	22.1	1.0	17.4	0.7	20.0	0.4	14.2	1.2
MAN	Cattle	Straw	13.6	0.1	8.8	0.1	52.0	1.9	5.3	1.9

Table A2. *Cont.*

Family	Type	Sub Type	Cellulose		Hemicelluloses		Lignin		Protein	COD
			Mean	SD	Mean	SD	Mean	SD	Calculated	Calculated
			(g/100g DM)		(g/100g DM)		(g/100g DM)		(% DM)	(g COD/g CxHyOz)
MAN	Horse	—	20.0	0.5	12.7	0.4	56.5	3.7	9.0	1.3
MAN	Cattle	Straw	29.2	1.4	16.9	0.2	30.4	0.3	7.5	1.3
MAN	Poultry	—	19.1	0.3	17.1	0.1	26.4	0.2	15.4	1.4
MAN	Horse	—	35.0	1.3	20.0	1.3	27.2	1.3	4.6	1.4
MAN	Horse	—	31.8	0.0	18.9	0.0	27.1	0.1	3.1	1.4
MAN	Zoo	—	21.9	0.2	15.2	0.1	38.6	1.6	7.8	1.4
MAN	Cattle	Straw	19.1	0.3	11.5	0.1	40.1	4.0	6.6	1.3
MAN	Mix	Straw	20.2	0.2	14.9	0.4	28.6	0.1	13.8	1.4
MAN	Mix	Straw	18.9	0.2	12.7	0.1	35.7	2.7	13.3	1.5
MAN	Horse	—	31.6	0.2	20.0	0.1	27.1	0.5	4.1	1.3
MAN	Horse	—	29.6	0.6	19.9	0.1	34.6	0.3	3.3	1.2
MAN	Cattle	Straw	28.5	0.2	15.7	0.1	36.4	0.0	12.8	1.3
MAN	Cattle	Straw	25.0	1.0	18.1	0.6	43.2	0.2	11.3	1.4
MAN	Horse	—	20.7	0.5	16.4	0.4	28.3	0.7	16.8	1.3
CER	Maize Residues	Follicle	49.0	1.2	10.8	0.5	12.0	0.0	8.2	1.3
CER	Wheat	Contaminated culture	59.3	1.1	6.0	0.1	5.5	0.5	11.3	1.2
CER	Mix	Cereals	50.2	0.6	7.9	0.3	14.2	0.0	9.1	1.3
CER	Mix	Cereal dust	29.7	0.4	21.3	0.3	21.4	1.2	6.1	1.3
CER	Mix	Cereal residue	33.3	1.7	14.7	1.2	16.1	0.1	22.3	1.4
CER	Maize	Fresh residue from sweet corn	29.5	1.5	16.1	1.5	15.4	0.4	5.1	1.2
CER	Maize	Fresh residue from sweet corn	25.8	0.4	18.9	0.2	16.4	0.0	6.3	1.2
CER	Maize	Fresh residue from sweet corn	29.5	0.1	19.0	0.2	16.2	0.1	6.3	1.3
CER	Maize	Fresh residue from sweet corn	29.5	0.2	20.5	0.0	12.2	0.1	5.6	1.3
CER	Maize	Fresh residue from sweet corn	29.0	0.3	19.9	0.0	12.9	0.3	7.6	1.3
CER	Maize	Fresh residue from sweet corn	27.6	0.4	16.8	0.6	16.3	0.2	4.8	1.3
CER	Maize	Fresh residue from sweet corn	28.4	0.1	18.9	0.1	14.0	0.1	4.8	1.3
CER	Mix	Cereals	29.3	0.9	10.9	0.3	18.7	0.4	22.3	1.5
CER	Maize	Flour	60.9	0.4	7.2	0.1	6.2	0.5	6.8	1.2
CER	Mix	Cereals	50.6	1.3	7.3	0.6	10.6	0.1	14.8	1.4
CER	Mix	Silo's lose	51.6	0.7	10.7	0.5	18.7	0.5	4.5	1.2
CER	Mix	Cereals	49.3	1.9	6.8	0.3	13.3	0.3	8.9	1.3
SLU	Cattle	—	8.7	0.0	7.7	0.3	38.6	1.1	16.1	1.8
SLU	Rabbit	—	20.6	0.7	13.5	0.1	28.6	0.0	10.8	1.4

Table A2. *Cont.*

Family	Type	Sub Type	Cellulose		Hemicelluloses		Lignin		Protein	COD
			Mean	SD	Mean	SD	Mean	SD	Calculated	Calculated
			(g/100g DM)		(g/100g DM)		(g/100g DM)		(% DM)	(g COD/g CxHyOz)
SLU	Cattle	—	26.4	0.7	20.0	0.9	28.8	0.1	14.0	1.0
SLU	Duck	—	13.2	0.1	23.5	1.3	19.1	0.1	18.6	1.6
SLU	Cattle	—	17.8	0.0	12.7	1.0	33.8	0.9	9.0	1.6
LCM	Maize Residue	Cob	29.0	0.6	26.0	0.4	19.8	0.1	0.6	1.2
LCM	Hemp	Dust	19.5	0.8	8.9	0.0	28.2	0.0	5.1	1.5
LCM	Straw	Plant residues	28.5	0.3	17.5	0.2	16.3	0.3	4.4	1.2
LCM	Straw	—	30.9	0.5	18.3	0.2	17.0	0.4	3.4	1.2
LCM	Maize	Beans	33.6	1.5	22.7	0.7	19.4	0.5	1.6	1.4
LCM	Bagasse and Straw	—	30.1	2.5	18.2	1.7	17.6	1.5	1.1	1.3
LCM	Bagasse	—	32.1	0.2	15.7	0.2	22.8	0.7	0.7	1.3
LCM	Straw	—	31.7	0.2	20.6	0.3	23.1	0.4	1.4	1.4
LCM	Bagasse	—	33.4	0.6	21.2	0.4	30.9	2.9	0.9	1.9
LCM	Straw	Waste	25.3	1.7	24.1	1.8	21.3	2.1	3.4	1.3
LCM	Green waste	—	31.9	0.7	13.8	0.4	17.9	1.3	2.0	1.3
LCM	Straw	—	30.5	0.8	18.5	0.3	18.8	0.5	4.2	1.2
LCM	Hay	Meadow	26.8	2.0	19.9	1.9	23.7	2.2	5.2	1.3
LCM	Straw	Plant residues	30.1	0.3	17.8	0.1	18.3	0.5	3.0	1.2
LCM	Straw	Plant residues	31.3	0.0	17.4	0.1	18.3	1.2	3.0	1.3
LCM	Straw	—	29.2	0.9	18.9	0.3	14.7	0.4	3.6	1.2
LCM	Straw	—	32.3	0.7	18.3	0.4	15.7	0.6	2.0	1.2
LCM	Straw	—	30.3	0.0	20.2	0.1	24.6	1.1	2.6	1.3
LCM	Straw	Waste	31.4	0.9	22.0	0.7	18.4	0.3	2.2	1.3
LCM	Flower residue	Lavender	20.8	0.7	12.1	0.5	30.8	0.1	6.8	1.4
LCM	Maize	Leaf	25.7	1.7	20.8	1.5	27.5	1.8	4.8	1.3
LCM	Straw	Plant residues	26.2	1.1	16.6	0.9	19.5	1.4	3.8	1.2
LCM	Straw	Waste	30.2	1.7	23.8	1.1	26.0	1.1	2.1	1.3
LCM	Straw	Rapeseed waste	24.5	1.5	11.7	0.8	23.7	2.0	8.1	1.3
LCM	Straw	—	29.8	0.2	19.7	0.3	20.9	1.1	1.7	1.2
LCM	Straw	Waste	27.8	2.1	21.5	1.4	23.7	1.3	2.3	1.3
LCM	Green waste	—	18.3	1.2	11.1	0.7	46.0	0.2	10.6	1.6
LCM	Mix	Green waste	16.8	1.6	12.5	1.1	50.2	4.2	10.7	1.5
LCM	Green waste	—	15.4	0.8	12.9	0.9	42.1	1.5	7.4	2.8
LCM	Flower residue	Pomace	17.1	0.8	7.8	0.0	17.1	0.6	13.6	1.3

Table A2. *Cont.*

Family	Type	Sub Type	Cellulose		Hemicelluloses		Lignin		Protein	COD
			Mean	SD	Mean	SD	Mean	SD	Calculated	Calculated
			(g/100g DM)		(g/100g DM)		(g/100g DM)		(% DM)	(g COD/g CxHyOz)
LCM	Straw	—	31.5	0.7	17.0	0.7	27.9	2.5	0.6	1.2
LCM	Flower residue	Lavender	24.5	0.9	10.7	0.5	30.0	0.1	4.0	1.4
LCM	Straw	Waste	29.2	0.3	20.1	0.0	26.6	0.1	1.0	1.2

Appendix B

Table A3. Literature references of BMP performed on large samples, biogas production and biochemical characterization are indicated for each families of substrates. DM: dry matter; VS: volatile solids, HCell: hemicellulose, Cell: cellulose, COD: chemical oxygen demand, Prot: proteins, and BMP: biochemical methane potential.

Reference	N.	Sample Familly	Sample Description	DM	VS	HCell	Cell	Lignin	COD	Prot	BMP (mL CH^4/g VS)
	2	Manures	Dairy and Separated liquid manure	58–124 91 g/kg	41–102 71 g/kg	10% VS	32% VS	14% VS	71–129 100 g/kg	6% VS	243–261 252
[14]	9	Food residue	Cheese whey, Plain pasta, Meat pasta, Used vegetable oil, Ice cream, Fresh dog food, Cola beverage, Cabbage, and Potatoes	71–991 274 g/kg	60–989 274 g/kg	0–0 0% VS	0–36 3% VS	0–0 0% VS	91–2880 642 g/kg	0–19 10% VS	216–649 390
	1	Switchgras	Switchgrass	930 g/kg	905 g/kg	42% VS	49% VS	8% VS	707 g/kg	1% VS	122
	1	Silage	Corn silage	217 g/kg	201 g/kg		12% VS		-	14% VS	296
[25]	20	Municipal solid wastes	Municipal solid wastes	94–99 97% RM	53–90 74% RM	-	-	ND–0.4 0.1 g/g VS	38–279 145 g/g VS	29–89 52 g/g VS	87–357 226
[26]	95	Grass	Meadow grass	51	288	-	-	-	-	-	406
[18]	204			295	329						355

Table A3. *Cont.*

Reference	N.	Sample Familly	Sample Description	DM	VS	HCell	Cell	Lignin	COD	Prot	BMP (mL CH^4/g VS)
[17]	9	Lawn cuttings	Meadow grass, Grass mixture, White clover, and Short bluegrass	-	-	22% VS	28% VS	6% VS	-	16% VS	298–404 333
	9	Hedge cuttings	Oval-leaved privet, Ivy, Beech hedge, Chokeberry, and Ground-elder	-	-	12% VS	28% VS	16% VS	-	12% VS	149–277 203
	16	Wood cuttings	Birch tree, Plane tree, Willow, and Cypress	-	-	12% VS	24% VS	24% VS	-	10% VS	138–245 177
	17	Wild plants	Northern bluegrass, Green foxtail, Bamboo, Common reed, Tufted hair-grass, Reed canary grass, Chrysanthemum, and Dandelion	-	-	24% VS	38% VS	10% VS	z	8% VS	106–319 227
	6	Crops	Maize, Wheat straw, and Sugar beet	-	-	30% VS	28% VS	4% VS	-	8% VS	223–479 404
[27]	58	Agro-industrial wastes	Solid food processing waste and non-conformed end products	-	4–99 52% DM	-	-	-	-	-	66–845 396
	1	Macroaglae	-	-	56% DM	-	-	-	-	-	238
	20	Biowaste	Household organic waste	-	3–88 42% DM	-	-	-	-	-	185–845 370
	4	Energy crops	Maize and switch grass	-	89–94 92% DM	-	-	-	-	-	211–370 264
	11	Fatty waste	Industrial sludge digester with fatty feedstock	-	0–29 13% DM	-	-	-	-	-	53–1321 475
	14	Meat waste	Slaughterhouse waste or stale meat	-	23–96 70% DM	-	-	-	-	-	172–594 475
	2	Co-digestion mix		-	83% DM	-	-	-	-	-	185
	66	Municipal solid wastes	Fresh wastes collected from different localisation and after different treatment	-	15–85 60% DM	-	-	-	-	-	26–423 211

Table A3. *Cont.*

Reference	N.	Sample Family	Sample Description	DM	VS	HCell	Cell	Lignin	COD	Prot	BMP (mL CH$_4$/g VS)
	42	Plant and Vegetable	Wheat and barley residues, Potatoes, Tomatoes, etc.	-	42–95 81% DM	-	-	-	-	-	0–449 264
	18	Agro-industrial sludges	Sludges produced from agro-industrial WWTP	-	2–80 18% DM	-	-	-	-	-	0–687 317
	30	Sewage sludge WWTP	Different WWTP at different process steps (pre-treated or not)	-	11–84 66% DM	-	-	-	-	-	13–343 172
	31	Stabilised municipal solid waste	Landfill drillings	-	14–66 40% DM	-	-	-	-	-	0–264 132
[20]	14	Leaf	Reed canary grass	-	-	22–36 31% DM	16–29 26% DM	1–5 3% DM	-	-	321–388 352
		Steam	Reed canary grass	-	-	24–34 30% DM	21–41 35% DM	1–10 7% DM	-	-	283–417 344
	3	Manures	Chicken, Dairy, and Swine manures	26–39 32% FM	20–29 23% FM	15–28 22% DM	11–20 17% DM	2–17 8% DM	-	13–20 17% DM	51–322 223
	3	Crops straws	Corn stover, Wheat straw, and Rice straw	85–93 89% FM	77–82 79% FM	25–30 27% DM	41–42 42% DM	8–11 10% DM	-	3–6 4% DM	241–281 256
[24]	5	Food and green wastes	Kitchen waste, Fruit and vegetable, Used animal/vegetable oil, and Yard waste	4–100 60% FM	3–100 57% FM	0–20 7% DM	0–21 10% DM	0–11 5% DM	-	0–21 9% DM	183–811 531
	2	Processing organic wastes	Vinegar residue and Rice husk	90–92 91% FM	74–85 80% FM	18–33 26% DM	23–41 32% DM	12–20 16% DM	-	3–12 7% DM	49–253 151
	1	Energy crops	Switchgrass	91% FM	87% FM	32% DM	43% DM	11% DM	-	3% DM	246
	2	Lignocellulosic biomass	Chenopodium album leaf, seed, and stalk	84–86 85% FM	78–83 81% FM	17–19 18% DM	20–39 30% DM	8–16 12% DM	-	3–17 10% DM	171–262 217
[28]	88	All		-	87–96 92% DM		9–76 57% DM		-	-	104–502 251

Table A3. *Cont.*

Reference	N.	Sample Family	Sample Description	DM	VS	HCell	Cell	Lignin	COD	Prot	BMP (mL CH^4/g VS)
[16]	18	Miscanthus	Miscanthus giganteus	-	-	25% DM	44% DM	9% DM	-	4% DM	263
	16	Switchgrass		-	-	33% DM	40% DM	7% DM	-	4% DM	213
	36	Spelt straw		-	-	31% DM	44% DM	7% DM	-	2% DM	275
	37	Fiber sorghum	Winter and Autumn	-	-	22–25 24% DM	33–42 37% DM	5–7 6% DM	-	4–7 5% DM	363–438 400
	369	Tall Fescue	Spring, Summer, and Autumn	-	-	22–25 24% DM	25–29 27% DM	4–4 4% DM	-	9–11 10% DM	400–425 408
	21	Immature rye		-	-	18% DM	22% DM	2% DM	-	9% DM	525
	73	Fiber corn	Winter and Autumn	-	-	2–4 3% DM	20–20 20% DM	18–18 18% DM	-	5–7 6% DM	313–400 356
[29]	23	Anaerobic sludges	Effluent from anaerobic digesters	-	-	-	-	-	-	-	32–214 73
	30	Standard compounds	Cellulose, Starch, and Gelatine	-	-	-	-	-	-	-	289–407 361
	50	Household wastes	Fruit and vegetable waste, Milk waste, Meat waste, and Co-digestion mixtures	-	-	-	-	-	-	-	214–900 461
	10	Agriculture wastes	Wheat straw, Bamboo waste, and Banana stem	-	-	-	-	-	-	-	139–300 224
	19	Sewage sludges	Primary and secondary Sludge and Co-digestion mixtures	-	-	-	-	-	-	-	171–429 353
[5]	6	Lipid rich wastes	Butter and Oil wastes	-	-	-	-	-	-	-	793–943 891
	6	Cereal crops	Barley, Wheat, Triticale, and Oats	54–69 62% FM	49–67 58% FM	-	-	-	-	-	281–366 336
	3	Oil seed rapes	Macerated, Whole crop, and Not macerated	88–93 91% FM	85–89 87% FM	-	-	-	-	-	215–646 393
	7	Root crops	Potatoes, Turnips, Sugar beet, Energy beet, and Fodder beet	11–26 19% FM	10–25 18% FM	-	-	-	-	-	306–399 349
	5	Grass silages	Grass silage and Fresh grass	12–29 19% FM	11–27 18% FM	-	-	-	-	-	368–400 385
	2	Baled silages	-	17–17 17% FM	15–16 15% FM	-	-	-	-	-	428–433 431

Table A3. Cont.

Reference	N.	Sample Family	Sample Description	DM	VS	HCell	Cell	Lignin	COD	Prot	BMP (mL CH^4/g VS)
	8	Other grass substrates	Silage, Hay, Savazi grass, Silage effluent, Grass digestate, Fresh maize, and Maize silage	6–87 29% FM	3–82 27% FM	-				-	127–394 324
	7	Dairy slurries	-	6–9 7% FM	4–7 6% FM	-	-	-	-	-	136–239 201
	4	Other agricultural wastes	Beef slurry, Pig slurry, Poultry manure, and Farm yard manure	5–51 21% FM	4–30 14% FM	-	-	-	-	-	99–311 194
	4	Milk processing wastes	Sludges with or without dissolved air floatation	4–16 9% FM	3–9 7% FM	-	-	-	-	-	189–787 473
	4	Abattoir wastes	Mix, paunch content, and Sludges	13–20 17% FM	11–18 15% FM	-	-	-	-	-	166–404 286
	7	Miscellaneous wastes	Bakery waste, Brewing stillage, Grocery waste, Fish offal mix, Bread waste, Park and grass waste, and WWTP	9–66 32% FM	7–64 29% FM	-	-	-	-	-	247–592 396
	10	Domestic and commercial food wastes	Rural and urban food waste, Food wastes from canteens and restaurants, and Centralised collection centre combining the two types or not	22–95 37% FM	19–88 32% FM	-				-	274–535 329
	3	Alternative wastes	Recycled paper, Used cooking oil, and Grease trap wastes	27–100 72% FM	26–99 68% FM	-	-	-	-	-	254–805 434
	12	Seaweeds	9 brown & 3 green Seaweeds	13–78 23% FM	8–46 15% FM	-	-	-	-	-	101–341 213
[19]	24	Main and secondary crops	Sugar beet, Barley/ryegrass, Maize, Triticale, Marrow stem kale, Rye/triticale, Potatoes, Oat/forage Pea/false flax, Rye, Sundangrass, Forage sorghum, Rye/fodder vetch, Barley/turnip rape, Oat, Amaranth, Quinoa, Rapeseed, Sunflower, Forage pea, and Buckwheat	9–59 33% FM	81–97 92% DM	2–25 15% DM	3–37 27% DM	1–13 6% DM	-	4–19 9% DM	210–399 294

Table A3. *Cont.*

Reference	N.	Sample Family	Sample Description	DM	VS	HCell	Cell	Lignin	COD	Prot	BMP (mL CH^4/g VS)
[30]	10	Catch crops	Triticale, Barley, Rye, Landsberger mix, Sudengrass hybrid, Forage sorgum, Ryegrass, Phacelia, Fodder radish, and Buckwheat/phacelia	9–58 24% FM	73–96 90% DM	0–24 17% DM	24–34 30% DM	2–9 5% DM	-	5–26 11% DM	235–376 311
	4	Annual grass and legume mix	Ryegrass, Clover, Alfalfa clover, and Alfalfa	15–48 28% FM	85–93 90% DM	11–18 14% DM	26–29 28% DM	4–7 5% DM	-	7–20 14% DM	240–388 307
	5	Perennial crops	Tall wheatgrass, Countru mallow, Jerusalem artichoke, Miscanthus, and Cup plant	14–40 28% FM	85–97 90% DM	5–24 16% DM	28–42 33% DM	7–13 10% DM	-	4–15 9% DM	179–259 228
	58	Solid manure				-	-	-	-	-	129–366 225
	7	Animal slurries				-	-	-	-	-	225–551 293
	3	Slaughterhouse waste				-	-	-	-	-	186–664 349
	16	Mix of AD feedstock				-	-	-	-	-	90–253 101
	6	AD digestats		2–99% FM	1–92% DM	-	-	-	-	-	214–405 304
	36	Grass and intermediate crops				-	-	-	-	-	191–444 304
	24	Cereals and crop residues				-	-	-	-	-	191–388 304
	26	Silages				-	-	-	-	-	186–495 338

Table A3. *Cont.*

Reference	N.	Sample Familly	Sample Description	DM	VS	HCell	Cell	Lignin	COD	Prot	BMP (mL CH^4/g VS)
[30]	38	Lignocellulosic plants				-	-	-	-	-	62–326 270
	15	Grape marcs				-	-	-	-	-	79–219 129
	3	Algae				-	-	-	-	-	146–169 165
	25	Food wastes and biowastes				-	-	-	-	-	96–518 338
	10	Sludges				-	-	-	-	-	56–776 259
	3	Effluents				-	-	-	-	-	225–281 276
	3	Fat and lipid wastes				-	-	-	-	-	596–878 630
	2	Products and wastes from meat				-	-	-	-	-	203–388 293
	2	Organic fraction of municipal waste				-	-	-	-	-	281
[21]	41	Energy crops	Barley, Clover, Cup plant, Grassland, Maize, Millet, Potatoes, Rye, Sugar beet, Sunflower, and Triticale	88–94 91% FM	79–89 85% FM	3–28 18% DM	5–39 27% DM	0–11 4% DM	-	4–20 9% DM	177–401 311
[22]	43	Grasses	Lolium perenne, Dactylis glomerata, Poa pratensis, and Fescuta pratensis	87–94 91% FM	78–88 84% FM	21–32 26% DM	20–36 29% DM	2–7 4% DM	-	6–20 11% DM	314–422 353
	18	Legumes	Trifolium pratense and Repens	88–93 90% FM	80–85 82% FM	3–22 11% DM	16–33 25% DM	5–9 7% DM	-	13–29 21% DM	265–346 301

Table A3. Cont.

Reference	N.	Sample Familly	Sample Description	DM	VS	HCell	Cell	Lignin	COD	Prot	BMP (mL CH^4/g VS)
	2	Biowaste	Banana peel waste, Tomato waste, and	11% FM	83% DM	-	-	-	2 g O2/g VS	-	329
	1	Effluent	Winery wastewater	3% FM	65% DM	-	-	-	3 g O2/g VS	-	251
	10	Plants		?	?	?	?	?	?	?	111–379 229
	21	Vegetables		?	?	?	?	?	?	?	186–443 314
	24	Fruits		?	?	?	?	?	?	?	185–529 314
[13]	7	Cereals		?	?	?	?	?	?	?	261–325 293
	12	Manures		?	?	?	?	?	?	?	154–325 211
	17	Diet		?	?	?	?	?	?	?	250–775 432
	10	Sludges		?	?	?	?	?	?	?	164–711 411
	4	Beverage wastewaters		?	?	?	?	?	?	?	250–593 411
	18	Organic fraction of municipal solid wastes		?	?	?	?	?	?	?	175–571 464
	8	Other		?	?	?	?	?	?	?	207–443 379
[23]	20	Sludges	10 primary and 10 bioglogical Sludges	5–46 21% FM	4–33 15% FM	-	-	-	1–2 2% VS	0–60 28 mg BSA/g VS	58–318 181

References

1. Brémond, U.; Bertrandias, A.; Steyer, J.-P.; Bernet, N.; Carrere, H. A vision of European biogas sector development towards 2030: Trends and challenges. *J. Clean. Prod.* **2021**, *287*, 125065. [CrossRef]
2. Scarlat, N.; Dallemand, J.-F.; Fahl, F. Biogas: Developments and perspectives in Europe. *Renew. Energy* **2018**, *129*, 457–472. [CrossRef]
3. Wang, X.; Lu, X.; Li, F.; Yang, G. Effects of Temperature and Carbon-Nitrogen (C/N) Ratio on the Performance of Anaerobic Co-Digestion of Dairy Manure, Chicken Manure and Rice Straw: Focusing on Ammonia Inhibition. *PLoS ONE* **2014**, *9*, e97265. [CrossRef] [PubMed]
4. Wang, X.; Yang, G.; Feng, Y.; Ren, G.; Han, X. Optimizing feeding composition and carbon–nitrogen ratios for improved methane yield during anaerobic co-digestion of dairy, chicken manure and wheat straw. *Bioresour. Technol.* **2012**, *120*, 78–83. [CrossRef]
5. Allen, E.; Wall, D.M.; Herrmann, C.; Murphy, J.D. A detailed assessment of resource of biomethane from first, second and third generation substrates. *Renew. Energy* **2016**, *87*, 656–665. [CrossRef]
6. Monlau, F.; Sambusiti, C.; Barakat, A.; Guo, X.M.; Latrille, E.; Trably, E.; Steyer, J.-P.; Carrere, H. Predictive Models of Biohydrogen and Biomethane Production Based on the Compositional and Structural Features of Lignocellulosic Materials. *Environ. Sci. Technol.* **2012**, *46*, 12217–12225. [CrossRef]
7. Achinas, S.; Euverink, G.J.W. Theoretical analysis of biogas potential prediction from agricultural waste. *Resour.-Effic. Technol.* **2016**, *2*, 143–147. [CrossRef]
8. Cresson, R.; Pommier, S.; Beline, F.; Bouchez, T.; Buffière, P.; Rivero, J.A.C.; Patricia, C.; Pauss, A.; Pouech, P.; Ribeiro, T. Etude Interlaboratoires Pour l'harmonisation Des Protocoles de Mesure Du Potentiel Méthanogène Des Matrices Solides Hétérogènes. In Proceedings of the Journées Recherche Industrie Biogaz et Méthanisation, Rennes, France, 3–5 February 2015.
9. Holliger, C.; Astals, S.; de Laclos, H.F.; Hafner, S.D.; Koch, K.; Weinrich, S. Towards a standardization of biomethane potential tests: A commentary. *Water Sci. Technol.* **2021**, *83*, 247–250. [CrossRef]
10. Holliger, C.; Alves, M.; Andrade, D.; Angelidaki, I.; Astals, S.; Baier, U.; Bougrier, C.; Buffière, P.; Carballa, M.; De Wilde, V.; et al. Towards a standardization of biomethane potential tests. *Water Sci. Technol.* **2016**, *74*, 2515–2522. [CrossRef]
11. Filer, J.; Ding, H.H.; Chang, S. Biochemical Methane Potential (BMP) Assay Method for Anaerobic Digestion Research. *Water* **2019**, *11*, 921. [CrossRef]
12. Bond, T.; Brouckaert, C.J.; Foxon, K.M.; Buckley, C. A critical review of experimental and predicted methane generation from anaerobic codigestion. *Water Sci. Technol.* **2012**, *65*, 183–189. [CrossRef]
13. Rodrigues, R.; Klepacz-Smolka, A.; Martins, R.; Quina, M. Comparative analysis of methods and models for predicting biochemical methane potential of various organic substrates. *Sci. Total. Environ.* **2019**, *649*, 1599–1608. [CrossRef]
14. Labatut, R.A.; Angenent, L.T.; Scott, N.R. Biochemical methane potential and biodegradability of complex organic substrates. *Bioresour. Technol.* **2011**, *102*, 2255–2264. [CrossRef]
15. Garcia, N.H.; Mattioli, A.; Gil, A.; Frison, N.; Battista, F.; Bolzonella, D. Evaluation of the methane potential of different agricultural and food processing substrates for improved biogas production in rural areas. *Renew. Sustain. Energy Rev.* **2019**, *112*, 1–10. [CrossRef]
16. Godin, B.; Mayer, F.; Agneessens, R.; Gerin, P.; Dardenne, P.; Delfosse, P.; Delcarte, J. Biochemical methane potential prediction of plant biomasses: Comparing chemical composition versus near infrared methods and linear versus non-linear models. *Bioresour. Technol.* **2015**, *175*, 382–390. [CrossRef]
17. Triolo, J.M.; Pedersen, L.; Qu, H.; Sommer, S.G. Biochemical methane potential and anaerobic biodegradability of non-herbaceous and herbaceous phytomass in biogas production. *Bioresour. Technol.* **2012**, *125*, 226–232. [CrossRef]
18. Grieder, C.; Mittweg, G.; Dhillon, B.S.; Montes, J.M.; Orsini, E.; Melchinger, A.E. Kinetics of methane fermentation yield in biogas reactors: Genetic variation and association with chemical composition in maize. *Biomass-Bioenergy* **2012**, *37*, 132–141. [CrossRef]
19. Herrmann, C.; Idler, C.; Heiermann, M. Biogas crops grown in energy crop rotations: Linking chemical composition and methane production characteristics. *Bioresour. Technol.* **2016**, *206*, 23–35. [CrossRef]
20. Kandel, T.P.; Sutaryo, S.; Møller, H.B.; Jørgensen, U.; Lærke, P.E. Chemical composition and methane yield of reed canary grass as influenced by harvesting time and harvest frequency. *Bioresour. Technol.* **2013**, *130*, 659–666. [CrossRef]
21. Dandikas, V.; Heuwinkel, H.; Lichti, F.; Drewes, J.; Koch, K. Correlation between biogas yield and chemical composition of energy crops. *Bioresour. Technol.* **2014**, *174*, 316–320. [CrossRef]
22. Dandikas, V.; Heuwinkel, H.; Lichti, F.; Drewes, J.E.; Koch, K. Correlation between Biogas Yield and Chemical Composition of Grassland Plant Species. *Energy Fuels* **2015**, *29*, 7221–7229. [CrossRef]
23. Catenacci, A.; Azzellino, A.; Malpei, F. Development of statistical predictive models for estimating the methane yield of Italian municipal sludges from chemical composition: A preliminary study. *Water Sci. Technol.* **2019**, *79*, 435–447. [CrossRef] [PubMed]
24. Li, Y.; Zhang, R.; Liu, G.; Chen, C.; He, Y.; Liu, X. Comparison of methane production potential, biodegradability, and kinetics of different organic substrates. *Bioresour. Technol.* **2013**, *149*, 565–569. [CrossRef]
25. Lesteur, M.; Latrille, E.; Maurel, V.B.; Roger, J.; Gonzalez, C.; Junqua, G.; Steyer, J. First step towards a fast analytical method for the determination of Biochemical Methane Potential of solid wastes by near infrared spectroscopy. *Bioresour. Technol.* **2011**, *102*, 2280–2288. [CrossRef] [PubMed]

26. Raju, C.S.; Ward, A.J.; Nielsen, L.; Møller, H.B. Comparison of near infra-red spectroscopy, neutral detergent fibre assay and in-vitro organic matter digestibility assay for rapid determination of the biochemical methane potential of meadow grasses. *Bioresour. Technol.* **2011**, *102*, 7835–7839. [CrossRef]

27. Doublet, J.; Boulanger, A.; Ponthieux, A.; Laroche, C.; Poitrenaud, M.; Rivero, J.C. Predicting the biochemical methane potential of wide range of organic substrates by near infrared spectroscopy. *Bioresour. Technol.* **2013**, *128*, 252–258. [CrossRef]

28. Triolo, J.M.; Ward, A.J.; Pedersen, L.; Løkke, M.M.; Qu, H.; Sommer, S.G. Near Infrared Reflectance Spectroscopy (NIRS) for rapid determination of biochemical methane potential of plant biomass. *Appl. Energy* **2014**, *116*, 52–57. [CrossRef]

29. Strömberg, S.; Nistor, M.; Liu, J. Early prediction of Biochemical Methane Potential through statistical and kinetic modelling of initial gas production. *Bioresour. Technol.* **2015**, *176*, 233–241. [CrossRef]

30. Mortreuil, P.; Baggio, S.; Lagnet, C.; Schraauwers, B.; Monlau, F. Fast prediction of organic wastes methane potential by near infrared reflectance spectroscopy: A successful tool for farm-scale biogas plant monitoring. *Waste Manag. Res.* **2018**, *36*, 800–809. [CrossRef]

31. Wei, Z.; Li, Y.; Hou, Y. Quick estimation for pollution load contributions of aromatic organics in wastewater from pulp and paper industry. *Nord. Pulp Pap. Res. J.* **2018**, *33*, 568–572. [CrossRef]

32. Jain, R.; Goomer, S. Evaluation of Food Nitrogen and Its Protein Quality Assessment Methods. *Int. J. Food Sci. Nutr.* **2019**, *6*, 68–74.

33. Sluiter, A.; Hames, B.; Ruiz, R.; Scarlata, C.; Sluiter, J.; Templeton, D.; Crocker, D. *Determination of Structural Carbohydrates and Lignin in Biomass*; Technical Report NREL/TP-510-42618; National Renewable Energy Laboratory: Golden, CO, USA, 2012.

34. Hafner, S.D.; De Laclos, H.F.; Koch, K.; Holliger, C. Improving Inter-Laboratory Reproducibility in Measurement of Biochemical Methane Potential (BMP). *Water* **2020**, *12*, 1752. [CrossRef]

35. ADEME. *Méthanisation de Fumiers Bovin et Volaille—Impact Du Stockage Du Fumier et Essais Pilote et Potentiel Énergétique*; ADEME Bourgogne: Dijon, France, 2013.

36. Teurki, R.; Agricultures & Territoires Chambre d'Agriculture Somme; Agricultures & Territoires Chambre d'Agriculture Nord-Pas de Calais; Agence de l'eau Picardie; Agence de l'eau Seine Normandie. Satege Les Effluents D'élevage: Mieux Les Connaître Pour Bien Les Valoriser. 2013.

37. Corno, L. *Arundo Donax L.(Giant Cane) as a Feedstock for Bioenergy and Green Chemistry*; University of Milano: Milano, Italy, 2016.

38. Hutňan, M. Maize Silage as Substrate for Biogas Production. *Adv. Silage Prod. Util.* **2016**, *16*, 173–196.

39. Doligez, P. Réussir Le Compostage de Fumier Équin. Available online: https://equipedia.ifce.fr/infrastructure-et-equipement/installation-et-environnement/effluents-delevage/reussir-le-compostage-de-fumier-equin?tx__%5Baction%5D=&tx__%5Bcontroller%5D=Standard&cHash=113657bc00a1d6a39f98a694daa686fb (accessed on 7 May 2021).

40. Luna-de Risco, M.; Normak, A.; Orupõld, K. Biochemical Methane Potential of Different Organic Wastes and Energy Crops from Estonia. *Agron. Res.* **2011**, *9*, 331–342.

41. Kafle, G.K.; Chen, L. Comparison on batch anaerobic digestion of five different livestock manures and prediction of biochemical methane potential (BMP) using different statistical models. *Waste Manag.* **2016**, *48*, 492–502. [CrossRef]

42. Cu, T.T.T.; Nguyen, T.X.; Triolo, J.M.; Pedersen, L.; Le, V.D.; Le, P.D.; Sommer, S.G. Biogas Production from Vietnamese Animal Manure, Plant Residues and Organic Waste: Influence of Biomass Composition on Methane Yield. *Asian-Australas. J. Anim. Sci.* **2015**, *28*, 280–289. [CrossRef]

43. Yang, G.; Li, Y.; Zhen, F.; Xu, Y.; Liu, J.; Li, N.; Sun, Y.; Luo, L.; Wang, M.; Zhang, L. Biochemical methane potential prediction for mixed feedstocks of straw and manure in anaerobic co-digestion. *Bioresour. Technol.* **2021**, *326*, 124745. [CrossRef]

44. Carabeo-Pérez, A.; Odales-Bernal, L.; López-Dávila, E.; Jiménez, J. Biomethane potential from herbivorous animal's manures: Cuban case study. *J. Mater. Cycles Waste Manag.* **2021**, *23*, 1404–1411. [CrossRef]

45. Barakat, A.; Monlau, F.; Steyer, J.-P.; Carrere, H. Effect of lignin-derived and furan compounds found in lignocellulosic hydrolysates on biomethane production. *Bioresour. Technol.* **2012**, *104*, 90–99. [CrossRef]

46. Dinuccio, E.; Balsari, P.; Gioelli, F.; Menardo, S. Evaluation of the biogas productivity potential of some Italian agro-industrial biomasses. *Bioresour. Technol.* **2010**, *101*, 3780–3783. [CrossRef]

47. Böske, J.; Wirth, B.; Garlipp, F.; Mumme, J.; Weghe, H.V.D. Anaerobic digestion of horse dung mixed with different bedding materials in an upflow solid-state (UASS) reactor at mesophilic conditions. *Bioresour. Technol.* **2014**, *158*, 111–118. [CrossRef] [PubMed]

48. Holliger, C.; De Laclos, H.F.; Hack, G. Methane Production of Full-Scale Anaerobic Digestion Plants Calculated from Substrate's Biomethane Potentials Compares Well with the One Measured On-Site. *Front. Energy Res.* **2017**, *5*, 12. [CrossRef]

waste

Article

Washing Methods for Remove Sodium Chloride from Oyster Shell Waste: A Comparative Study

Jung Eun Park, Sang Eun Lee and Seokhwi Kim *

Center for Bio Resource Recycling, Institute for Advanced Engineering, Yongin-si 17180, Republic of Korea
* Correspondence: shkim5526@iae.re.kr; Tel.: +82-31-330-7203

Abstract: The oyster shell is a valuable calcium resource; however, its application is limited by its high NaCl content. Therefore, to establish the use of oyster shells as a viable resource, conditional experiments were conducted to select optimum parameters for NaCl removal. For this purpose, we compared leaching methods with batch and sequential procedures, determined the volume of water used for washing, and evaluated the mixing speed. The batch system removed NaCl when washed for >24 h over a shell to water ratio of 1:5. Results from the batch experiments confirmed that washing twice can completely remove NaCl from the shells on a like-for-like basis. Additionally, the efficiency of washing was sequentially evaluated in terms of the number of washing cycles. Compared to batch experiments, continuous washing could remove NaCl in approximately 10 min at a shell to water ratio of 1:4. We found that regardless of the washing methods, the volume of water used for washing is key for enhancing NaCl removal. Consequently, increasing the volume of water used for washing coupled with a proper sorting of fine particles can help enhance the purity of calcium, which will enable the use of oyster shell as an alternate Ca-resource.

Keywords: oyster shell waste; washing method; sodium chloride; impurities; particle size

1. Introduction

Oysters are widely used as part of a healthy diet, as they help improve immunity and in the rapid absorption of zinc and iron; moreover, increased awareness of their health benefits has led to an increase in the consumption of oysters [1]. As of 2017, 639 tons per year of Pacific oysters are produced in Asia, which constitutes approximately 92% of the global oyster market [2]. An oyster comprises about 14% oyster meat and 86% shell; during oyster processing, the shells are disposed as waste into the sea or buried into the ground [3,4]. Shells buried into the ground pose problems, as they produce leachates and odors. Furthermore, the organic matter in the shells discarded into the sea may pollute the ocean ecosystem. Hence, there is a need for a rapid establishment of policy support to address the substantial increase in the number of shells generated at present and in the future. Moreover, it is desirable to utilize shells as resources without disposing them as waste, as they can contribute to an environmental and economic circular economy [5,6]. Both Korea and Southern Brazil have suggested recycling of shells for environmental benefits; as of 2018 [6,7], Japan has recycled 140,000 tons of oyster shells to improve the environment of fishing farms and to enhance the sediment quality by establishing underwater plants [8]. Taiwan generates 34,000 tons of oyster shells, whose disposal into the environment causes sanitation, pollution, and protection issues [9].

Oyster shells have been disposed as waste along the coast of Korea [10]; however, after the amendment of the Framework Act on Resources Circulation in 2022, the status of oyster shells has been changed from waste to resources to help control the influx of impurities such as organic residues. Nevertheless, there are two challenges associated with the application of oyster shells as an alternative to industrial material such as limestone: (i) the levels of impurities equivalent to those of limestone and (ii) the demand for a massive quantity [11–13].

Citation: Park, J.E.; Lee, S.E.; Kim, S. Washing Methods for Remove Sodium Chloride from Oyster Shell Waste: A Comparative Study. *Waste* **2023**, *1*, 166–175. https://doi.org/10.3390/waste1010012

Academic Editors: Vassilis Athanasiadis and Dimitris P. Makris

Received: 23 November 2022
Revised: 2 January 2023
Accepted: 4 January 2023
Published: 10 January 2023

Oyster shells are subjected to physical and chemical treatments before they can be used as high value-added resources [4]. However, their consumption as a calcium substitute is minimal, because only a meagre percentage is applied during the dark fermentation of cement waste [14,15]. Previous studies indicate that the oyster shell content in concrete cannot exceed 15% [4]; the suggested concentration in dark fermentation is less than 8% [15]. Thus, the limited application of oyster shells in the construction industry cannot fundamentally help in the recycling of the neglected oyster shells (NOSs). Oyster shells have also been used in water treatment, as they are considered as absorbents that help remove phosphorus and nitrogen compounds [11,15,16]. However, their absorption capacity is minimal at ~228.15 mg-P/kg-oyster shell, and the recovered absorptive materials are generated as secondary waste [17].

There is a massive demand for calcium compounds such as $CaCO_3$ and $Ca(OH)_2$ as they are commonly used for acid gas treatment; moreover, several studies have been conducted on SO_x removal using $CaCO_3$ [18,19]. This process involves two stages where $CaCO_3$ reacts with SO_4 to produce gypsum ($CaSO_4 \cdot 2H_2O$) as the final product, as shown in Equation (1). Oyster shells can be used as a substitute for limestone, which would effectively address the over-exploitation of limestone in the environment.

$$H_2SO_4 + CaCO_3 + 0.5O_2 + H_2O \leftrightarrow CO_2 + CaSO_4 + 2H_2O \tag{1}$$

Interestingly, Lim et al. [18] found that the efficiency of flue gas desulfurization (FGD) increased with the addition of NaCl. SO_2 removal increased up to 52% under Ca/S = 2 with the addition of 3.2% of NaCl, whereas it was relatively low at 32% in the absence of NaCl [20]. The sulfidation of CaO with alkali NaCl was much higher than that of CaO without alkali NaCl, as shown in Equation (2) [21]:

$$H_2SO_4 + CaCO_3 + 2NaCl \leftrightarrow Na_2SO_4 + CaCl_2 + H_2O + CO_2 \tag{2}$$

However, when the sulfidation temperature was over 750 °C with the inclusion of sodium, the higher sodium concentration led to a greater spreading resistance ratio for sulfidation. As $NaCl\text{-}CaSO_4$ partially melts under high temperature, the liquid phase present locally blocks the pores. In other words, a higher sodium concentration increases gas spreading resistance in the pores [22]. Furthermore, the gypsum ($CaSO_4 \cdot 2H_2O$) recovered after FGD is high in sodium, which limits its conversion into viable products. Additionally, previous studies have reported that NaCl impurities affect the degree of limestone calcination [23–26]. Therefore, the use of oyster shells as a viable resource requires the removal of NaCl.

To overcome the existing limitations, in the present study, we investigated washing processes of oyster shells to convert them into useful resources.

2. Materials and Methods

2.1. Oyster Shell Sample Collection

Oyster shells were collected from an open-air yard in Tongyeong, South Korea, where shells are stored outdoors for over six months to generate shell fertilizer; this facilitates the natural washing out of NaCl from the shells. The shell separates the luggage rope using a cylindrical centrifuge, and in this process, some of the shell is crushed. The speed of the centrifuge is about 250 rpm, and the size of the crushed shell is about 5 cm or less. The NaCl content in these shells is expected to be lower than in those generated freshly at the shucking place.

2.2. Oyster Shell Samples Subjected to Washing Process

2.2.1. Batch Washing

To identify the characteristics of NaCl removal from the oyster shells, an elution experiment was conducted using batch as well as sequential reactions. The batch experiment was conducted with 500 g of shells in a container by washing them with tap water, varying the ratio of the volume of tap water from one to six times the weight of the shells; the different ratios of 1:1, 1:3, 1:4, 1:5, and 1:6 were denoted as B-OWR1 to B-OWR6 (B-OWR, batch-oyster shell to tap water ratio). Each oyster shell sample was agitated with the corresponding volume of tap water at 100 rpm at room temperature. Most of the experiments were conducted at 100 rpm, but the effect of rpm was investigated by changing the rpm to 200.

2.2.2. Continuous Washing

In the continuous washing process, oyster shells were washed once, and the tap water used for elution of NaCl from the shells was removed, followed by the addition of new tap water. The washing and elution steps were repeated six times per sample. The samples were agitated for 1 min; the reacted solution was recovered after solid–liquid separation by filtration. The ratios of tap water used for the washing steps were maintained the same as those used for batch washing; the different ratios were denoted as C-OWR1 to C-OWR6 (C-OWR, continuous-oyster shell to tap water ratio). The number of elution processes was determined based on the concentration of NaCl obtained with each wash.

2.3. Estimation of NaCl Leachates

Considering that the oyster shells were not washed with tap water during shucking, the concentrations of leached dissolved ions from the shell could be associated with those of seawater components. The NaCl content of the tap water eluted after batch or continuous washing was estimated indirectly using electrical conductivity (EC) (Pro 30, YSI, Yellow Springs, Ohio, USA). From the measured EC values, the NaCl content in the shells was calculated according to Equation (3) [27]:

$$\text{Total dissolved solids (mg/L)} = k \times \text{Electrical conductivity (liquid} - \text{tap water)} \atop (\mu S/cm) \times SSR \tag{3}$$

where k is conversion of NaCl from the observed TDS value, that is 0.064, and SSR is the solution to solid ratio.

2.4. Analysis of Oyster Shell Particle Size and Chemical Composition

Shells recovered after washing were classified based on their particle size and chemical composition. The particle-size of the shells that ranged from 0.2 mm to 5 mm were classified using standard sieves. Chemical compositions of the shells before and after washing were identified using X-ray fluorescence spectroscopy (XRF-1800, Shimadzu, Kyoto, Japan), and an X-ray diffractometer (XRD-6100, Shimadzu, Kyoto, Japan) was used to analyze the mineral phases.

3. Results and Discussion

3.1. Physical Properties of Neglected Oyster Shells

This study evaluated the basic characteristics of NOSs near Tongyeong, South Korea. The NaCl content of NOSs was measured considering the number of repeated experiments, which were repeated more than six times at B-OWR6 for 24 h. As shown in Figure 1, the NaCl content of shells varied depending on the duration. From the result values, the average of the measured values repeated six times was used and the moisture content of sampled shells was measured as 1.22% ± 0.03 on average.

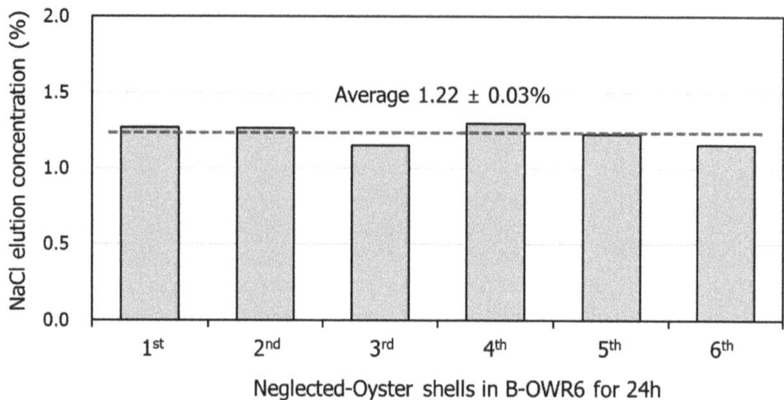

Figure 1. NaCl content of neglected oyster shells in B-OWR6 for 24 h.

Chemical analysis of the oyster shell samples using XRF indicated CaO, SiO_2, Na_2O, SO_3, and MgO as the major constituents, and the average of the measured values repeated six times was used. As shown in Table 1, the presence of Na, Mg, and Mg in the shells suggests that major elements of the seawater could be captured when oysters grow in their habitat. Meanwhile, Al, Fe, and Si were believed to have originated from marine sediments [28,29].

Table 1. Chemical composition of neglected oyster shells (NOSs).

Chemical Composition	NOSs (%)
CaO	81.2 ± 1.6
C	14.1 ± 0.7
SiO_2	1.4 ± 0.3
Na_2O	0.7 ± 0.2
SO_3	0.6 ± 0.1
MgO	0.5 ± 0.1
Al_2O_3	0.2 ± 0.1
P_2O_5	0.2 ± 0.1
Fe_2O_3	0.2 ± 0.1
SrO	0.2 ± 0.0
K_2O	0.1 ± 0.0
TiO_2	0.0 ± 0.0

However, Cl was not detected in the XRF analysis. Because NaCl was regarded as the major salt, the XRF results suggest that NaCl was possibly washed out owing to its high water solubility (up to 36%). In fact, oyster shells in this study were obtained from a fertilizer production facility, where shells were left outdoors for more than six months, indicating that the high level of Cl in the shells had been washed away naturally, and hence, it was not detected.

3.2. Removal of NaCl Using Washing Methods

Since washing efficiencies could be attributed to the physical properties of the oyster shells, the washing method was tested based on the volume of water used, washing duration, and mixing speeds.

3.2.1. Batch Washing

As previously mentioned, NaCl is highly soluble in water; therefore, its leaching would depend on the volume of water used in the washing step. Figure 2 illustrates the leaching properties of NaCl from the shells with elapsed time depending on the volume of water, and the average of the measured values repeated three times was used.

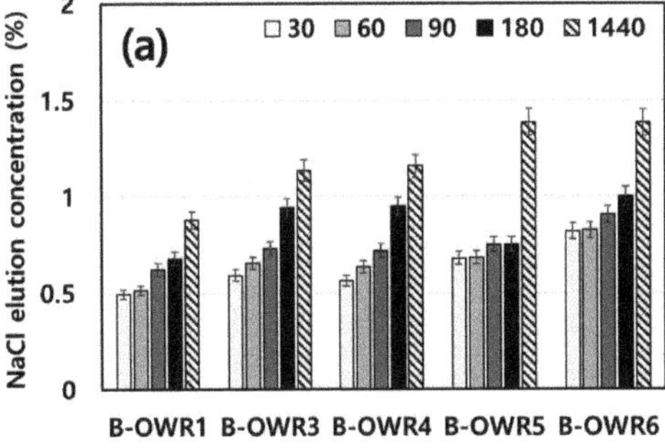
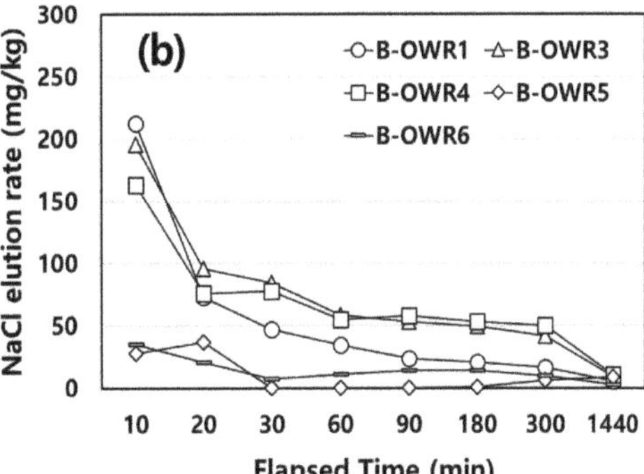

Figure 2. Neglected-oyster shell to water ratio (OWR) depending on elution time. (a) NaCl elution concentration and (b) NaCl elution rate. B-OWR, batch-oyster shell to water ratio.

With respect to the volume of water, the absolute value of eluted NaCl was relatively higher when the water to shell ratio was higher (Figure 2a). However, it was observed that the elution volume of NaCl increased among samples with a smaller ratio of water used for washing (OWR < 4.0). In addition, the leaching of NaCl increased with an increase in elution time in general, except for the OWR of 5. However, no substantial effect was observed when a large ratio of water (OWRs of 5 and 6) was used. We observed a relatively higher absolute elution value when shells were washed for a relatively longer time at OWRs > 5 than that when washed at lower OWRs (Figure 2b). NaCl was eluted into the water during the initial reaction, which indicates that the concentration of eluted NaCl depends on the volume of water used for washing. Hence, it could be inferred that continuous washing for shorter periods was more effective as it enhances NaCl removal from the shells. The average NaCl content in the shells was 1.22 wt.%, and the removal efficiencies are shown in Table 2. The average of the measured values repeated three times was used. The batch system sufficiently removed NaCl residues from the shells under a B-OWR of $\geq$5. However, some differences persisted depending on the volume of washing water used.

Table 2. Maximum NaCl elution volume depending on water.

Samples	Elution NaCl (%)	Washing Efficiency (%)
B-OWR1	0.76 ± 0.07	63 ± 0.06
B-OWR3	0.99 ± 0.02	81 ± 0.03
B-OWR4	1.00 ± 0.03	82 ± 0.04
B-OWR5	1.20 ± 0.07	99 ± 0.06
B-OWR6	1.20 ± 0.03	99 ± 0.02

Figure 3 shows leaching of NaCl from the shells based on the mixing speed and the average of the measured values repeated three times was used. Leaching patterns of NaCl were very similar regardless of mixing speeds of 100 and 200 rpm. However, the amount of NaCl leached from the shell at 200 rpm was approximately 20% higher than that at 100 rpm. Leaching concentrations of NaCl at 200 rpm increased from the start of the reaction and eventually reached up to 80%. This indicates that over 90% of NaCl removal efficiency relies on the volume of water, even though leaching properties primarily rely on the mixing speed.

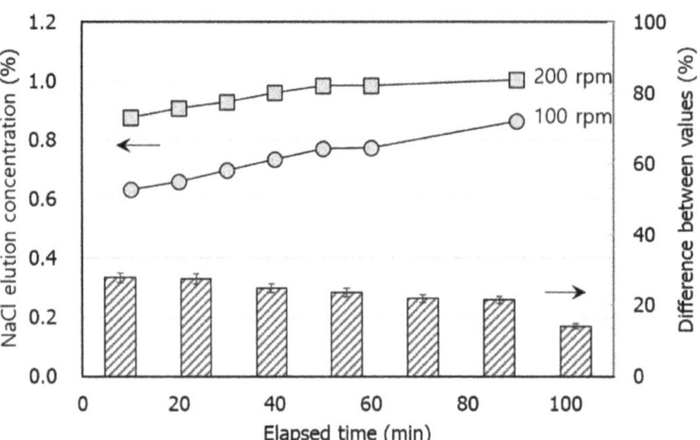

Figure 3. NaCl elution concentration (**left arrow**) with elapsed time at B-OWR3 with different mixing speeds. The bar chart is difference between values 100 rpm and 200 rpm (**right arrow**) with elapsed time at B-OWR3.

These results indicate that the volume of water used for washing is key for effective removal of NaCl from the surface of oyster shells. The removal efficiency of NaCl from the shells reached over 90% when the OWR was >5.0 (Table 2).

3.2.2. Continuous Washing

Figure 4 shows the washing efficiencies for each washing step by continuous washing of oyster shell samples and the average of the measured values repeated three times was used. Leaching of NaCl from the shell was proportionate to the number of washing steps. The amount of leached NaCl at the first step was over 63–96% depending on the OWR. As shown in Figure 2 and Table 2, over 90% leaching was observed when the OWR was >5. Considering that the NaCl removal efficiency reached 90% by sequentially washing the shells twice even at a first OWR, a continuous washing method would help reduce the volume of water used for washing.

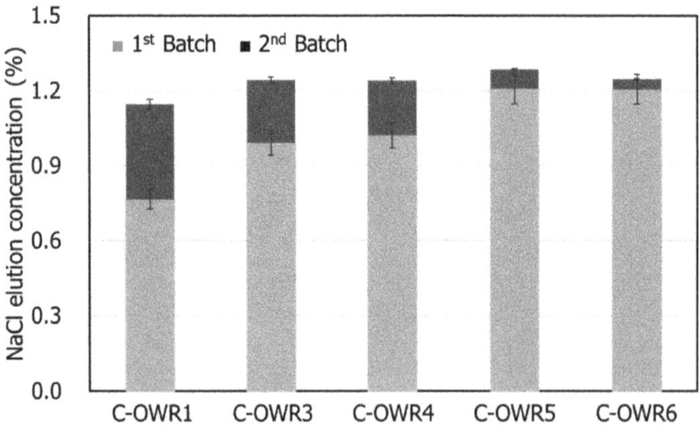

Figure 4. NaCl elution concentration via first and second batch system with various OWRs.

In the present study, we comparatively performed shell washing under different OWRs to reduce the reaction time. OWRs were chosen at two, three, and four times water volumes. The washing steps were repeated 10 times sequentially, as shown in Figure 5a, which also shows the unit mass of leached NaCl from the shells at every single step. As confirmed by batch experiments (Figure 2), a high concentration of eluted NaCl was observed during the early stages of washing steps (up to three times) and reduced gradually reaching nearly zero after six washes. The leaching patterns observed were similar regardless of OWRs; however, the total amount of leached NaCl differed largely under varied OWRs. The average of measured values repeated three times was used, as shown in Figure 5b. C-OWR4 had the highest value of 11.1 g/kg-OS among the three samples, which is 2.2-fold higher than that of C-OWR2, implying that volume of water is the key factor that controls washing properties.

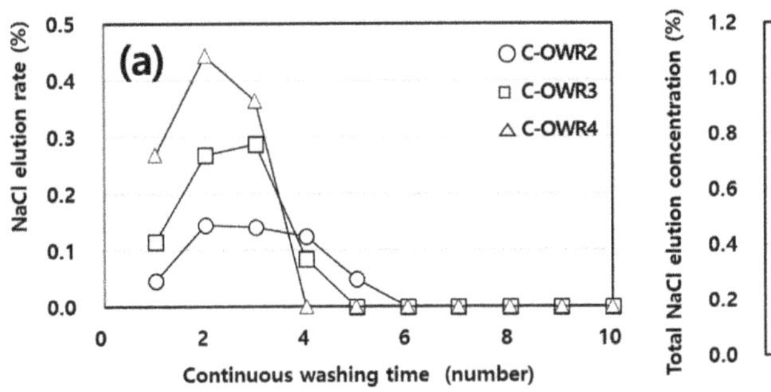

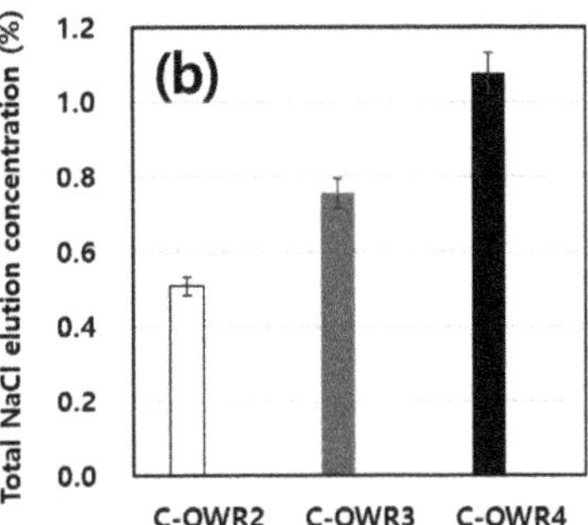

Figure 5. Amount of NaCl elution with continuous washing steps at different OWRs. (**a**) Changes in NaCl concentration as a function of washing steps. (**b**) Total NaCl elution concentration after 10 times of sequential washing.

We observed a substantial difference in NaCl leaching properties between the batch and continuous experiments. Assuming that the percentage of maximum eluted NaCl could be approximately 1.2%, leaching rates of NaCl at B-OWR5, B-OWR6, and C-OWR4 are repeated three time and indicated in Table 3. The maximum values (1.2%) of C-OWR4 were reached at 10 min, whereas batch experiments (B-OWR5 and B-OWR6) were 15 times more time consuming and utilized larger volumes of water to achieve the maximum elution value. As a result, the leaching rates of NaCl from the shells seem to be more affected by the volume of water than the duration of washing periods.

Table 3. Comparison of the optimum conditions between the batch and continuous systems.

Samples	Elution Time (min)	NaCl Elution (%)	NaCl Elution Rate (%/h)
B-OWR5	1440	1.20 ± 0.07	0.05 ± 0.02
B-OWR6	1440	1.20 ± 0.03	0.05 ± 0.01
C-OWR4	10	1.14 ± 0.02	6.73 ± 0.01

3.3. Particle Size of Washed Shells and Changes in Mineral Composition

The particle size distributions are depicted in a parabolic shape, accounting for 70% of coarse grains over 5.0 mm, and the average of measured values repeated three times was used in Figure 6a. The particle size for raw shells indicated that particles ≥1 mm accounted for 78% and those ≥5 mm in the coarse-grains accounted for 70%.

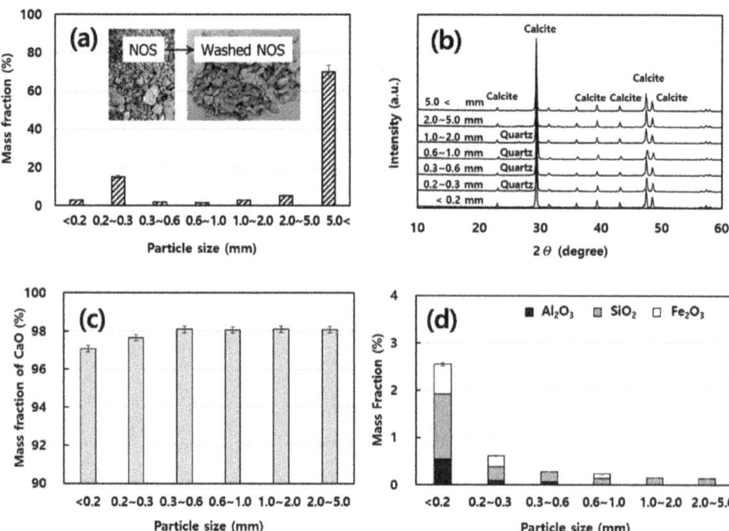

Figure 6. Properties of particle size of washed neglected oyster shells (NOSs). (**a**) Particle size distribution. (**b**) Results of X-ray diffractometer (XRD) analysis of washed NOSs with different particle sizes. (**c**) CaO contents of washed NOSs with different particle sizes. (**d**) Al_2O_3, SiO_2, and Fe_2O_3 contents of washed NOS with different particle sizes.

The NOSs were assessed using an XRD, and the patterns are represented in Figure 6b. Mineral composition of separated shells analyzed by XRD analysis confirmed that the main mineral was $CaCO_3$ regardless of the particle size; some quartz was observed in the fine-grained particles [30]. These results were averaged three times and agreed with those of XRF; the main element found in particle sizes of over 0.2 mm was Ca, which accounted for >98% (Figure 6c). The content of CaO increased as the particles became coarser, and Si, Fe, and Al were relatively more enriched in the finer particles. Generally, these elements are the main components (Al_2O_3, SiO_2, and Fe_2O_3) of the soil, suggesting that marine sediments could be recovered together with oyster shells from oyster aquacultural points. In Figure 6d, these results were averaged three times, the composition of Si, Fe, and Al were relatively more concentrated in the finer particles, and the content of these impurities clearly increased in particles ≤2 mm.

3.4. Leaching Experiment

Total component analysis results by elution test were compared for the washed and unwashed shells (Table 4). Although some differences were observed depending on the particle size, Na and Cl contents in washed shells clearly decreased by 77–93%. With the exception of Zn, heavy metals were not detected and decreased with increasing particle size. Some Na content remained after washing, which is likely because some seawater components are captured when oysters grow. Nonetheless, given that impurities in shells were relatively concentrated in particles, it is important to apply washing conditions considering the particle size to improve washing efficiency. Therefore, the sorting of particle size helps to remove impurities and NaCl simultaneously, as most of the impurities are confined to the finer particles. These results indicate that it is important to remove the impurities along with salts to use shells as Ca resources and selectively utilize fine-grained shells of ≥1.0 mm.

Table 4. NOSs and washed NOS components under different particle size by elution test.

Components	As-Received	Washed Shells		
		0.4–0.5 mm	1.0–2.0 mm	>5.0 mm
Na (%)	0.017	0.004	0.003	0.001
Cl (%)	0.007	0.000	0.000	0.000
As (mg/kg)	0.00	0.00	0.00	0.00
Cd (mg/kg)	0.000	0.00	0.00	0.00
Hg (mg/kg)	0.00	0.00	0.00	0.00
Pb (mg/kg)	0.00	0.00	0.00	0.00
Cu (mg/kg)	0.00	0.00	0.00	0.00
Zn (mg/kg)	9.49	12.23	7.55	0.00

4. Conclusions

The results of our study indicate the elution efficiency of NaCl with varied OWRs under the batch and continuous systems and different washing speeds and particle sizes. To completely elute out NaCl in one wash, at least five times as much water and 24 h mixing is required under the batch system. The NaCl elution is effective under a higher OWR where elution time can be minimized. The optimum NaCl elution efficiency of NOSs was predicted at an OWR of over 5, with an agitation time of 10 min and under a continuous washing system. Furthermore, because the NaCl proportion mainly constitutes fine particles, selective washing of particles ≥ 1.0 mm would increase washing efficiency. Larger particles of NOSs may benefit and be enhanced by the elution efficiency of NaCl. The washing speed is proportional to washing efficiency; otherwise, the elution efficiency according to the washing speed has a limitation in that the elution time would be longer in a batch system than in a continuous system. In conclusion, the washing process is essential for use neglected-oyster shells as a resource and materialize, and the use of the neglected oyster shells is a very important consideration for solve environmental problems.

Author Contributions: Conceptualization, S.K. and J.E.P.; formal analysis, data curation, S.E.L. and J.E.P.; writing—original draft preparation, writing—review and editing, J.E.P.; supervision, project administration, S.K. All authors have read and agreed to the published version of the manuscript.

Funding: This work was supported by Environment Industry & Technology Institute (KEITI) through the Project to develop eco-friendly new materials and processing technology derived from wildlife Project, funded by Korea Ministry of Environment (MOE) (2021003280003).

Conflicts of Interest: The authors declare no conflict of interest.

References

1. Song, Q.; Wang, Q.; Xu, S.; Mao, J.; Li, X.; Zhao, Y. Properties of water-repellent concrete mortar containing superhydrophobic oyster shell powder. *Constr. Build. Mater.* **2022**, *337*, 127423. [CrossRef]
2. Chairopoulou, M.A.; Garcia-Trinanes, P.; Teipel, U. Oyster shell reuse: A particle engineering perspective for the use as emulsion stabilizers. *Powder Technol.* **2022**, *408*, 117721. [CrossRef]
3. Yoon, G.-L.; Kim, B.-T.; Kim, B.-O.; Han, S.-H. Chemical-mechanical characteristics of crushed oyster-shell. *Waste Manag.* **2003**, *23*, 825–834. [CrossRef]
4. Ruslan, H.N.; Muthusamy, K.; Mohsin, S.M.S.; Jose, R.; Omar, R. Oyster shell waste as a concrete ingredient: A review. *Mater. Today Proc.* **2022**, *48*, 713–719. [CrossRef]
5. Ubachukwu, O.A.; Okafor, F.O. Towards green concrete: Response of oyster shell powder-cement concrete to splitting tensile load. *Niger. J. Technol.* **2020**, *39*, 363–368. [CrossRef]
6. Alvarenga, R.A.F.; Galindro, B.M.; Helpa, C.F.; Soares, S.R. The recycling of oyster shells: An environmental analysis using Life Cycle Assessment. *J. Environ. Manag.* **2012**, *106*, 102–109. [CrossRef]
7. Baek, E.-Y.; Lee, W.-G. A study on the Rational Recycling of Oyster-shell. *J. Fish. Bus. Adm.* **2020**, *51*, 71–87. [CrossRef]
8. Baek, E.-Y. Oyster Shell Recycling and Marine Ecosystems: A Comparative Analysis in the Republic of Korea and Japan. *J. Coast. Res.* **2021**, *114*, 350–354. [CrossRef]
9. Wang, H.-Y.; Kuo, W.-T.; Lin, C.-C.; Chen, P.-Y. Study of the material properties of fly ash added to oyster cement mortar. *Constr. Build. Mater.* **2013**, *41*, 532–537. [CrossRef]

10. Framework Act on Resource Circulation. Available online: https://elaw.klri.re.kr/kor_mobile/viewer.do?hseq=51210&type=sogan&key=16 (accessed on 4 October 2022).
11. Lu, M.; Shi, X.; Feng, Q.; Li, X.; Lian, S.; Zhang, M.; Guo, R. Effects of humic acid modified oyster shell addition on lignocellulose degradation and nitrogen transformation during digestate composting. *Bioresour. Technol.* **2021**, *329*, 124834. [CrossRef]
12. Yen, L.-T.; Wang, C.-H.; Than, N.A.T.; Tzeng, J.-H.; Jacobson, A.R.; Iamsaard, K.; Dang, V.D.; Lin, Y.-T. Mode of inactivation of Staphylococcus aureus and Escherichia coli by heated oyster-shell powder. *Chem. Eng. J.* **2022**, *432*, 134386. [CrossRef]
13. Park, K.; Sadeghi, K.; Panda, P.K.; Seo, J.; Seo, J. Ethylene vinyl acetate/low-density polyethylene/oyster shell powder composite films: Preparation, characterization, and antimicrobial properties for biomedical applications. *J. Taiwan Inst. Chem. Eng.* **2022**, *134*, 104301. [CrossRef]
14. Sun, Q.; Zhao, C.; Qiu, Q.; Guo, S.; Zhang, Y.; Mu, H. Oyster shell waste as potential co-substrate for enhancing methanogenesis of starch wastewater at low inoculation ratio. *Bioresour. Technol.* **2022**, *361*, 127689. [CrossRef] [PubMed]
15. Shi, Z.; Zhang, L.; Yuan, H.; Li, X.; Chang, Y.; Zuo, X. Oyster shells improve anaerobic dark fermentation performances of food waste: Hydrogen production, acidification performances, and microbial community characteristics. *Bioresour. Technol.* **2021**, *335*, 125268. [CrossRef] [PubMed]
16. Shih, P.-K.; Chang, W.-L. The effect of water purification by oyster shell contact bed. *Ecol. Eng.* **2015**, *77*, 382–390. [CrossRef]
17. Lee, J.-I.; Kang, J.-K.; Oh, J.-S.; Yoo, S.-C.; Lee, C.-G.; Jho, E.H.; Park, S.-J. New insight to the use of oyster shell for removing phosphorus from aqueous solutions and fertilizing rice growth. *J. Clean. Prod.* **2021**, *328*, 129536. [CrossRef]
18. Lim, J.; Cho, H.; Kim, J. Optimization of wet flue gas desulfurization system using recycled waste oyster shell as high-grade limestone substitutes. *J. Clean. Prod.* **2021**, *318*, 128492. [CrossRef]
19. Lim, J.; Kim, J. Optimization of a wet flue gas desulfurization system considering low-grade limestone and waste oyster shell. *J. Korea Soc. Waste Manag.* **2020**, *37*, 263–274. [CrossRef]
20. Liu, Y.; Che, D.; Xu, T. Effects of NaCl on the capture of SO_2 by $CaCO_3$ during coal combustion. *Fuel* **2006**, *85*, 524–531. [CrossRef]
21. Cho, S.; Lim, J.; Cho, H.; Yoo, Y.; Kang, D.; Kim, H. Novel process design of desalination wastewater recovery for CO_2 and SO_x utilization. *Chem. Eng. J.* **2022**, *433*, 133602. [CrossRef]
22. Koralegedara, N.H.; Pinto, P.X.; Dionysiou, D.D.; Al-Abed, S.R. Recent advances in flue gas desulfurization gypsum processes and applications—A review. *J. Environ. Manag.* **2019**, *251*, 109572. [CrossRef] [PubMed]
23. Lee, J.W.; Choi, S.-H.; Kim, S.-H.; Cha, W.-S.; Kim, K.; Moon, B.-K. Mineralogical Changes of Oyster Shells by Calcination: A Comparative Study with Limestone. *Econ. Environ. Geol.* **2018**, *51*, 485–492. [CrossRef]
24. Ha, S.H.; Kim, K.; Kim, S.-H.; Kim, Y. The effects of marine sediments and NaCl as Impurities on the Calcination of Oyster Shells. *Econ. Environ. Geol.* **2019**, *52*, 223–230. [CrossRef]
25. Laursen, K.; Grace, J.R.; Lim, C.J. Enhancement of the Sulfur Capture Capacity of Limestones by the Addition of Na_2CO_3 and NaCl. *Environ. Sci. Technol.* **2001**, *35*, 4384–4389. [CrossRef] [PubMed]
26. Shearer, J.A.; Johnson, L.; Turner, C.B. Effects of sodium chloride on limestone calcination and sulfation in fluidized-bed combustion. *Environ. Sci. Technol.* **1979**, *13*, 1113–1118. [CrossRef]
27. Rusydi, A.F. Correlation between conductivity and total dissolved solid in various type of water: A review. *Earth Environ. Sci.* **2018**, *118*, 012019. [CrossRef]
28. Ramakrishna, C.; Thenepalli, T.; Nam, S.Y.; Kim, C.; Ahn, J.W. Oyster Shell Waste is alternative source for Calcium carbonate ($CaCO_3$) instead of Natural limestone. *J. Energy Eng.* **2018**, *27*, 59–64. [CrossRef]
29. Prusty, J.K.; Patro, S.K.; Basarkar, S.S. Concrete using agro-waste as fine aggregate for sustainable built environment—A review. *Int. J. Sustain. Built Environ.* **2016**, *5*, 312–333. [CrossRef]
30. Kontoyannis, C.G.; Vagenas, N.V. Calcium carbonate phase analysis using XRD and FT-Raman spectroscopy. *Analyst* **2000**, *125*, 251–255. [CrossRef]

MDPI

St. Alban-Anlage 66

4052 Basel

Switzerland

www.mdpi.com

Waste Editorial Office

E-mail: waste@mdpi.com

www.mdpi.com/journal/waste